ANNALS OF THE NEW YORK ACADEMY OF SCIENCES

Volume 708

TYPES OF ALCOHOLICS
EVIDENCE FROM CLINICAL, EXPERIMENTAL, AND GENETIC RESEARCH

Edited by Thomas F. Babor, Victor Hesselbrock, Roger E. Meyer, and William Shoemaker

The New York Academy of Sciences
New York, New York
1994

Cover: Degas, "Au café, dit l'Absinthe," 1876, Musée du Jeu de Paume.

Library of Congress Cataloging-in-Publication Data

Types of alcoholics : evidence from clinical, experimental, and
   genetic research / edited by Thomas F. Babor . . . [et al.].
         p.   cm. — (Annals of the New York Academy of Sciences ; v.
708)
      "This volume is the result of a conference . . . held on October
22–23, 1992, in Farmington, Connecticut"—Contents p.
      Includes bibliographical references and index.
      ISBN 0-89766-799-9. — ISBN 0-89766-800-6 (pbk.)
      1. Alcoholism—Classification—Congresses.   2. Alcoholism—Genetic
aspects—Congresses.   3. Biochemical markers—Congresses.
I. Babor, Thomas F.   II. Series.
Q11.N5   vol. 708
[RC565]
500 s—dc20
[616.86'1'0012]   94-3082
                                                                                              CIP

AMP/PC
Printed in the United States of America
ISBN 0-89766-799-9 (cloth)
ISBN 0-89766-800-6 (paper)
ISSN 0077-8923

ANNALS OF THE NEW YORK ACADEMY OF SCIENCES

Volume 708
February 28, 1994

TYPES OF ALCOHOLICS

EVIDENCE FROM CLINICAL, EXPERIMENTAL, AND GENETIC RESEARCH[a]

Editors and Conference Organizers
Thomas F. Babor, Victor Hesselbrock, Roger E. Meyer,
and William Shoemaker

CONTENTS

[a]This volume is the result of a conference entitled Genetic Susceptibility, Biological Markers of Vulnerability and Alcoholic Subtypes, which was sponsored by the National Institute on Alcohol Abuse and Alcoholism and held on October 22–23, 1992 in Farmington, Connecticut.

Biological Markers of Vulnerability

Childhood Vulnerability to Alcoholism and Developmental Evidence for Alcoholic Subtypes

Implications for Research, Theory, and Clinical Practice

Conclusion

Financial assistance was received from:

• NATIONAL INSTITUTE ON ALCOHOL ABUSE AND ALCOHOLISM

Introduction

Method and Theory in the
Classification of Alcoholics[a]

THOMAS F. BABOR

Department of Psychiatry
University of Connecticut Health Center
Farmington, Connecticut 06030

The papers collected in this volume were commissioned for a scholarly symposium held in Farmington, Connecticut, on October 22 and 23, 1992. Sponsored by the National Institute on Alcohol Abuse and Alcoholism, the conference proceedings describe recent developments in the classification of types of alcoholics. By bringing together a diverse group of clinical, epidemiological, and experimental researchers, the conference proceedings provide an opportunity to take stock of new research findings, explore commonalities across diverse methodological approaches, and consider the implications of subtyping for theory, diagnosis, and treatment.

The topical areas for this symposium were derived from the conference organizers' conviction that typology research is likely to show continued progress to the extent that methodological innovation is integrated with theory development and clinical application. Because no professional discipline, theoretical perspective, or methodological approach has achieved dominance in the recent revival of interest in alcoholic subtyping, we have chosen to follow an eclectic course in setting the conference agenda. To this end, special emphasis was given to the findings emerging from different methodological approaches, such as genetic epidemiology, clinical assessment, longitudinal investigations, laboratory research, and follow-up studies of treatment response.

THE EVOLUTION OF TYPOLOGICAL THEORY

A type is an abstract category that brings together a set of associated defining characteristics that give the category a conceptual, theoretical, and clinical meaning. A typology of alcoholics is a classification system and a set of decision rules used to differentiate relatively homogeneous groups.

Classification has long been used as a strategy to deal with the complexity of physical disease and mental disorder. Etiological factors, morphological characteristics, functional disabilities, and therapeutic considerations are just a few of the criteria that have been used to guide the prediction of course and the selection of treatment for diseases such as cancer, diabetes, and hypertension. In the field of alcohol studies, multidimensional typology research is capable of making similar contributions not only to nomenclature and classification, but also to the development of better theories and more effective treatments.

The notion of distinguishing different subtypes of alcoholics has varied according to different historical periods and national traditions.[1] Typological thinking about

[a]The writing of this paper, and the conference proceedings it introduces, were supported by a grant from the National Institute on Alcohol Abuse and Alcoholism (No. 1R13 AA09017-01).

1

alcoholics had its origins in the pre-scientific speculations of 19th century asylum physicians who attempted to differentiate alcoholics on the basis of family history, drinking patterns, medical consequences, and sociocultural factors. Much of this literature was synthesized by Bowman and Jellinek[2] in an article published in the first volume of the *Quarterly Journal of Studies on Alcohol.* Their compendium provided a basis for Jellinek's[3] later classification of alcoholics into gamma and delta types. Although this typology has dominated the popular thinking about types of alcoholics, it has stimulated little empirical research.[4]

Despite a general conviction that alcoholism is a unitary disorder befitting a single disease label, the clinical and epidemiological research that flowered in the 1960s and 1970s provided ample evidence of the heterogeneity that characterizes persons diagnosed as alcoholics. This research reinforced the conclusion that alcoholics differ with respect to a wide variety of defining characteristics, such as genetic predisposition, antecedent personality characteristics, presenting symptoms, drinking patterns, severity of dependence, co-morbid psychopathology, the age at which problem drinking begins, the rapidity with which alcohol dependence develops, and the severity of alcohol-related consequences.[4–12]

Although there is ample evidence that the heterogeneity among alcoholics is not random, the search for the optimal classification system has led to a confusing array of single domain typologies (e.g., familial alcoholics, antisocial alcoholics). With the advent of multivariate statistical procedures and computer technology, typological theorists began to incorporate this complexity into their theories by postulating subtypes that encompass multiple defining characteristics. Examples of these newer multidimensional typologies include Cloninger's[13] neurobiological learning model, Morey and Skinner's[14] hybrid model, Zucker's[15] developmental model, and Babor *et al.*'s[16] vulnerability-severity classification.

Despite the differences in methodology and terminology, the single-domain and multidimensional typologies developed during the last century share a number of commonalities. (See, for example, reviews by Babor and Lauerman[1] and Babor and Dolinsky.[4]) The repeated finding that multiple defining characteristics tend to overlap in relatively homogeneous clusters of alcoholics has created the possibility that most of the typology theories proposed in the literature are referring to two basic types of alcoholics. One type is characterized by later onset, slower course, fewer complications, less psychological impairment, and better prognosis. Another type is characterized by genetic precursors, early onset, a more rapid course, more severe symptoms, greater psychological vulnerability, and poorer prognosis. If this hypothesis is correct, it could help to guide theory development, particularly in light of further advances in research methods, assessment technology, and statistical analysis.

METHODOLOGICAL APPROACHES

Among the methodological approaches that have been employed in typology research are clinical and statistical description, genetic epidemiology, experimental studies of high-risk groups, and treatment matching. In the last few years these approaches have advanced to the point where, alone and in combination, they offer great potential to the study of alcoholic subtypes. For example, a major limitation of previous descriptive research has been the general lack of comprehensive assessment to measure important defining characteristics. With the development of new verbal report measurement techniques, alcoholics can now be described according to a vast array of theoretically relevant dimensions, such as psychopathology, personality, drinking patterns, family history, and alcohol-related consequences. Similarly, to the extent that subtypes identified through clinical description can be discriminated according to biological markers and other putative indicators of vulnerability, these new methodological approaches can significantly advance our understanding of the etiology of different types of alcoholism.

Descriptive Approaches

Clinical description has advanced tremendously from the pre-scientific intuitive approaches employed up to the 1950s. Traditionally, descriptive typologies grew out of clinical observation. With the advent of structured interview schedules, criterion-based diagnostic procedures, personality inventories, and family history assessments, researchers have complemented and often replaced clinicians by introducing quantitative procedures to the identification of homogeneous groups. At the same time, prospective, longitudinal research designs have been employed to trace the natural history of clinical samples and at-risk groups. Although armchair speculation has given way to numerically based taxonomies, too often, as Skinner[17] has noted, statistical clustering algorithms "have been applied to a convenient data set as an end in itself. Few attempts have been made to determine whether the types have prognostic value with respect to treatment outcome or to integrate the types with previous research."

A recent study that combines comprehensive assessment with empirical clustering techniques and prospective prognostic evaluation is one conducted by Babor and colleagues.[16] An empirical clustering technique was applied to extensive diagnostic information collected from a sample of 321 treated alcoholics. A statistical clustering procedure identified two "types" of alcoholics who differed consistently across 17 defining characteristics in both male and female samples. One group, designated Type A alcoholics, is characterized by later onset, fewer childhood risk factors, less severe dependence, fewer alcohol-related problems, and less psychopathology. The other group, termed Type B alcoholics, is characterized by childhood risk factors, familial alcoholism, early onset of alcohol problems, greater severity of dependence, polydrug use, a more chronic treatment history (despite their younger age), greater psychopathology, and more life stress. The two types also differed with respect to treatment outcome assessed prospectively at 12 and 36 months.

Genetic Epidemiology

Another method that has contributed to the development of current typology theory is genetic epidemiology. Evidence from adoption and twin studies has identified clinical subtypes of alcoholism with distinct genetic characteristics. For example, on the basis of prospective adoption studies, Cloninger[13] proposed a neurobiological learning model of alcoholism that distinguishes two genetic subtypes, termed milieu-limited (Type I) and male-limited (Type II). Type I alcoholics have a later onset of alcohol problems, develop psychological rather than physical dependence, and experience guilt feelings about their alcohol use. In contrast, Type II alcoholics manifest alcohol problems at an early age, exhibit spontaneous alcohol-seeking behavior, and are socially disruptive when drinking. Dimensions of personality are hypothesized to account for these different types of alcoholism. Other researchers[18,19] have reported that age of onset provides a convenient way to classify patients who resemble Type I and Type II alcoholics.

Experimental Research

More recently, laboratory-based experimental methods and research on biological markers have added scientific depth to typology theory. Laboratory-based methods have been developed to study high-risk groups that might have characteristics suggesting different genetic vulnerabilities or typological patterns. Much of this effort has focused on the identification of biological markers of susceptibility to alcoholism in general and to specific types of alcoholism in particular. These include tests for biochemical markers (e.g., cortisol, prolactin) that suggest susceptibility to alcoholism,[20] studies of event-related potentials[21] that might serve as trait markers, and research on behavioral indicators of tolerance.[20] Some evidence indicates that event-related potentials in the sons[21] of alcoholic fathers are abnormal even before the onset of drinking. Abstinent alcoholics may also possess this defect,[22] suggesting that the neurophysiological measurements are "trait" markers of susceptibility to alcoholism. In a related line of research, Tabakoff *et al.*[23] identified abnormalities in platelet monoamine oxidase and adenylate cyclase activity that may represent another kind of genetic susceptibility. Animal models of alcohol preference have been developed in selected lines of rats over several generations, suggesting that genes influencing alcohol drinking display heterogeneity and can be manipulated even in rodent populations. Some of these rodent lines may prove useful models for some types of alcoholism.[24] Finally, studies of college students considered at high risk of alcoholism demonstrate that the sons of alcoholics are differentially resistant to the effects of a challenge dose of ethanol.[25]

Treatment Response

A final method that has been applied to the study of alcoholic subtypes is the comparison to treatment response in different therapeutic modalities. To be valid, not to mention useful, typological classification systems should permit the rational assignment of patients to the most appropriate treatment. Kadden *et al.*,[26] for example, randomly assigned different types of alcoholics to cognitive-behavioral and interactive therapy to evaluate the treatment-matching implications of typological classifications. They found that sociopathic alcoholics responded better to cognitive-behavioral therapy, whereas patients without this kind of personality disorder responded best to interactive therapy.

TOWARD AN INTEGRATION OF THEORY AND METHOD

If the history of typology research is a prologue to the future, it is clear that classification theory should be grounded in sound methodology and research design. As just indicated, a variety of creative methodological approaches have been employed during the last decade to identify genetic and biological markers, describe clinical subtypes, trace familial relationships, identify defining characteristics, follow natural histories, and evaluate treatment response. These approaches include family history methods; twin and adoption studies; psychological, physiological, and behavioral testing of at-risk adolescents; prospective studies of clinical samples; and random assignment studies of treatment response. A major impetus to recent work in typology research has been the availability of new assessment technologies, new biological monitoring procedures, and better diagnostic criteria. Because recent typology theory has become organized largely around these different methodological approaches, it is important to look for commonalities across these different disciplinary and scientific orientations. This, in brief, is the purpose of the papers collected in this volume.

ACKNOWLEDGMENTS

The editors would like to thank John Allen, Shirley Crall, and Debbie Talamini for their assistance in organizing the conference.

REFERENCES

1. BABOR, T. F. & R. LAUERMAN. 1986. Classification and forms of inebriety: Historical antecedents of alcoholic typologies. *In* Recent Developments in Alcoholism. M. Galanter, ed. Vol. **5:** 113–114. Plenum Press. New York.
2. BOWMAN, K. M. & E. M. JELLINEK. 1941. Alcohol addiction and its treatment. Q. J. Stud. Alcohol **2:** 98–176.
3. JELLINEK, E. M. 1960. Alcoholism: A genus and some of its species. Can. Med. Assoc. J. **83:** 1341–1345.
4. BABOR, T. F. & Z. S. DOLINSKY. 1988. Alcoholic typologies: Historical evolution and empirical evaluation of some common classification schemes. *In* Alcoholism: Origins and Outcome. R. M. Rose & J. Barrett, eds.: 245–266. Raven Press. New York.
5. WANBERG, K. W. & J. L. HORN. 1983. Assessment of alcohol use with multidimensional concepts and measures. Am. Psychol. **38:** 1055–1069.
6. COTTON, N. S. 1979. The familial incidence of alcoholism. J. Stud. Alcohol **40:** 89–116.
7. HESSELBROCK, M. N., R. E. MEYER & J. J. KEENER. 1985. Psychopathology in hospitalized alcoholics. Arch. Gen. Psychiatry **42:** 1050–1055.
8. HESSELBROCK, V. M., J. R. STABENAU, M. N. HESSELBROCK, R. E. MEYER & T. F. BABOR. 1982. The nature of alcoholism in patients with different family histories for alcoholism. Prog. Neuropsychopharmacol. Biol. Psychiatry **6:** 607–614.
9. POWELL, B. J., M. R. READ, E. C. PENICK, N. S. MILLER & S. F. BINGHAM. 1987. Primary and secondary depression in alcoholic men: An important distinction? J. Clin. Psychiatry **48:** 98–101.
10. BABOR, T. F. & R. E. MEYER. 1986. Typologies of Alcoholics: Overview. *In* Recent Developments in Alcoholism. M. Galanter, ed. Vol. **5:** 105–111. Plenum Press. New York.
11. NERVIANO, V. J. & H. W. GROSS. 1983. Personality types of alcoholics on objective inventories. J. Stud. Alcohol **44:** 837–851.
12. SCHUCKIT, M. 1985. The clinical implications of primary diagnostic groups among alcoholics. Arch. Gen. Psychiatry **42:** 1043–1049.
13. CLONINGER, C. R. 1987. Neurogenetic adaptive mechanisms in alcoholism. Science **236:** 410–416.
14. MOREY, L. C. & H. A. SKINNER. 1986. Empirically derived classifications of alcohol-related problems. *In* Recent Developments in Alcoholism. M. Galanter, ed. Vol. **5:** 45–168. Plenum Press. New York.
15. ZUCKER, R. A. 1987. The four alcoholisms: A developmental account of the etiologic process. *In Alcohol and Addictive Behavior.* P. C. Rivers, ed.: 27–83. University of Nebraska Press. Lincoln, Nebraska.
16. BABOR, T. F., M. HOFMANN, F. DELBOCA, V. HESSELBROCK, R. MEYER, Z. S. DOLINSKY & B. ROUNSAVILLE. 1992. Types of alcoholics, I. Evidence for an empirically derived

typology based on indicators of vulnerability and severity. Arch. Gen. Psychiatry **49:** 599–608.

17. SKINNER, H. A. 1981. Toward the integration of classification theory and methods. J. Abn. Psychol. **90:** 69.

18. VON KNORRING, L., V. PALM & H. ANDERSON. 1985. Relationship between treatment outcome and subtype of alcoholism in men. J. Stud. Alcohol **46:** 388–391.

19. BUYDENS-BRANCHEY, L., M. H. BRANCHEY & D. NOUMAIR. 1989. Age of alcoholism onset. I. Relationship to psychopathology. Arch. Gen. Psychiatry **46:** 231–236.

20.. SCHUCKIT, M., E. GOLD & S. C. RISCH. 1987. Changes in blood prolactin levels in sons of alcoholics and controls. Am. J. Psychiatry **144:** 854–859.

21. BEGLEITER, H., B. PORJESZ, B. BIHARI & B. KISSIN. 1984. Event-related brain potential in boys at risk for alcoholism. Science **225:** 1493–1496.

22. PORJESZ, B. & H. BEGLEITER. 1991. Neurophysiological factors in individuals at risk for alcoholism. *In* Recent Developments in Alcoholism. 9. Children of Alcoholics. M. Galanter, ed.: 53–64. Plenum Press. New York.

23. TABAKOFF, B., P. L. HOFFMAN, J. M. LEE, T. SAITO, B. WILLARD & F. DELEON-JONES. 1988. Differences in platelet enzyme activity between alcoholics and controls. N. Engl. J. Med. **318:** 134–139.

24. LI, T. K., L. LUMENG, D. P. DOOLITTLE, W. J. MCBRIDGE, J. M. MURPHY, J. C. FROELICH & S. MOZORATI. 1988. Behavioral and neurochemical association of alcohol seeking behavior. *In* Biomedical and Social Aspects of Alcohol and Alcoholism. K. Kuriyama, A. Takada & H. Ishii, eds.: 435–439. Exerpta Medica. New York.

25. SCHUCKIT, M., M. IRWIN & M. G. MONTERO. 1988. Differences in intensity of reaction to ethanol in children of alcoholics and controls. *In* Biomedical and Social Aspects of Alcohol and Alcoholism. K. Kuriyama, A. Takada & H. Ishii, eds.: 453–459. Exerpta Medica. New York.

26. KADDEN, R. M., N. L. COONEY, H. GETTER & M. D. LITT. 1989. Matching alcoholics to coping skills or interactional therapies: Posttreatment results. J. Consulting & Clin. Psychol. **57:** 698–704.

Personality-Based Subtypes of Chemically Dependent Patients

JOHN P. ALLEN,[a] JOANNE B. FERTIG,
AND MARGARET E. MATTSON

National Institute on Alcohol Abuse and Alcoholism
Room 14C-20, 5600 Fishers Lane
Rockville, Maryland 20857

Interest has recently focused on assigning chemically dependent patients to treatments based on their specific needs.[1,2] In fact, the intriguing potential of effective "patient-treatment matching" as a means of improving outcome has prompted the National Institute on Alcohol Abuse and Alcoholism to further explore the hypothesis by supporting the largest and most complex clinical trial ever undertaken on treatment of alcoholism.[3] The National Institute on Drug Abuse has also initiated a cooperative agreement to test the validity of the "matching" hypothesis as it might relate to drug abuse.

Prerequisite to identifying positive interactions between patients and interventions is derivation of a patient classification system which results in specific recommendations for choice of treatment. Such a taxonomy might be based on chemical use variables (such as pattern of use, high risk situations, etiology, or consequences of chemical use), treatment-oriented variables (such as motivation or needs for specific treatment services), demographic variables, personality characteristics, or combinations of variables "cutting across" domains. Regardless of the basis of the taxonomy, the system needs to be easy for clinicians to employ, should permit assignment of patients early enough in the rehabilitation process that specific treatment plans can be implemented, and, ideally, should allow categorization of all patients into a relatively small number of discrete types for which treatment options can be tailored.

Attempts to derive typologies to classify substance abusers are not new. One reviewer documents such efforts at least as far back as the Ebers Papyrus.[4] The more recent history of the development of alcoholism typologies is particularly well detailed by Babor and Dolinsky.[5]

Because most current therapies for substance abuse are of a verbal nature, use of personality measures for patient assignment to treatment seems particularly relevant. To date, personality variables including field dependence, anxiety, sociopathy, autonomy, and locus of control have served as bases for subtyping chemically dependent patients and evaluating the effects of differing interventions.

Objective personality instruments have served as the most common means of assessing such variables[6,7] and the Minnesota Multiphasic Personality Inventory (MMPI) has proven the most frequent choice of instrument. Nevertheless, the heavy emphasis of the MMPI on psychopathology, absence of scales to measure inter-

[a]Address for correspondence: John P. Allen, Division of Treatment Research, National Institute on Alcohol Abuse and Alcoholism, Room 14C-20, 5600 Fishers Lane, Rockville, Maryland 20857.

personal style variables,[7] and factorial complexity of its clinical scales appear to limit usefulness of the MMPI in this regard. Other personality inventories may well prove more helpful. The Personality Research Form (PRF)[8] is a particularly promising candidate. The PRF was developed to assess traits deriving from Murray's[9] conceptually rich theory of personality, employed multi-trait, multi-method factor analysis in scale derivation, includes scales to evaluate response styles of social desirability and infrequency, keys half of the items as true and half as false to minimize response styles of acquiescence and criticalness, and measures personality needs in a bipolar fashion. Form E of the PRF is written at a low reading level, approximately fifth or sixth grade,[10] and hence is applicable to a broader clientele than are many other personality measures.

The PRF has been subjected to extensive research and has played an especially important role in chemical dependency studies, including research on risk factors[11] personality correlates of substance use and abuse,[12–19] and therapy-related characteristics of chemically dependent patients.[20–27] Scores on the PRF have also served as dependent variables for several previous taxonomies of alcoholism[28–32] and have themselves served as the basis for such taxonomic research on alcoholism.[32–34]

The latter three studies will be described in some detail inasmuch as the current project was partially designed to replicate and validate their findings. Nerviano[32] performed a correlational cluster analysis of 12 PRF Form AA subscales which loaded saliently on 5 factors. Subjects were male inpatients with a primary diagnosis of alcoholism. Forty-nine percent of the sample was classified into seven subtypes. The subtypes and percentage of cases assigned to each were: Passive-Ambivalent (14.5%), Impulsive (8.5%), Explosive (8%), Passive-Dependent (6%), Avoidant (6%), Asocial Schizoid (3.1%), and Narcissistic (3%). Meaningful correlates of subtypes were found with the 16PF and with the remaining 9 PRF trait scales.

Nerviano[33] also evaluated PRF Form AA personality profiles among a similar group of alcoholic male inpatients in a V.A. hospital. The same clustering method as in the 1976 study was employed. Subtypes discovered in the previous study were closely replicated, and an additional group, "Hostile Withdrawn," was also distinguished. Patients assigned to this subtype scored high on the PRF traits of Defendence, Aggression, and Autonomy and low on traits of Affiliation and Succorance. Fifty-five percent of the sample was classified. The subtypes were projected into a two-space defined by orthogonal vectors of MMPI scales, distress-withdrawal and acting-out tendencies. The report suggests a range of treatment variations that may be appropriate for the patient types.

Again employing the same clustering method, but using Form E of the PRF, Zivich[34] replicated the five most commonly occurring Nerviano types and thereby classified 41% of his cases. Additionally, he disaggregated "no type" profiles into those truly "no type" (i.e., all scores near the mean on the 12 PRF scales), a small "mixed type" group, and a group of profiles with distinctive, but differing, scale elevations that could not be readily classified. For the most part, the eight types varied in treatment outcome as measured by self-report. As with Nerviano's samples, Zivich's group was entirely male. Zivich's sample, however, was younger, included a much higher percentage of blacks, and was drawn from a municipal treatment center rather than from a V.A. hospital.

The present investigation was partially designed to evaluate the robustness of this classification system by employing an iterative partitioning technique to classify all cases and to determine if patient types differ as a function of diagnosis, age, education, gender, MMPI scales, and "high risk" situations for abuse of alcohol or drugs.

METHOD

Subjects. The sample consisted of 407 successive admissions to a private, for-profit adult substance abuse inpatient treatment facility. Following clinical evaluation and interviews with them and, when possible, with family members and employers, subjects were diagnosed according to DSM III-R criteria as alcohol dependent, drug dependent, or both alcohol and drug dependent. Demographic characteristics of the three diagnostic groups are presented in TABLE 1.

Procedure and Instruments. On the sixth day after admission or as soon thereafter as they had completed detoxification, patients were administered an assessment battery consisting of the PRF (Form E), the MMPI, and the "Chemical Use Survey" (CUS). The CUS is a 23-item self-report inventory dealing with internal and external cues that may elicit drinking or use of drugs. Seventeen items reflect Miller and Mastria's[35] conceptual schema of five classes of stimuli: social, situational, emotional, physiological, and cognitive. The additional items were based on comments from the rehabilitation staff about other stimuli that they reported having frequently heard from patients. The CUS was scored on a five-point Likert scale with "1" indicating that the stimulus was "not at all likely" to prompt usage and "5" that it was "very likely" to lead to use of alcohol or drugs.

Data Analysis. Data on individual tests were screened for univariate and multivariate outliers. Separate factor analyses were then performed on PRF scales and CUS items. Factor scores on six retained PRF factors were used in an iterative partitioning procedure to generate five subtypes of patients. These five clusters were then used as predictor variables in two separate discriminant functions. In the first, CUS factor scores were used as classification variables. In the second, nine MMPI clinical scales and scales L, F, and K served as classification variables. Demographic covariates of age, gender, and education were used in both discriminant analyses.

TABLE 1. Sample Characteristics

Characteristic	Group		
	Alcohol	Drug	Dual Diagnosis
Education	12.1 ± 2.4	11.8 ± 1.9	11.9 ± 1.7
Age	38.5 ± 10.1	29.7 ± 7.5	27.7 ± 6.7
Gender			
Male	135	52	139
Female	24	16	20
Religion			
Protestant	71%	63%	63%
Catholic	12%	9%	13%
Jewish	0%	0%	0%
No religion	15%	21%	20%
Other	2%	7%	4%
Race			
White	95%	67%	91%
Black	5%	33%	8%
Hispanic	0%	0%	1%
Other	0%	0%	0%

Last, separate contrast analyses were performed to characterize each subtype on the CUS and MMPI variables.

RESULTS

Following exclusion of seven subjects with PRF Infrequency scores of 3 or higher, PRF scale scores were standardized separately for the male and female patients using the most recently published norms.[36] FIGURES 1 and 2 show PRF scale means for men and women in this study compared to these norms. Visual inspection of the figures suggests that, as a group, study subjects are more hostile, resentful, and risk-taking and less reflective and cognitively oriented than normative subjects.

MMPI scale scores were also standardized separately for male and female patients using norms from Dahlstrom, Welsh, and Dahlstrom.[37] FIGURE 3 shows MMPI scale scores for male and female study subjects. Visual inspection of FIGURE 3 suggests that study subjects tend to be more depressed, asocial, hypomanic, and disturbed than MMPI normative cases.

Subjects' profiles for the PRF, CUS, and MMPI were evaluated individually for multivariate outliers. Using a regression approach, Mahalanobis distances, that is, the distance between each case and the mean of all other cases for the independent variables in the regression equation, were calculated. These distances were tested according to a chi-square sampling distribution with degrees of freedom equal to the

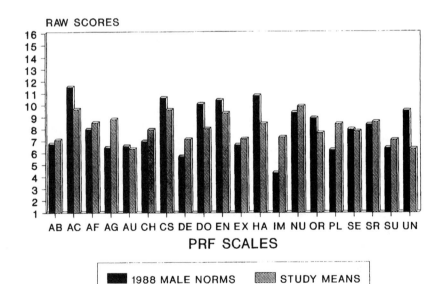

FIGURE 1. Personality Research Form raw scores for male study subjects compared to Costa and McCrae,[36] 1988 norms. PRF = Personality Research Form[8]; AB = Abasement; AC = Achievement; AF = Affiliation; AG = Aggression; AU = Autonomy; CH = Change; CS = Cognitive Structure; DE = Defendence; DO = Dominance; EN = Endurance; EX = Exhibition; HA = Harm Avoidance; IM = Impulsivity; NU = Nurturance; OR = Order; PL = Play; SE = Sentience; SR = Social Recognition; SU = Succorance; UN = Understanding.

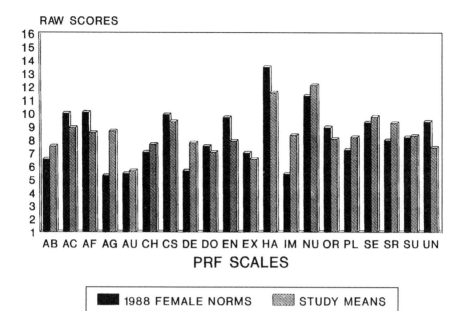

FIGURE 2. Personality Research Form raw scores for female study subjects compared to Costa and McCrae,[36] 1988 norms. PRF = Personality Research Form[8]; AB = Abasement; AC = Achievement; AF = Affiliation; AG = Aggression; AU = Autonomy; CH = Change; CS = Cognitive Structure; DE = Defendence; DO = Dominance; EN = Endurance; EX = Exhibition; HA = Harm Avoidance; IM = Impulsivity; NU = Nurturance; OR = Order; PL = Play; SE = Sentience; SR = Social Recognition; SU = Succorance; UN = Understanding.

number of variables for each test. Cases with Mahalanobis distance values exceeding the chi-square critical value for alpha = 0.001 were adjudged multivariate outliers and were excluded from further analyses. One case on the PRF, two on the MMPI, and four on the CUS were thus removed from the sample prior to subsequent analyses.

PRF scales (with the exception of Infrequency and Social Desirability) were submitted to a principal components extraction. Retained factors were subsequently rotated to a varimax solution and exact method factor scores computed. Based primarily on interpretability, a six-factor solution was chosen. The solution accounted for 65% of the variance, with the lowest eigenvalue for the unrotated solution being 1.01. TABLE 2 depicts loadings of PRF traits on rotated factors. Factor I is a bipolar dimension appearing to deal with cognitive control *versus* impulsivity. Factor II seems to tap social involvement. Factor III is a bipolar dimension that appears to involve hostility and interpersonal control *versus* self-blame. Factor IV contrasts the need for emotional support with the need for autonomy and is termed dependency. Factor V seems to assess personal agreeableness. Factor VI appears to gauge cognitive reflectiveness.

Five patient clusters were derived from the previously calculated PRF factor scores using the BMDP KM clustering procedure.[38] This algorithm establishes a

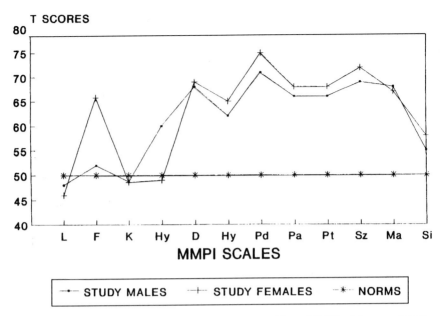

FIGURE 3. MMPI T scores for male and female study subjects. MMPI = Minnesota Multiphasic Personality Inventory[37]; HS = Hysteria; D = Depression; HY = Hypochondriasis; PD = Psychopathic Deviance; PA = Paranoia; PT = Psychasthenia; SZ = Schizophrenia; MA = Hypomania; SI = Social Introversion.

TABLE 2. Personality Research Form Factor Loadings

Factors	Factors					
	1	2	3	4	5	6
Impulsivity	−77	10	33	03	02	08
Order	76	15	02	06	05	09
Cognitive structure	75	06	11	23	01	14
Endurance	60	18	18	−31	27	27
Achievement	56	13	06	21	42	34
Affiliation	15	82	09	18	13	08
Exhibition	05	74	29	12	00	16
Play	−47	63	01	12	13	17
Change	06	52	00	−35	08	16
Aggression	20	07	77	15	16	14
Defendence	10	08	72	00	−34	00
Social recognition	00	25	52	42	22	06
Dominance	27	47	50	25	14	18
Succorance	06	14	08	72	00	11
Autonomy	15	01	19	−68	−25	15
Harm avoidance	15	−37	13	59	15	19
Abasement	08	16	−35	01	75	03
Nurturance	17	33	01	16	68	09
Understanding	26	02	12	15	15	78
Sentience	09	34	−27	09	12	69

Note: Decimals eliminated. Loadings of .40 or greater are underlined.

fixed number of homogeneous groups of cases based on minimizing euclidean distances from cluster centroids. The procedure begins with all cases in a single cluster and repeatedly splits the cluster until the specified number of clusters is reached. Cases are iteratively reallocated into the cluster whose midpoint is closest in euclidean distance. Various numbers of clusters from four to seven were considered. Based primarily on interpretability and comparability with previous PRF clustering research, a five-cluster solution was chosen. TABLE 3 displays cluster membership as a function of gender and diagnosis. The five types derived from the clustering procedure may be characterized in terms of the underlying PRF factors as follows: Cluster 1 subjects seem socially oriented but hostile and dependent on others. Cluster 2 patients appear somewhat nonreflective, quite socially invested, and agreeable; Cluster 3 cases are generally the least socially oriented; Cluster 4 subjects are quite impulsive and little interested in relating to others; Cluster 5 cases may be described as hostile and overcontrolled. Means of the clusters on PRF factors are displayed in FIGURE 4.

CUS items were also intercorrelated, factored, and rotated using the same procedures. Five factors were retained, accounting for 49% of the variance, the lowest eigenvalue for the unrotated solution being 1.13. Factor loadings for each of the CUS items are provided in TABLE 4. Factor I seems to reflect depression and lack of self-esteem as a precipitant to use. Factor II is interpreted as peer influence to use. Factor III is characterized as "self-medicational" use to reduce physical or emotional stress. Factor IV seems to involve self-reinforcing, hedonic reasons for use. Factor V seems to deal with using to in some way punish oneself or others. CUS factor scores for the clusters are displayed in FIGURE 5.

A discriminant function analysis was performed using the patient clusters as the predictor variable and CUS factor scores as the classification variables. Age, education, and gender were employed as covariates. (In preliminary analyses, diagnosis did not contribute to the classification function and was thus deleted from the discriminant function analysis.) Gender played a highly significant role in the discrimination between groups (F $[4,382]$ = 35.34, $p < 0.0001$) as did age (F $[4,382]$ = 10.13, $p < 0.001$). Education played a significant, but smaller, role (F $[4,382]$ = 3.43, $p < 0.01$). From the classification functions derived from the CUS factor scores and covariates, 44% of the cases were accurately classified, with 53% correctly

TABLE 3. Gender-Diagnosis Composition of Clusters: Number of Cases

| Cluster | Gender | Diagnosis | | |
		Alcohol	Drug	Dual Diagnosis
1	Male	21	20	49
	Female	2	0	2
2	Male	35	8	35
	Female	5	3	4
3	Male	33	14	24
	Female	1	1	1
4	Male	37	8	23
	Female	5	1	3
5	Male	9	2	8
	Female	11	11	10

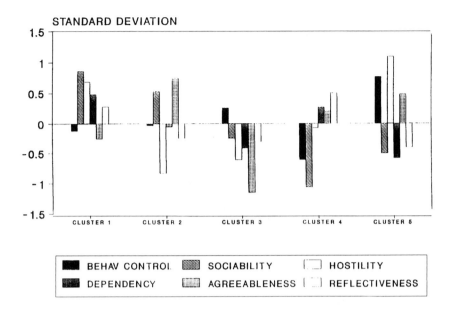

FIGURE 4. Characterization of clusters derived from PRF factor scores.

TABLE 4. Factor Loadings of Self-Reported Stimuli for Substance Abuse Subjects

	Factors				
Stimuli	1	2	3	4	5
Feeling ashamed	74	17	08	05	18
Feeling guilty	72	22	06	08	03
Feeling frustrated at self	71	05	13	04	06
Feeling badly about life	67	07	24	11	05
Lacking hope	57	01	30	01	06
Feelings of loss	56	−21	06	19	00
Seeing others use	08	78	11	12	04
Being encouraged by others	12	70	−10	15	09
Being where used before	03	60	02	13	09
Using to sleep	10	−09	75	08	00
Using to relieve pain	11	−03	63	19	20
Feeling physical need to use	22	32	57	−01	05
Liking taste/smell	01	10	07	70	17
Using to reward self	14	21	16	65	08
Feeling happy	−10	26	17	55	06
Believing use was safe	08	08	−05	55	31
Wanting to hurt self	15	00	21	−15	74
Using to "get back" at others	16	25	−06	20	68
Feeling sexually frustrated	47	06	02	29	23
Using to feel calmer with others	38	33	33	19	−21
Feeling bored	19	38	36	20	−14
Remembering past enjoyable use	10	45	35	20	09
Feeling tense for no reason	39	15	47	−03	−06

Note: Decimals eliminated. Loadings of .40 or greater are underlined.

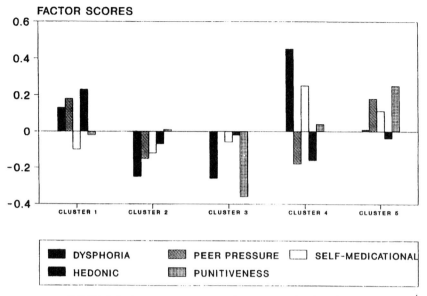

FIGURE 5. Chemical Use Survey factor scores for five clusters.

classified into Cluster 1 and 64% into Cluster 5. The canonical correlations for the first (0.53) and second (0.39) discriminant functions between cluster membership and CUS factors were determined to be significant according to Bartlett's test. The first canonical variable assesses differences according to a dimension of depression, punitiveness, and self-medication and reflects primarily contributions of CUS Factor 1 (−0.85), Factor 5 (−0.40), and Factor 3 (0.38). It appears to assess dysphoria. The second canonical variable distinguishes clusters according to tendency to use for hedonic reasons (Factor 4 [0.66]) and peer pressure stimuli (Factor 2 [0.66]). It appears to assess conviviality as a basis for substance usage. Fifty-six percent of the between-groups variance was accounted for by the first discriminant function and 29% by the second discriminant function. The multivariate discrimination between groups was highly significant using all demographic variables and CUS factor predictor variables (Wilks' lambda = 0.54; F [32,1388] = 7.83, p < 0.001). FIGURE 6 indicates locations of the centroids (multivariate means) of the clusters plotted in the space of the two CUS canonical variables.

A second discriminant function analysis was conducted, again using PRF subtypes as predictor variables, but with the nine MMPI clinical scales and the F, L, and K scales as the classification variables. (Scale 5, masculinity-femininity, was not included.) As before, age, education, and gender were introduced as covariates and they significantly discriminated the clusters. Using the classification functions derived from MMPI scores and demographic covariates, 54% of cases were correctly classified. The highest percentage (73%) of correct classification was again to Cluster 5. The canonical correlations for the first (0.58) and second (0.52) discriminant functions relating group membership to MMPI variables were significant by Bartlett's test. After removing the demographic covariates, the first canonical variable assessed differences according to a dimension of mood and social involvement with

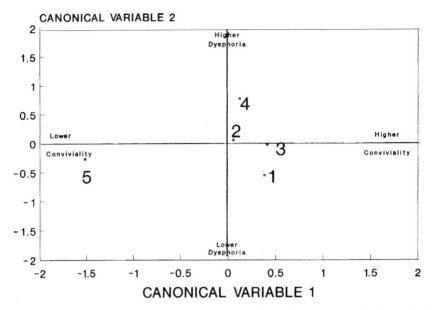

FIGURE 6. Personality Research Form cluster centroids (multivariate means) in the space of the discriminant functions relating Chemical Use Survey variables to cluster membership.

salient canonical variable loadings from Si (−0.82), D (−0.54), Ma (0.42), and F (0.23). This dimension seems to tap emotional responsiveness. The second canonical variable aligned types according to degree of behavioral control and was related to canonical variable loadings by Pd (0.76) and L (−0.39). Forty-five percent of the between-groups variance was accounted for by the first discriminant function and 33% by the second. The multivariate discrimination between groups was highly significant when all demographic and MMPI predictor variables were included in the equation. (Wilks' lambda = 0.37; F [52, 1435] = 8.02, p < 0.0001). FIGURE 7 shows the centroids of the five clusters plotted in the space of the canonical variables.

DISCUSSION

Cluster analyses of PRF factor scale scores established five subtypes of alcoholics. Consideration of factor scores most distinctive of the groups suggests the following descriptive labels: Cluster 1: Hostile-dependent; Cluster 2: Cooperative-nonreflective; Cluster 3: Socially uninvolved; Cluster 4: Impulsive-unsociable; and Cluster 5: Hostile-overcontrolled.

After statistically equating groups for age, education, and gender differences, relationships were found between personality types and stimuli for chemical use. The hostile-dependent group appears to use primarily for hedonic and peer pressure reasons. The cooperative-nonreflective and socially uninvolved subtypes seem prompted primarily because of dysphoria. The impulsive-unsociable group reports using substances especially as a means of self-medication and to reduce dysphoria. The hostile-overcontrolled subtype differs from the other clusters in using substances

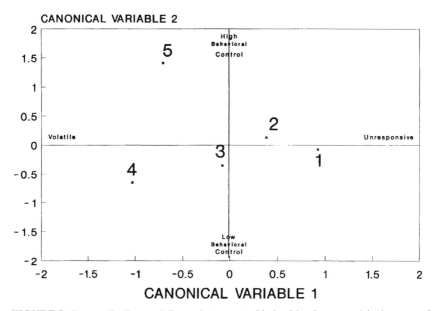

FIGURE 7. Personality Research Form cluster centroids (multivariate means) in the space of the discriminant functions relating Minnesota Multiphasic Personality Inventory clinical scales to cluster membership.

to in some way punish themselves or others as well as for peer pressure reasons.

These results have considerable intuitive appeal. For example, while social relationships seem valued by Cluster 1 patients, their competing tendencies for hostility and the need for emotional support likely keep them from developing satisfying interpersonal relationships. They may, therefore, derive alternative reinforcers in life from intrapersonal, hedonic activities to include drug and alcohol use. Clusters 2 and 3 likely suffer considerable dissatisfaction, and substance use may provide a temporary respite from depression. The sources of the unhappiness probably differ for the two clusters. Cluster 2 patients probably derive few rewards from intellectual or aesthetic pursuits. Cluster 3 patients probably lack a social support system to reduce depression. Cluster 4 patients seem socially aloof and may use substances to assuage emotional discomfort arising from alienation from peers. The impulsivity of Cluster 4 patients may be driven by hyperexcitability, and substances may dampen this heightened and uncomfortable state of arousal. Cluster 5 patients appear quite angry but seem unable to directly express such feelings. They are likely passive aggressive and report using substances to punish and hurt others.

The subtypes also differ on the two canonical variables derived from the MMPI. The first canonical variable seems to assess the degree of emotional stability, with higher scores representing greater stability. Nerviano[32] labeled his comparable MMPI dimension "distress-withdrawal" *versus* "integration." The second canonical variable seems to deal with degree of socialization and behavioral control and seems similar to Nerviano's second MMPI vector termed "tendency toward acting out."

The mean group MMPI configuration for Cluster 1 is 4–9, a classic characterologic profile generally associated with substance abuse and acting out. The MMPI configuration associated with Cluster 2 membership is within normal limits. Cluster 3 membership is associated with a slight 4–8 profile suggesting social nonconformity. Most striking is the MMPI configuration of 2–7–8 for Cluster 4. This combination of scale elevations is generally associated with emotional turmoil, problems in concentration, and social withdrawal. Cluster 5 patients reveal a moderately high 6–8–9 profile, often associated with resentment, social skill deficits, and emotional lability.[39]

Cluster 1 patients seem to be the most emotionally stable with Clusters 2, 3, 5, and 4 being successively less so. Cluster 5 patients seem the most alienated from social norms and Cluster 4 the most socialized. Clusters 1, 2, and 3 patients seem to differ little from each other and score near the total sample midpoint on the dimension.

Despite our ability to delineate five meaningful PRF-based subtypes of chemically dependent patients with significant correlates on the MMPI and the CUS, the subtypes do not generally resemble the Nerviano and Zivich[32–34] clusters. There are several probable reasons for this. Our subject population was of considerably higher social status, being largely employed, included drug-only and dually dependent patients not simply pure alcoholics, was considerably younger, and included female patients. Our analytic method was also different. With the BMDP KM clustering program we classified all but 14 of our patients (98%), whereas the TYPOL procedure employed in the earlier studies classified only about half of the cases. We also used all of the PRF trait scales, whereas the other studies employed only 11 of them and Social Desirability. We believe that our approach is preferable because it included women as well as alcoholics with collateral drug dependence (an increasingly common treatment population) and was less discretionary in setting the PRF trait cut-points to establish clusters.

Consideration of PRF factors underlying subtype membership and CUS correlates of subtypes leads us to believe that the subtypes would differentially benefit from alternative chemical dependency interventions.

Cluster 1 patients would be expected to do well with treatments that encourage personal goal setting and appropriate assertion skills to include training on how to resist peer pressure to abuse substances. Training in self-reinforcement activities that do not involve substance abuse may also be beneficial.

Clusters 2 and 3 are lowest on self-reported stimuli of abuse and are nearer the midpoints of two MMPI dimensions. Neither do the PRF factor scores underlying membership to group 2 suggest particular treatment needs. These patients may do well with less intensive interventions such as outpatient treatment or involvement in peer-support groups such as AA or NA. Cluster 3 patients, however, are probably socially isolated and might well benefit from social skills training and interactional therapies.

Cluster 4 patients would likely profit from techniques to diminish depression. These might include cognitive behavioral therapy, cognitive restructuring, and rational emotive therapy. Social skills training might also assist them in developing an emotionally supportive peer group.

Cluster 5 patients might be expected to do best in treatment regimens that encourage spontaneity. Gestalt techniques might be considered. At the same time it is important to help these patients learn to curb their desires and attempts to harm others.

This is the first study of PRF substance abuser profiles to include female patients. Even though PRF scores were standardized separately for male and female subjects, 53% of the female subjects were assigned to Cluster 5. It is therefore not surprising that gender played a significant role in discriminating the five clusters for both CUS and MMPI correlates. Although the small number of female subjects in the present sample precludes a separate analysis for female subjects, future studies will need to consider gender effects in formulating clinically useful typologies.

As noted earlier, a variety of patient characteristics might serve as the basis for taxonomies for chemically dependent patients, and determination of the most useful domain of patient characteristics is ultimately a function of how effective the resulting taxonomy is in specifying choice of intervention. Personality variables likely contribute to the patient's ability to establish rapport with the therapist, proclivity to affiliate with extra-treatment peer support groups such as AA, motivation to persevere in treatment, tendencies to seek long-term *versus* short-term changes in behavior, willingness to objectively analyze one's own behavior, and so forth. So too, considerably more research has been done on personality traits than on other variables that would appear to be reasonable contenders for taxonomizing such as patient motivation, high risk stimuli, or needs for treatment services.

A potential difficulty with most bases of chemical-dependent patient taxonomies, including personality measures, is influence of recent chemical use or detoxification. Research is needed to determine the extent of these influences on patient responses and to specify optimal points in treatment for assessment.

Secondly, it should be borne in mind that the three instruments employed in the current investigation all involve structured written self-report and, hence, share considerable methods variance. Subjects' needs for consistency in description of internal states may underlie some of the relationship across the three measures as well as reflect common patterns of attribution. Unfortunately, other assessment methods such as physiological or performance measures to assess the dimensions are yet unavailable. If the predicted interactions between personality subtypes of abusers and alternative interventions occur, this issue would become largely moot.

Research is needed on a variety of other issues that concern the use of personality measures as bases for treatment-relevant patient taxonomies. For example, it is important to understand the relationship of personality variables to substance abuse dynamics, use patterns, treatment prognosis, and threats to maintaining abstinence. The work of Cloninger and von Knorring represents an important step in this regard. These researchers have posited thought-provoking parallels between personality needs, neurophysiological response patterns, genetic vulnerability to alcohol dependence, severity of drinking consequences, and treatment prognosis. Unfortunately studies have yet to be conducted linking their proposed typology to choice of treatment.

Inasmuch as temperament and basic personality need patterns are established early in life, subtyping of individuals before development of chemical dependence might have important implications for prevention programs as well as treatment programs.

It would also be of interest to explore whether changes in substance abuse are reflected in changes in personality needs as well as in the measures of those needs.

Although the topic has as yet received little research attention, substance abuse counselors differ greatly in effectiveness and ability to retain patients in treatment. Extending research in personality subtyping to counselors might further clarify the bases for differential success and also suggest favorable interaction effects between types of patients and types of therapists.

Finally, although the current investigation was performed with individuals suffering alcohol and/or drug problems, it is possible that the PRF-based subtypes are not unique to such individuals either in structure or in prevalence and that personality traits might cluster similarly in other clinical samples or in normal individuals. It would be of interest to determine if the taxonomic scheme transcends the current sample and might have implications for types of behavior change outside of substance abuse remission.

The challenges in developing a clinically useful typology of chemical-dependent patients are considerable. The typology should be based on characteristics of patients that offer relevance to specific choice of treatment. The typology should be comprehensive, ideally assigning all patients to subtypes. At the same time, subtypes should have a high degree of internal homogeneity and considerable cross-group variability. Granted the variety of algorithms for clustering and the sometimes inconsistent results they yield, typologies need to be cross-validated in new samples using differing agglomerative procedures. Finally, the number of subtypes should be relatively small so that their differing needs can be satisfied within a single facility.

We believe that our proposed typology satisfies several of these goals. Nevertheless, the typology must be viewed as speculative until it has been validated by rigorous scientific testing. We believe that the ultimate value of the typology can be demonstrated only by determining that interventions selected on the basis of types yield more favorable outcomes than does employing the same treatment with all patients regardless of cluster membership and that assignment of patients to interventions based on their particular subtype is clearly more beneficial than is intentionally mismatching interventions and subtypes.

SUMMARY

Jackson's Personality Research Form E (PRF)[8] was administered to a large, heterogeneous group of adult inpatients with diagnoses of alcohol and/or drug dependence. Cluster analysis of PRF factor scores yielded five distinct patient subtypes: hostile-dependent, cooperative-nonreflective, socially uninvolved, impulsive-unsociable, and hostile-overcontrolled. The subtypes differed on two MMPI canonical variables and on self-reported high risk factors for substance usage. Implications of the findings for more appropriate matching of patients to chemical dependency interventions are offered.

REFERENCES

1. INSTITUTE OF MEDICINE. 1990. Broadening the base of treatment for alcohol problems. National Academy Press. Washington, DC.
2.. NATIONAL INSTITUTE ON ALCOHOL ABUSE AND ALCOHOLISM. 1990. Alcohol and Health. Seventh special report to the US Congress. DHHS Publication No. ADM 90-1656. U.S. Government Printing Office. Washington, DC.
3. MATTSON, M. E. & J. P. ALLEN. 1991. Research on matching alcoholic patients to treatment: Findings, issues and implications. J. Addict. Dis. 11: 33–49.
4. MCKENLAY, A. P. 1949. Ancient experience with intoxicating drink: Non-classical people. Q. J. Stud. Alcohol 9: 388–414.
5. BABOR, T. F. & Z. S. DOLINSKY. 1988. Alcoholic typologies: Historical evaluation and empirical evaluation of some common classification schemes. In Alcoholism: Origins and Outcome. R. M. Rose & J. E. Barrett, eds.: 245–266. Raven Press. New York.
6. SKINNER, H. A. 1982. Comparison of clients assigned to in-patient and out-patient treatment for alcoholism and drug addition. Br. J. Psychiatry 138: 312–320.
7. MOREY, L. C. & R. K. BLASHFIELD. 1981. Empirical classifications of alcoholism: A review. J. Stud. Alcohol 42: 925–937.

8. JACKSON, D. N. 1984. Personality Research Form Manual, 3rd Ed. Research Psychologists Press. Port Huron, MI.
9. MURRAY, H. A. 1938. Explorations in Personality. Harvard University Press. Cambridge.
10. REDDEN, J. R. & D. N. JACKSON. 1989. Readability of three adult personality tests: Basic Personality Inventory, Jackson Personality Inventory, and Personality Research Form-E. J. Pers. Assess. **53:** 180–183.
11. LABOUVIE, E. W. & C. R. McGEE. 1986. Relation of personality to alcohol and drug abuse in adolescence. J. Consult. Clin. Psychol. **54:** 289–293.
12. BRAASH, R. W. 1989. College student alcohol use: A validation of naturally occurring quantity and frequency of alcohol use subgroups and cluster analytically derived typologies for males and females. Diss. Abstr. Int. **50:** 340B.
13. SCOUFIS, P. & M. WALKER. 1982. Heavy drinking and the need for power. J. Stud. Alcohol **43:** 1010–1019.
14. CARROLL, J. F. X. 1980. Similarities and differences of personality and psychopathology between alcoholics and addicts. Am. J. Drug Alcohol Abuse **7:** 219–236.
15. STARK-ADAMEK, C. & R. O. PIHL. 1980. Personality and non-medical use of drugs. Psychol. Rep. **46:** 103–110.
16. HUBA, G. J., B. SEGAL & J. L. SINGER. 1977. Organization of needs in male and female drug and alcohol users. J. Consult. Clin. Psychol. **45:** 34–44.
17. KAHN, M. 1974. Personality factors in student drug use. J. Consult. Clin. Psychol. **42:** 236–243.
18. GROSS, W. F. & V. J. NERVIANO. 1973a. The use of the Personality Research Form with alcoholics: Effects of age and IQ. J. Clin. Psychol. **29:** 378–379.
19. HOFFMAN, H. 1970. Personality characteristics of alcoholics in relation to age. Psychol. Rep. **27:** 167–171.
20. ALLEN, J. P., V. FADEN & R. R. RAWLINGS. 1992. Relationship of diagnostic, demographic, and personality variables to self-reported stimuli for chemical use. Addict. Behav. **17:** 359–366.
21. LEIGH, G., A. C. OGBORNE & P. CLELAND. 1984. Factors associated with patient dropout from an outpatient alcoholism treatment service. J. Stud. Alcohol **45:** 359–362.
22. SKINNER, H. A. 1981. Statistical approaches to the classification of alcohol and drug addiction. Br. J. Addict. **77:** 259–273.
23. BONYNGE, E. R. & H. HOFFMAN. 1977. Personality measurements in selection of applicants for an alcohol counselor training program. Psychol. Rep. **41:** 493–494.
24. NERVIANO, V. J. & W. F. GROSS. 1976. Loneliness and locus of control for alcoholics: Validity against Murray need and Cattell trait dimensions. J. Clin. Psychol. **32:** 479–484.
25. NERVIANO, V. J. 1973. Cross-validation of some personality correlates of the A-B Therapist Scale among male alcoholics. Psychol. Rep. **32:** 1338.
26. GROSS, W. F. & V. J. NERVIANO. 1973. The prediction of dropouts from an inpatient alcoholism program by objective personality inventories. Q. J. Stud. Alcohol. **34:** 514–515.
27. HOFFMAN, H. 1971. Personality changes of hospitalized alcoholics after treatment. Psychol. Rep. **29:** 948–950.
28. ALLEN, J. P., V. B. FADEN, A. MILLER & R. R. RAWLINGS. 1991. Personality correlates of chemically dependent patients scoring high versus low on the MacAndrew Scale. Psychol. Assess. **3:** 273–276.
29. CARROLL, J. F. X., T. E. MALLOY, D. L. ROSCIOLI & D. R. GODARD. 1981. Personality similarities and differences in four diagnostic groups of women alcoholics and drug addicts. J. Stud. Alcohol **42:** 432–440.
30. MOREY, L. C., H. A. SKINNER & R. K. BLASHFIELD. 1984. A typology of alcohol abusers: Correlates and implications. J. Abnorm. Psychol. **93:** 408–417.
31. WILLIAMS, R. L., K. V. GUTSCH, R. KAZELSKIS, J. P. VERSTEGEN & J. SCANLON. 1980. An investigation of relationships between level of alcohol use impairment and personality characteristics. Addict. Behav. **5:** 107–112.
32. NERVIANO, V. J. 1976. Common personality patterns among alcoholic males. J. Consult. Clin. Psychol. **44:** 104–110.

33. NERVIANO, V. J. 1981. Personality patterns of alcoholics revisited: Delineation against the MMPI and clinical implications. Int. J. Addict. **16:** 723–729.
34. ZIVICH, J. M. 1981. Alcoholic subtypes and treatment effectiveness. J. Counsel. Clin. Psychol. **49:** 72–80.
35. MILLER, P. M. & M. A. MASTRIA. 1978. Alternatives to Alcohol Abuse: A Social Learning Model. Research Press. Champaign, IL.
36. COSTA, P. T. & R. R. McCRAE. 1988. From catalog to classification: Murray's needs and the five-factor model. J. Pers. Soc. Psychol. **55:** 258–265.
37. DAHLSTROM, W. G., G. WELSH & L. E. DAHLSTROM. 1972. An MMPI Handbook. Volume I. Clinical Interpretation. University of Minnesota Press. Minneapolis.
38. BMDP STATISTICAL SOFTWARE, INC. 1990. BMDP Statistical Software Manual, Vol. 2. University of California Press. Berkeley, CA.
39. GRAHAM, J. R. 1987. The MMPI: A Practical Guide, 2nd Ed. Oxford University Press. New York, NY.

The Type A/Type B Distinction

Subtyping Alcoholics According to Indicators of Vulnerability and Severity[a]

JOSEPH BROWN, THOMAS F. BABOR, MARK D. LITT,
AND HENRY R. KRANZLER

University of Connecticut School of Medicine
Alcohol Research Center
Farmington, Connecticut 06030

The ability to assimilate and accommodate new information is said by the famous Swiss psychologist Jean Piaget to be the basis of human learning.[1] Babor and Lauerman,[2] in a review of typologies of alcoholics dating from 1850 through 1941, identified four major criteria that account for the content of almost all of the 39 typologies: dependence or addiction severity, consumption pattern, chronicity, and hypothesized etiology.

The relative importance assigned to these different characteristics will likely vary with the purpose of the typology. One purpose for attempting to classify the population of alcoholics is to arrive at more homogeneous groups suitable for scientific study. False assumptions of homogeneity wreak havoc on the scientific method's ability to test hypotheses by introducing unwanted error variance.[3] Thus, typologies developed solely for scientific inquiry may be of limited clinical value because they usually define a smaller proportion of the population.

Typologies developed for clinical use have the more pragmatic goal of defining subtypes of alcoholics who might benefit from specific forms of treatment. The treatment-matching hypothesis in alcoholism has a long history and has spawned a number of typological schemes.[4–9] The grosser distinctions made by these typologies are unlikely to meet the rigorous needs of the scientific method. Most typologies of alcoholics fall somewhere between these two extremes.

More recent typologies have sought to bridge the gap between scientific rigor and clinical utility, by allowing for the definition of highly homogeneous groups for research purposes while also defining groups that research indicates might be helpful in differentially prescribing treatments. These typologies are notable in their use of multidimensional conceptions of alcoholism, thereby postulating subtypes that encompass multiple defining characteristics. Examples of these newer multidimensional typologies include Cloninger's[10] neurobiological learning model, Morey and Skinner's[11] hybrid model, and Zucker's[12] developmental model.

On the basis of prospective adoption studies, Cloninger[10] and Cloninger *et al.*[13] proposed a neurobiological learning model of alcoholism that distinguishes two genetic subtypes, termed milieu-limited (Type I) and male-limited (Type II). Type I alcoholics have a later onset of alcohol problems, develop psychological rather than physical dependence, and experience guilt feelings about their alcohol use. In contrast, Type II alcoholics manifest alcohol problems at an early age, exhibit spontaneous alcohol-seeking behavior, and are socially disruptive when drinking. Dimensions of personality are hypothesized to account for these different types of alcoholism.

[a]This research was supported in part by grants from the National Institute on Alcohol Abuse and Alcoholism (No. P50AA03510 and AA00143).

23

Morey and Skinner[11] applied cluster analysis to data obtained from an extensive alcohol use questionnaire. Three types of drinkers were identified: (1) early-stage problem drinkers, a fairly heterogeneous group experiencing drinking problems but still free of major symptoms of alcohol dependence; (2) affiliative drinkers, being more socially oriented with a tendency toward daily consumption and moderate alcohol dependence; and (3) schizoid drinkers who were more socially isolated, tended to drink in binges, and reported the most severe symptoms of alcoholism.

Zucker's[12] developmental model postulates four kinds of alcoholism: (1) anti-social alcoholism, hypothesized to have a genetic diathesis, poor prognosis, and early onset of both alcohol-related problems and antisocial behavior; (2) developmentally cumulative alcoholism, in which alcohol dependence is considered primary to any comorbid psychopathology and in which, over the life course, culturally induced drinking processes become sufficiently cumulative to produce alcohol dependence; (3) negative affect alcoholism, considered to occur primarily in women and characterized by the use of alcohol for mood regulation or for enhancing social relationships; and (4) developmentally limited alcoholism, characterized primarily by frequent heavy drinking with a tendency toward remission to social drinking with the successful assumption of adult career and family roles.

More recently another multidimensional typology was proffered by Babor and colleagues.[14] It is derived by cluster analyses of measures representing several dimensions: premorbid risk and vulnerability, severity of dependence and alcohol-related problems, chronicity and negative alcohol-related consequences, and comorbid psychopathology. It identifies two types of alcoholics. Type A alcoholics are characterized by less premorbid risk and vulnerability, later onset, less severe dependence, fewer alcohol-related problems, a less chronic course with fewer negative consequences of alcohol consumption, and less comorbid psychopathology. Conversely, Type B alcoholics present with more premorbid risk and vulnerability, earlier onset, more severe dependence and alcohol-related problems, a more chronic course with more negative consequences of alcohol consumption, and more comorbid psychopathology.

Further analyses by Babor et al.[14] provide good evidence of both discriminative and predictive validity for the typology. Comparison of the two types on measures not included in the original cluster analyses shows significant differences in the expected directions. Specifically, both the male and female Type B alcoholics were significantly more experimental (vs conservative), less controlled, and more tense than their Type A counterparts. A measure of childhood aggressive behavior also discriminated significantly between subtypes, with Type B alcoholics reporting more symptoms of childhood aggression. The two male groups differed significantly with respect to most of the drinking history variables (specifically, daily consumption and times in alcohol treatment), as did the two female groups (specifically, lifetime consumption and daily consumption).

Regarding predictive validity, Babor et al.[14] compared the two types at 12 and 36 months following treatment. At the 12-month follow-up evaluation, the two male cluster groups differed significantly on measures of alcohol consumed on drinking days, symptoms of pathological drinking, alcohol-related social and occupational problems, average Minnesota Multiphasic Personality Inventory (MMPI) scale elevation, average distress experienced in association with different life events, and frequency of drug use. In all comparisons, Type B alcoholics indicated more drinking problems and greater alcohol-related impairment. On a graded measure of global outcome, 64% of Type B males had a relapse and required additional treatment in the 12 months after discharge compared with 45% of Type A males. Results in the 36-month follow-up are similar, with the two male groups again differing on the same measures except for intensity of drinking (ounces consumed per drinking day). By this time, 9% (n = 8) of the Type B group had died compared with 3% (n = 3) of the Type A group. Similar results for the female subtypes were found at the two follow-ups, but the differences were not as pronounced.

The Babor *et al.*[14] typology was replicated in an attempt to ascertain its generalizability as well as to test its utility in treatment matching.[5] Subjects were 79 male alcoholic inpatients. Regarding replication, one cluster of patients indicated more familial alcoholism, a younger age of onset of problem drinking, greater dependence on alcohol, and more symptoms of antisocial personality than did the other cluster. This group closely resembled patients labeled Type B alcoholics by Babor *et al.*[14] The second group, similar to the Type A subtype identified by Babor *et al.*,[14] was characterized by lower scores on vulnerability and risk indicators, less severe alcohol problems, a more benign course, and less psychopathology.

The Litt *et al.*[15] study also demonstrated the clinical utility of the typology for treatment matching. Subjects were assigned in cohorts to one of two types of aftercare groups: coping skills training or interactional therapy. Aftercare treatment in both conditions consisted of 90-minute group sessions scheduled weekly for 6 months. Coping skills training[16] consisted of didactic presentations, behavioral rehearsal within group sessions, and homework practice exercises. Interactional group therapy was based on the work of Brown and Yalom.[17] The goal of these groups was to explore the participants' relationships with one another, concentrating on immediate feelings and issues within the group.

It was hypothesized that more severely affected Type B alcoholics would have better outcomes with coping skills treatment and that less severe Type As would do better with interactional group therapy. One dependent variable was the number of heavy drinking days, collected posttreatment, 12 and 24 months after inpatient discharge. No main effect was determined for treatment approach or patient type; however, a significant time effect was found in that drinking levels generally increased over the follow-up periods. A significant patient type X treatment approach interaction was also found, whereby Type B alcoholics who received interactional treatment fared more poorly than did the rest of the patients.

Survival analysis was also employed to test the hypothesis that an effective patient-treatment match would also prolong the time to relapse. When outcome was defined as time to first drinking day, survival analysis showed no significant main effects attributable to either treatment approach or patient type. The interaction of treatment approach and patient type was significant whereby those patients considered "unmatched" (Type A, coping skills; Type B, interactional) tended to relapse much more quickly than did those considered matched to treatment.

Both the original Babor *et al.*[14] typology study and the Litt *et al.*[15] replication and extension employed alcoholic inpatients. Whereas the original publication studied both males and females, the later study examined only males. This study seeks to replicate the typology using outpatient samples of alcohol-dependent subjects of both genders. Furthermore, both previous studies used an extensive battery of assessment measures that would be too inefficient for routine clinical use. The time to collect all the data required for each subject is conservatively estimated at 2.5–3.0 hours. Although this study also collected the same data, the possibility of increasing clinical efficiency is explored by a discriminant function analysis conducted to identify those variables most likely to result in correct cluster membership using only a subset of the most discriminating variables.

METHOD

Alcohol-dependent persons (n = 193) responded to public advertisements seeking research subjects at two study sites. These subjects were used in placebo-controlled studies of buspirone, fluoxetine, and naltrexone as adjuncts to relapse prevention treatment. Inasmuch as data used in this analysis were collected prior to randomization to active *versus* placebo treatments, eventual group membership is not an issue. Mean subject age was 39.7 (SD = 8.9) years for study site I and 40.5 (SD = 9.7) years for study site II. Subjects were predominantly male (148, 76.7%). Inclusion criteria were: (1) alcohol dependence, as defined by the DSM-III-R; (2) 7–30 days of self-reported abstinence from alcohol; and (3) age between 18 and 65 years. Exclusion criteria were: (1) current DSM-III-R dependence diagnosis on other substances, save nicotine; (2) history of psychosis; (3) current risk to self or others; (4) current psychiatric symptoms requiring medication; (5) current use of disulfiram; (6) evidence of significant disease, especially cerebral, thyroid, renal, or cardiac diseases; (7) history of cirrhosis; and (8) laboratory evidence of hepatocellular injury. Additionally, women were excluded if they were pregnant, nursing, or not using a reliable form of birth control. Subjects for the naltrexone study were also excluded for opioid abuse history.

Measures

Indicators of Vulnerability and Risk. (a) Familial alcoholism. This continuously distributed variable represents the proportion of all first degree relatives, excluding offspring, who had a serious alcohol problem. The buspirone and fluoxetine samples utilized a 45-minute family history interview.[18] The naltrexone sample used a similar interview. Only comparable data for all three samples were used. (b) Childhood behavior problems were collected by self-report questionnaire using 26 items. (c) Onset of problem drinking is computed as the average age for three "milestone" events: age at which regular drinking began, age at which getting drunk regularly began, and age at beginning of heaviest drinking, all collected by self-report questionnaire.

Severity of Dependence and Other Related Behaviors. (a) Ounces of alcohol consumed per day. An estimate of the average daily amount of pure ethanol consumed during the 6-month period before treatment was computed from a self-report questionnaire[19] on the quantity and frequency of drinking beer, wine, and spirits. (b) Self-medication. A measure of alcohol consumption to relieve withdrawal and psychological distress was computed from nine self-report items rating the importance of relief drinking during the previous 6 months. (c) Dependence syndrome. Severity of alcohol dependence was measured on a 17-item scale representing the behavioral, cognitive, and physiological elements of the alcohol dependence syndrome experienced within the preceding 6 months, as reported via self-report questionnaire. (d) Frequency of benzodiazepine use during the preceding 6 months, as reported by a retrospective questionnaire. (e) Polydrug use. A measure of drug use was constructed from questionnaire items assessing amphetamines, barbiturates, marijuana, opiates, and cocaine. Each substance was coded in terms of its frequency of use (1 = never, 5 = daily) in the prior 6-month period. Item scores were summed to provide an estimate of the total frequency of other substance use.

Chronicity and Consequences of Drinking. (a) Medical problems. A summary of alcohol-related medical symptoms was obtained from 26 self-report items. (b) Physical consequences of alcohol consumption were determined by the frequency of five alcohol-related physical consequences (e.g., cognitive confusion and impotence) during the previous 6 months.[20] (c) Social consequences. The frequency of alcohol-related social problems (e.g., job trouble, family complaints, arrests, violence, and accidents) occurring during the previous 6 months was assessed using a six-item scale.[20] (d) Lifetime severity. Cumulative alcohol-related symptoms over the course of an individual's drinking career were estimated by the total (weighted) score obtained from the Michigan Alcoholism Screening Test (MAST).[21]

Psychopathology. (a) Depressive symptom count. For the buspirone and fluoxetine samples the number of all depressive symptoms as measured by the computerized Diagnostic Interview Schedule (C-DIS)[22] was computed to estimate the lifetime severity of depressive symptoms. For the naltrexone sample the DIS was administered by trained interviewers, and the number of DSM-III depression criteria met was recorded. Because of this minor discrepancy between samples, these measures were standardized within samples before collapsing across samples, where the locally standardized measures were again standardized to control any site-specific variance. (b) Antisocial personality (ASP) symptom count. A frequency count of the number of ASP symptoms was measured by the DIS (for the naltrexone sample) and the C-DIS (for the buspirone and fluoxetine samples). Alcohol abuse items were excluded from these symptom counts. (c) Anxiety severity. The total score obtained using the Taylor Manifest Anxiety Scale derived from the MMPI was employed as a measure of general anxiety. (d) Dependence liability was measured by the MacAndrew Alcoholism Scale[23] derived from the MMPI.

RESULTS

Replication

Gender-specific K-means cluster analyses were performed using the variables defined above to sort cases into two groups. TABLE 1 shows the number of males and females assigned to each of the two clusters. Most of the females ($n = 43$, 95.6%) were assigned to one cluster, which is very similar to Type A alcoholics defined by Babor *et al.*,[14] whereas only two (4.4%) were assigned to the cluster similar to Type B alcoholics. Males were more equitably distributed between the two clusters. Still, the majority ($n = 108$, 73.0%) were assigned to the cluster that closely resembles Babor *et al.*'s[14] Type A group. The smaller male cluster closely resembles Babor *et al.*'s[14] Type B cluster and accounted for the remaining 40 (27.0%) males.

The two clusters were differentiated by measures in each of the four categories of defining characteristics for both genders. TABLE 2 presents means and standard deviations for each of the clustering variables for the 148 males. Results of one-way analysis of variance for each variable are shown in the table. As in the Babor *et al.*[14] Type As, the larger male cluster reported less familial alcoholism and fewer child-

TABLE 1. Gender and Cluster Membership

Gender	Type A	Type B	Total
Male	108	40	148
Female	43	2	45
Total	151	42	

TABLE 2. Profiles of Empirically Derived Male Alcoholic Subtypes

	Cluster Membership				
	Type A (n = 108)		Type B (n = 40)		
	73%		27%		
Defining Characteristics	Mean	SD	Mean	SD	F Value
Vulnerability and Risk Factors					
Familial alcoholism	0.21	(0.23)	0.31	(0.25)	5.13*
Childhood behavior problems	0.39	(0.20)	0.58	(0.19)	28.67***
Onset of problem drinking	24.26	(6.71)	22.48	(6.18)	2.15
Dependence Severity and Other Related Behaviors					
Oz. ETOH per day	9.18	(9.62)	9.83	(9.02)	0.13
Self-medication	8.49	(4.95)	7.24	(6.20)	1.61
Dependence severity	1.32	(0.49)	2.17	(0.66)	73.77***
Benzodiazepine use	0.10	(0.31)	0.28	(0.45)	7.02**
Polydrug use	0.57	(0.94)	0.73	(0.99)	0.73
Chronicity and Consequences					
Medical problems	0.36	(0.22)	0.69	(0.25)	58.29***
Physical consequences	0.55	(0.47)	1.21	(0.68)	45.10***
Social consequences	0.35	(0.37)	1.21	(0.88)	68.79***
Lifetime severity	19.39	(8.78)	36.50	(7.85)	114.01***
Psychopathology					
Depressive symptoms	−0.28	(0.80)	0.47	(1.10)	18.80***
Antisocial personality symptoms	−0.21	(0.77)	0.81	(1.26)	30.87***
Anxiety	18.46	(9.10)	27.80	(8.25)	31.56***
Dependence liability	22.27	(3.77)	27.03	(3.27)	49.59***

*p < 0.05; **p < 0.01; ***p < 0.001.

hood behavior problems than did their Type B counterparts. This cluster also demonstrated less dependence severity and other related problems. They reported fewer dependence symptoms in the preceding 6 months than did the Type B cluster. The two male clusters did not differ on reported non-benzodiazepine drug use, but they did differ on reported benzodiazepine use. The direction of this difference indicates that the Type A cluster reported fewer benzodiazepine use than did the Type B cluster. The Type A cluster also exhibited less chronicity and consequences of alcohol consumption. They reported less alcohol-related medical problems, physical consequences, and social consequences as well as less lifetime symptoms as measured by the MAST than did the Type B cluster. Additionally, the Type A cluster also manifested fewer symptoms of psychopathology. They reported significantly fewer depressive symptoms, less anxiety, fewer antisocial personality disorder symptoms, and less dependence liability than did the Type B cluster. Results of the cluster comparison for females are not interpretable given the exceedingly low number of subjects assigned to the Type B cluster (n = 2).

Clinical Efficiency

With only the larger and more equitably distributed male sample, a stepwise discriminant function analysis was performed using cluster membership as the dependent variable. The 16 variables defining the four general concepts underlying the Babor *et al.*[14] typology were used as potential classifying variables. The analysis sought to minimize Wilks' Lambda, using a criterion F value of 3.84 for entry into the discriminative formula. Subjects having missing values on any of the potential discriminating variables were deleted from the analysis. This left 104 subjects, 75 (72%) being classified as Type A by the cluster analysis and the remaining 29 (28%) as Type B. Please note that these proportions are very similar to those discovered by cluster analysis using all subjects regardless of missing values, thus indicating no differential loss of subjects between the two cluster groups. These proportions also served as prior probabilities of group membership in the discriminant function analysis.

TABLE 3 presents the variables entered into the discriminant function analysis before termination as well as their associated Wilks' Lambda and F values. Only five steps were completed before the termination criterion was reached. The variables chosen during these five steps represent each of the four general concepts in the Babor *et al.*[14] typology. The first variable to enter, representing Chronicity and Consequences, was lifetime severity as measured by MAST ($F[1,102] = 76.63$, $p < 0.0001$). The second variable to enter, also representing Chronicity and Consequences, was alcohol-related medical problems ($F[1,102] = 20.49$, $p < 0.001$). The third variable entered, representing Psychopathology, was antisocial personality disorder symptoms ($F[1,102] = 14.57$, $p < 0.001$). Childhood behavior problems, representing Vulnerability and Risk Factors, entered in the fourth step ($F[1,102] = 5.13$, $p < 0.05$). The last variable to enter was severity of dependence symptoms within the previous 6 months, representing Dependence Severity and Other Related Behaviors ($F[1,102] = 4.94$, $p < 0.05$). Analysis ceased at this point because none of the remaining nine variables could muster the F-criterion value of 3.84 for inclusion.

Comparison of group membership, as predicted by discriminant function analysis to group membership as defined by cluster analysis, showed a 94.3% correct classification rate for discriminant function analysis. Of the 122 subjects who had no missing values on any of the five identified discriminating variables, 115 were correctly classified. Of the 87 Type As identified by cluster analysis, 85 (97.7%)

TABLE 3. Results of Discriminant Function Analysis

Variables in Order of Entry	Wilks' Lambda	F Value
Chronicity and Consequences		
Life severity	0.57	76.63***
Medical problems	0.47	20.49**
Psychopathology		
Antisocial personality disorder symptoms	0.41	14.57**
Vulnerability and Risk Factors		
Childhood behavior problems	0.39	5.13*
Dependence Severity and Other Related Problems		
Dependence severity	0.38	4.94*

*$p < 0.05$; **$p < 0.001$; ***$p < 0.0001$.

were also so classified by discriminant function analysis. Of the 35 Type Bs identified by cluster analysis, 30 (85.7%) were also so classified by discriminant function analysis.

DISCUSSION

Replication of the Type A/Type B typology of Babor et al.[14] was successful, at least for male subjects. Owing to the disproportionately low number of female subjects identified as Type B, the success of this replication is still in doubt for female subjects. TABLE 4 shows the results of the present study and the two previous studies[14,15] in terms of significant differences found on the clustering variables. Variables representing each of the four concepts in the Babor et al.[14] typology showed significant differences between the two clusters in each of the three studies. Review of this table shows the cluster structures to be similar across the three studies. In the Vulnerability and Risk Factor measures, only age at onset of problem drinking

TABLE 4. Comparison of Three Typology Studies

	Babor et al.[14]		Litt et al.[15]		This Replication	
	Type A ($n = 107$)	Type B ($n = 121$)	Type A ($n = 38$)	Type B ($n = 41$)	Type A ($n = 108$)	Type B ($n = 40$)
Vulnerability and Risk Factors						
Familial alcoholism	−	+	−	+	−	+
Childhood behavior problems	−	+	−	+	−	+
Onset of problem drinking	+	−	+	−	=	=
Dependence Severity and Other Related Behaviors						
Alcohol consumption	=	=	−	+	=	=
Self-medication	−	+	−	+	=	=
Dependence severity	−	+	−	+	−	+
Benzodiazepine use	−	+	?	?	−	+
Other drug use	−	+	=	=	=	=
Chronicity and Consequences						
Medical problems	−	+	−	+	−	+
Physical consequences	−	+	?	?	−	+
Social consequences	−	+	−	+	−	+
Life severity	−	+	−	+	−	+
Psychopathology						
Depressive symptoms	−	+	−	+	−	+
Antisocial personality symptoms	−	+	−	+	−	+
Anxiety	−	+	−	+	−	+
Dependence liability	=	=	−	+	−	+

+Mean value for the variable is significantly greater than that for the companion cluster.
−Mean value for the variable is significantly less than that for the companion cluster.
=Mean values for the variable are not significantly different.
?A comparable variable was not utilized in the study.

did not differentiate between outpatient clusters. In both inpatient studies, Type Bs had a significantly earlier onset of problem drinking than did Type As. The single other case in which the outpatient sample differed from both inpatient samples was in the tendency toward alcohol consumption for self-medication from withdrawal symptoms. Although both inpatient studies found that Type Bs report more self-medicating consumption, the outpatient study found no such difference between the alcoholic subtypes. It is obvious that the Type A/Type B typology is meaningful in this outpatient sample, at least for males.

As expected, a larger proportion of Type A alcoholics was identified in this replication than in either the original research[14] or the first replication.[15] In the Babor *et al.*[14] study, 46.9% (107) of the 228 male subjects were classified as Type A alcoholics, with the remaining being classified as Type B alcoholics. The Litt *et al.*[15] research classified 48.1% (38) of the 79 male subjects as Type A alcoholics. This replication classified 73% (108) of the males as Type A alcoholics and 27% (40) as Type B alcoholics. The Babor *et al.*[14] study also identified 62.4% (53) of the females as Type A and 37.6% (32) as Type B. Thus, even using an inpatient sample results in a decreased proportion of women who were identified as Type B alcoholics. This replication identified only 2 (4.4%) of 45 females as Type B alcoholics, which replicates the finding of less Type Bs in female samples. This difference in relative proportion of Type A and Type B alcoholics, although expected from an outpatient sample, is probably an overestimate of the general treatment-seeking population. It is expected that some Type B subjects were excluded on the basis of exclusion criteria which were specific to pharmacologic research, especially those excluded on the basis of a positive history of cirrhosis or evidence of severe hepatocellular injury. For females, an additional exclusion criterion of refusal to use a reliable birth control method may have added to the low number of Type Bs identified.

Regarding the attempt to define a more efficient subset of measures for clinical use, the five variables selected by discriminant function analysis resulted in a correct classification rate of almost 95%. The time savings from this reduction in data collection are estimated to be about 2 hours. This time savings is expected to be clinically significant in both outpatient settings as well as inpatient units where the pressure to shorten lengths of stay continues to result in shorter and shorter stays with increasingly less time available for assessment. This is a preliminary finding, however, and needs to be replicated with a more heterogeneous sample containing both inpatients and outpatients of both genders. Furthermore, some shrinkage in the correct classification rate would be expected in other independent samples.

Only one aspect of the clinical efficiency issue is examined in this paper. Although data collection is time consuming, other clinical efficiency issues that need to be explored are questions of scoring the typology so as to permit classification with a minimum of mathematics. The use of a discriminant function formula is probably beyond the scope of many clinics. Unit weighting should be explored for both clinical efficiency and the possibility of more reliable and valid classifications.[24] Notably, the classification function coefficients derived from this sample resulted in all positive values for the Type B group, ranging from a low of 0.725 to a high of 1.463. Conversely, classification function coefficients for the Type A group were all negative, ranging from −0.177 to −0.669. These values suggest that a unit weighting scheme is likely to be effective. Again, further research will be necessary to determine the actual usefulness of unit weighting.

Other possible research directions include examining the typology for convergent validity. In part this was done in the original Babor *et al.*[14] study, where other measures thought to be related to the clustering variables were found to indicate similar between-group differences. More testing of predictive validity vis-à-vis treatment matching is also necessary before widespread clinical use of the typology can be suggested. Given that these samples were collected for testing medications, one obvious research direction is to evaluate the potential for matching subtypes of alcoholics with specific medications. This possibility relates directly to neuropharmacologic hypotheses developed by Cloninger[10] to explain dimensional differences between alcoholic subtypes. Alternative means to scoring and classifying subjects should also be investigated, including the potential to replace unit weighting or more complicated classification schemes with the use of Bayes' Diagnostic Inference Equation.

In conclusion, this replication and extension suggest that although progress has been made in developing a useful typology for both the researcher and the clinician, further research is required.

SUMMARY

Multidimensional typologies of alcoholics are reviewed, including Cloninger's[10] neurobiological learning model, Morey and Skinner's[11] hybrid model, and Zucker's[12] developmental model. The more recent Type A/Type B typology proposed by Babor and colleagues[14] is reviewed in more depth, as is a previous replication and extension by Litt and colleagues.[15] Both the original study and the replication indicate this typology is a useful tool in classifying alcoholic inpatients into two groups and in matching alcoholics to the most suitable treatment. The present study replicates the typology using outpatient samples of male alcoholics. The resulting two clusters are very similar to those identified by the two earlier studies. As expected, the relative proportion of Type A alcoholics is higher in the outpatient samples than in the previously studied inpatient samples. Preliminary analysis of the typology's clinical efficiency suggests that the variables used to classify subjects might be appreciably reduced, thus effecting a considerable time savings in assessment. A discriminant function analysis indicates that using only 5 of the original 16 clustering variables results in a correct classification rate of almost 95%. Future research directions are addressed.

ACKNOWLEDGMENT

The authors thank Drs. Stephanie O'Malley, Bruce Rounsaville, and Adam Jaffe for their assistance with data collection.

REFERENCES

1. PHILLIPS, J. L. 1969. The Origins of Intellect: Piaget's Theory. W. H. Freeman & Co. San Francisco.
2. BABOR, T. F. & R. LAUERMAN. 1986. Classification and forms of inebriety: Historical antecedents of alcoholic typologies. In Recent Developments in Alcoholism, Vol. V. M. Galanter, ed.: 113–114. Plenum Press. New York.
3. BUCHSBAUM, M. S. & R. O. RIEDER. 1979. Biologic heterogeneity and psychiatric research. Arch. Gen. Psychiatry 36: 1163–1169.
4. BROWN, J. & J. P. LYONS. 1981. A progressive diagnostic schema for alcoholism with evidence of clinical efficacy. Alcoholism: Clin. & Exp. Res. 5: 17–25.

5. EWING, J. A. 1977. Matching therapy and patients. Br. J. Addict. **42:** 13–18.
6. GIBBS, L. E. 1980. A classification of alcoholics relevant to type-specific treatment. Int. J. Addict. **15:** 461–488.
7. GLASER, F. B. 1980. Anybody got a match? Treatment research and the matching hypothesis. *In* Alcoholism Treatment in Transition. G. Edwards & M. Grant, eds.: 178–196. University Park Press. Baltimore, MD.
8. GOTTHEIL, E., A. T. MCLELLAN & K. A. DRULEY. 1981. Matching Patient Needs and Treatment Methods in Alcoholism and Drug Abuse. Charles C Thomas. Springfield, IL.
9. O'LEARY, M. R., D. M. DONOVAN, E. F. CHANEY & D. E. O'LEARY. 1980. Relationship of alcoholic personality subtypes to treatment follow-up measures. J. Nervous & Mental Dis. **168:** 475–480.
10. CLONINGER, C. R. 1987. Neurogenetic adaptive mechanisms in alcoholism. Science **236:** 410–416.
11. MOREY, L. C. & H. A. SKINNER. 1986. Empirically derived classifications of alcohol-related problems. *In* Recent Developments in Alcoholism, Vol. 5. M. Galanter, ed. : 145–168. Plenum Press. New York.
12. ZUCKER, R. A. 1987. The four alcoholisms: A developmental account of the etiologic process. *In* Alcohol and Addictive Behavior. P. C. Rivers, ed.: 27–83. University of Nebraska Press. Lincoln, Nebraska.
13. CLONINGER, C. R., M. BOHMAN & S. SIVARDSSON. 1981. Inheritance of alcohol abuse. Cross-fostering analysis of adopted men. Arch. Gen. Psychiatry **38:** 861–868.
14. BABOR, T. F., F. K. DEL BOCA V. HESSELBROCK, R. E. MEYER, Z. S. DOLINSKY & B. ROUNSAVILLE. 1992. Types of Alcoholics I: Evidence for an empirically derived typology based on indicators of vulnerability and severity. Arch. Gen. Psychiatry **49:** 599–608.
15. LITT, M. D., T. F. BABOR, F. K. DEL BOCA, R. M. KADDEN & N. L. COONEY. 1992. Types of alcoholics II: Application of an empirically derived typology to treatment matching. Arch. Gen. Psychiatry **49:** 609–614.
16. MONTI, P. M., D. B. ABRAMS, R. M. KADDEN & N. L. COONEY. 1989. Treating Alcohol Dependence: A Coping Skills Training Guide. Guilford. New York.
17. BROWN, S. & I. D. YALOM. 1977. Interactional group therapy with alcoholics. J. Studies on Alcohol **38:** 426–456.
18. HESSELBROCK, V. M., J. R. STABENAU, M. N. HESSELBROCK, R. E. MEYER & T. F. BABOR. 1982. The nature of alcoholism in patients with different family histories for alcoholism. Progr. Neuro-psychopharmacol. Biol. Psychiatry **6:** 607–614.
19. HESSELBROCK, M. N., T. F. BABOR, V. M. HESSELBROCK, V. M. MEYER & K. WORKMAN. 1983. 'Never believe an alcoholic?' On the validity of self-report measures of alcohol dependence and related constructs. Int. J. Addiction **18:** 593–609.
20. BABOR, T. F., N. L. COONEY & R. J. LAUERMAN. 1987. The dependence syndrome concept as a psychological theory of relapse behavior: An empirical evaluation of alcoholic and opiate addicts. Brit. J. Addiction **82:** 393–405.
21. SELZER, M. L. 1971. The Michigan Alcoholism Screening Test: The quest for a new diagnostic instrument. Am. J. Psychiatry **12:** 1653–1658.
22. ROBINS, L. N., J. E. HELZER, J. CROUGHAN & K. S. RATCLIFF. 1981. The NIMH Diagnostic Interview Schedule: Its history, characteristics, and validity. Arch. Gen. Psychiatry **38:** 381–389.
23. MACANDREW, C. 1981. What the MAC scale tells us about men alcoholics. An interpretive review. J. Stud. Alcohol **42:** 604–625.
24. COHEN, J. 1990. Things I have learned (so far). Am. Psychol. **45:** 1304–1312.

Sex, Gender, and Alcoholic Typologies[a]

FRANCES K. DEL BOCA

Department of Psychiatry
University of Connecticut School of Medicine
Farmington, Connecticut 06030

Classification systems for subtyping alcoholics have a long history in alcohol research. The typological approach promises parsimonious theoretical explanations for the etiology of alcoholism in the large and diverse population of affected individuals. Moreover, the prospect of treatment-matching effects, that is, improved outcomes as a result of matching subsets of clients to appropriate types and intensities of care, is another potential benefit. Although less frequently mentioned, an enhanced understanding of the different pathways that lead to alcoholism may promote the development of prevention and intervention efforts tailored to particular subgroups with different vulnerabilities for the development of alcohol problems.

Although females have been understudied in alcohol research, sex and gender have been explicitly or implicitly implicated in typological schemes since investigators began to conceptualize alcoholics in these terms. The present paper selectively reviews research findings that may be relevant to understanding the role of sex and gender in alcoholic subtyping and offers a heuristic framework for integrating existing findings and guiding future research. Although the topics of sex and gender concern males as well as females, research that pertains more directly to the development of alcohol-related problems in women is emphasized. New data analyses are reported; however, the major aims of this paper are to highlight issues and to formulate hypotheses. Particular attention is focused on research methods and analytic procedures that are appropriate for addressing the questions and hypotheses presented.

THE EXISTING KNOWLEDGE BASE

Although research on women and alcohol has lagged behind that on men, there is a large literature on the characteristics and clinical needs of female alcoholics and a growing literature that deals with the processes through which women and men develop alcohol problems. Several recent literature reviews provide excellent summaries of the research data and identify key areas where additional work is needed.[1-4] For the purposes of the present paper, five areas of study are pertinent: (1) sex differences relating to alcohol consumption, concomitant problems, and consequences; (2) alcohol subtyping in relation to sex and gender; (3) the role of stress, negative mood states, and psychopathology as antecedents to drinking; (4) the role of sociocultural and situational factors in the etiology of alcohol problems; and (5) the effect of "gendered" psychological variables (e.g., sex-role attitudes) on alcoholism. The first two areas are relevant to explicating the structure and organization of attributes that may be differentially associated with each sex or with particular alcoholic subtypes. Findings in each of the five areas are presented selectively to illustrate the

[a]Support for this paper was provided by a grant from the National Institute on Alcohol Abuse and Alcoholism (AA03510).

limitations and strengths of the existing knowledge base and to inform the development of hypotheses that relate sex and gender to alcoholic subtypes.

SEX DIFFERENCES IN ALCOHOL USE, CONCOMITANT PROBLEMS, AND CONSEQUENCES

The following are frequently cited findings regarding sex differences in alcohol consumption, concomitant problems, help seeking, and treatment outcome:

1. Although females are believed to have a later age of onset for alcohol disorders than males, they appear to have a greater biological vulnerability to alcohol effects. The progression of negative physical, medical, and social consequences stemming from alcohol consumption is more rapid and more severe for females than for males, a phenomenon referred to as "telescoping.[5,6]

2. Women who seek treatment have more ancillary problems than do men, including financial, family, and medical problems. Treatment is sought more for a problem that may be related to alcohol consumption than for alcohol or substance use disorders *per se*.[7,8]

3. Female alcoholics frequently evidence psychiatric and emotional problems. A high proportion have primary affective disorders, with depression particularly common.[9] Psychologically, alcoholic women report feeling more guilty and anxious than do their male counterparts, and they tend to report lower feelings of self-worth.[10]

4. Major reviews of treatment outcome research on alcoholic women agree that there is little *direct* evidence regarding the differential effectiveness of various therapeutic modalities or treatment components.[3,11,12] This conclusion is generally not the result of failure to demonstrate differences between therapies in treatment outcome research. Rather, it is more the consequence of a lack of systematic research on treatment efficacy for female alcoholics. Without differentiating among therapeutic modalities, reviewers conclude, however, that female alcoholics clearly benefit from treatment, despite their greater concomitant psychological, social, and medical problems. In some studies, they fare better at follow-up than do their male counterparts.[11,13]

5. Although there is little direct evidence regarding gender as a treatment-matching characteristic, there is agreement that alcoholic women tend to have specific material needs and psychological problems that might be best addressed with particular treatment programs. Many of these concern the circumstances of women (e.g., the need for child care), but others (e.g., lack of self-assertion skills) relate more directly to targets of therapeutic intervention.

In sum, sex differences have been documented across a variety of significant dimensions. However, several limitations to existing data should be noted. First, specific studies typically compare the sexes in terms of a relatively small number of variables within a particular domain (e.g., psychopathology). A second set of problems concerns research subjects. Sample sizes, particularly for women, tend to be small. Additional problems of interpretation arise from the preponderance of clinical samples in the literature; observed differences between the sexes in treatment samples may not generalize to population samples. This seems particularly likely for female alcoholics, given the existing data regarding treatment-seeking patterns. Third, in many studies client sex is confounded with other variables that may offer alternative explanations for observed sex differences. A small number of studies have attempted to deal with this problem by comparing samples of alcoholic subjects to "control" groups of nonalcoholic women and men. Although there are excellent studies that attempt to untangle the unique contribution of sex to alcohol use,[14,15] most comparisons of males and females have not included statistical controls for age, socioeconomic status, or other factors potentially confounded with sex within a particular sample.

In part because relatively little work has focused on women, investigators have tended to devote more attention to identifying sex differences between male and female alcoholics than to explaining differences that have been observed. In some instances, the implicit assumption exists that "sex" itself constitutes an explanation for observed differences, especially when findings parallel those found in non-alcoholic samples. Whether "sex" is construed as a biological input, a socially defined status, or some combination of these is often unclear.

ALCOHOL SUBTYPING IN RELATION TO SEX AND GENDER

Since the seminal work of Jellinek,[16] a large number of studies concerned with alcoholic subtypes have appeared in the literature.[17–19] Most empirical studies, however, were based on samples of male alcoholics. In a review published only a decade ago, only 6 of 26 studies cited included women in their samples, and 4 of these relied primarily on male subjects (i.e., men comprised 78% or more of the study sample).[17]

Several studies focus exclusively on women. These studies generally have attempted to identify two or more female-specific subtypes based on measures of personality or psychopathology. For example, Hart and Stueland[20] used Cattell's 16 Personality Factors instrument to classify alcoholic women, and Eshbaugh and colleagues[21] used the Minnesota Multiphasic Personality Inventory (MMPI) for this purpose. Perhaps the most influential work relating personality to alcoholic subtypes in women was reported by Schuckit,[22,23] who proposed at least two female subtypes: primary alcoholics and those for whom alcoholism was secondary to affect disorder. This conclusion is consistent with numerous studies documenting sex differences in psychopathology in both population and clinical samples of alcoholics.[9,24] Nevertheless, the focus on personality and psychopathology to the exclusion of other potentially important factors indicates clearly that more research is needed.

Taking a somewhat different tack, Cloninger[25] proposed two major types of alcoholics based on a theory of neurological learning: a "male-limited" subtype (Type II) that includes alcoholics with a specific genetically determined temperament profile, and a "milieu-limited" subtype (Type I) that includes members of both sexes. Like other writers, Cloninger views the "true" genetically based form of alcoholism as a disorder confined primarily to men. Other forms of alcoholism are thought to arise from an extended history of drinking, with levels of alcohol consumption determined by a variety of social and situational factors.

Babor and colleagues[14] investigated the predictive utility of five different single-dimension typology schemes, including gender. Patient sex had considerable prognostic value; however, this attribute tended to overlap with characteristics in other typology systems, and those other factors were also predictive of outcome. On the basis of these findings, the authors proposed the development of a synthetic typology system in which subtypes would be empirically derived and based on a broader range of defining attributes. In a subsequent study,[26] 17 variables including risk factors for alcoholism (e.g., parental alcoholism, childhood conduct problems), patterns of alcohol and other drug use (e.g., ounces of alcohol consumed per day), measures of the chronicity and consequences of drinking (e.g., medical problems related to alcohol use, years of heavy drinking), and psychiatric symptoms (e.g., symptom counts for antisocial personality disorder and depression) were used in K-means cluster analyses to empirically derive alcoholic subtypes. Separate analyses were performed for male and female subjects. The results for both sexes were best represented by a two-cluster solution that partitioned cases into subtype categories. One group, labeled Type A alcoholics, was characterized by later onset, fewer childhood risk factors, less severe dependence, fewer alcohol-related problems, and less psychopathology. The second group, Type B alcoholics, was characterized by familial alcoholism, childhood risk factors, early onset of alcohol-related problems, greater dependence severity, polydrug use, a more chronic treatment history, more life stress, and greater psychiatric symptoms.

This typological structure was consistent with most previous formulations. Type B alcoholics are similar to the primary and male-limited alcoholic subtypes, whereas Type A corresponds more closely to subtypes in which alcoholism is secondary to other forms of psychopathology or determined by social (e.g., the "milieu") factors. With a similar set of input variables, the two-cluster scheme has been replicated in two additional samples of alcoholic men.[27,28]

In sum, there is a growing consensus that there are two major alcoholic subtypes. Although there is clear overlap in the "content" of these subtypes in classification schemes, agreement regarding the key dimensions that differentiate subtypes — psychopathology, neurobiology, or a synthetic composite of factors associated with risk and problem severity — is lacking. Furthermore, relatively little work has been directly concerned with the issues of sex and gender in relation to alcoholic typologies. Research on psychopathology suggests two forms of alcoholism among women (primary and secondary). On the other hand, there is also the suggestion that biologically based primary alcoholism predominates in men. And Babor's empirically derived typology includes two categories of alcoholics that are similar for males and females. Thus, important questions remain unanswered: Are there primary alcoholic females? To what extent is "secondary" alcoholism confined to women? If both sexes evidence both forms of alcoholism, why are the relative proportions of each sex within each subtype different? To what extent do women and men within each subtype tend to differ along important dimensions? Are there different etiologies for each subtype that are somehow related to sex and gender?

STRESS, NEGATIVE MOOD, AND PSYCHOPATHOLOGY AS ANTECEDENTS TO DRINKING PROBLEMS

In part because alcoholic women more often exhibit symptoms of anxiety and depression than do their male counterparts, the notion that women drink to "self-medicate," that is, to cope with stress and to relieve negative affect, has considerable currency among clinicians and investigators.[12] If women and men drink for different reasons, perhaps different pathways to alcoholism account for the differential prevalence of each form of alcoholism within each sex.

Consistent with the self-medication hypothesis, survey research indicates that women often attribute their drinking and other substance use to stressful or traumatic events and that they do so more frequently than do men.[3] Although some evidence suggests that female alcoholics do drink in response to stress, particularly difficult major life transitions,[29] it is important to acknowledge that self-reported attributions for behavior are subject to numerous biases and these are known to differ as a function of sex.[30] Furthermore, several studies have failed to find the expected sex differences in alcohol consumption in response to laboratory-induced stress or anxiety.[15] Thus, as concluded in two reviews of this literature, the findings in this area are more equivocal than is often reported.[1,31]

In clarifying the role that stress and negative affect may play in the development of alcohol problems in women, several considerations need to addressed. Although women and men may report different levels of depression and anxiety, these may in fact bear no relation to alcohol use. Even if negative affect is associated with alcohol *use* in one or both sexes, it may not be related to the development of problem drinking. Furthermore, if a relationship does exist, negative affect may be as much a consequence as a cause of drinking. As already noted, women tend to experience more guilt as a function of drinking than do men, and it may be that drinking produces more stress in their lives.

In investigating these issues, care must be taken in conceptualizing and operationalizing key variables. Drinking behavior and alcoholism are very different criterion measures. Similarly, negative affect is not equivalent to psychopathology, and stress does not always produce anxiety. Laboratory-induced stress may be a poor surrogate for the types of stressful occurrences in everyday life. Investigators also need to acknowledge that the impact of stressful events is known to depend on a variety of other factors, including the coping resources of the individual and the social support available in the immediate environment. Finally, stressful events vary on a number of dimensions (e.g., predictability, controllability), and the relative importance of these for the two sexes has yet to be delineated. Thus, although there are strong hints in the literature that depression and other affective disturbances may be a distinguishing feature of a female alcoholic subtype, considerable research is needed to explicate the nature of the associations between variables and the processes that account for relationships that are observed.

SOCIOCULTURAL AND SITUATIONAL FACTORS
IN THE ETIOLOGY OF ALCOHOL PROBLEMS

Biological sex is a major social stratification variable within virtually all cultures. Different roles and behaviors are prescribed and proscribed on the basis of sex, and it seems probable that sociocultural factors are related to sex differences among alcoholics. Virtually no research has examined the effects of sociocultural and situational factors on the prevalence of different forms of alcoholism within each sex. Although a biological or genetic basis for sex differences in the occurrence of "primary" alcoholism may exist, it seems unlikely that these explanations are adequate for understanding secondary alcoholism in women and men.

The most obvious sociocultural effects would appear to be the most direct. Traditionally, there have been very different social norms regarding male and female alcohol use, particularly heavy drinking. Women report experiencing more social disapproval for alcohol consumption than do men.[32] Perhaps as a result, female use is less public, and the problems associated with alcoholism in women tend to be less visible. In turn, it may become easier for others to ignore or deny drinking problems in women. These factors may partially explain why female alcoholics report that they are discouraged from seeking treatment by their families and friends and why they are referred less frequently for treatment by primary care physicians or their employers.[7,33]

Sociocultural factors may influence alcohol use in other ways. The gap between women and men in levels of reported drinking has narrowed over time. Some investigators have attempted to link the liberalization of traditional sex roles with this broad social trend. Two different mechanisms have been proposed. On the one hand, it is argued that females now have more freedom to engage in "male" behaviors such as heavy drinking; on the other, it is suggested that increases in alcohol consumption among women are a consequence of the added stress associated with filling both work and family roles.[34] The results of the few studies that have attempted to address these possibilities have been equivocal,[35] perhaps because of the difficulty of using cross-sectional data to address questions regarding the effects of social change over time. Furthermore, it seems likely that the effects of sociocultural factors are mediated by a wide range of individual-level variables. One set of potentially important mediators concerns the differential circumstances of women and men that are conditioned by traditional gender roles and by the relative status of the two sexes within society.

GENDERED PSYCHOLOGICAL VARIABLES AND ALCOHOLISM

In addition to their impact on individual circumstances, social expectations regarding sex-appropriate behavior are likely to influence psychological variables that may be differentially related to alcohol problems in women and men. Although a number of such factors exist, the present focus is on variables that are gender-related in that they have obvious connections with sex. Among others, these "gendered" variables include sex-role attitudes (attitudes toward traditional sex roles) and psychological masculinity and femininity (the extent to which the individual possesses attributes that are traditionally associated with males and females, respectively). Despite the repeated observation that sex differences among alcoholics are often consistent with those found more generally, little has been done to examine the association between gender-related variables and alcoholism.

Clear parallels are noted between social definitions of masculinity and femininity, on the one hand, and antisocial personality and depression, on the other. These latter variables have been described as exaggerations of traditional male and female roles.[36] To the extent that these forms of psychopathology are causally related to different forms of alcoholism, gendered personality and attitudinal variables may also have some etiologic significance.

Other sex differences observed in alcoholics may also have some basis in gendered personality. For example, women are more likely than men to have a spouse or significant other who abuses alcohol or other drugs. They more frequently report that they initiated alcohol or drug use at the urging of their partners and that they continue to use as a consequence of their partner's influence.[37–40] Alcoholic women who are well socialized into the traditional female role may lack such traditional masculine traits as self-assertion and be more susceptible to the influence of their partners as a consequence. Thus, there are good reasons to suspect that gendered psychological variables are important mediators of sociocultural factors, but currently little empirical data are available to test this proposition.

Most alcohol research dealing with sex and gender (or with women specifically) has not adopted a typological approach. And with few exceptions, investigators interested in alcoholic typologies have not explicitly considered the role of sex and gender in the development of alcoholism. In the interest of theory and clinical practice, these lines of research need to be integrated. In so doing, investigators should clearly distinguish between differences that are attributable to biological sex and those that are conditioned by gender-related sociocultural and situational factors. Furthermore, in broadening the domain of variables studied, more attention needs to be focused on gendered psychological variables (e.g., sex-role orientation).

INTEGRATING AND EXPANDING THE KNOWLEDGE BASE:
TOWARD A HEURISTIC FRAMEWORK

As the foregoing selective review illustrates, there are already empirical findings that can inform theoretical development regarding the relation of sex and gender to alcoholic subtyping. At the same time, significant gaps in knowledge exist that can only be addressed with further research. The remainder of this paper offers suggestions for approaches to building the knowledge base in a manner that will enhance our understanding of the role that sex and gender play in the etiology of different forms of alcoholism. These recommendations fall into two broad categories. The first concerns the structural representation of alcoholic subtypes; the second concerns the etiologic processes involved in the development of alcoholism in males and females.

ISSUES OF STRUCTURE AND ORGANIZATION

Existing studies suggest that two primary approaches to thinking about sex, gender, and alcohol typologies have been used, approaches that are often more implicit than explicit. The first considers women and men in relation to each other in an effort to determine if female alcoholics constitute a relatively homogeneous alcoholic subtype in their own right. Methodologically, this approach typically involves a simple comparison of female and male alcoholics along one or more dimensions.

The second approach considers the heterogeneity within female or male alcoholic groups. This approach is typically sex-specific; single-sex samples are investigated, or males and females are analyzed separately to examine within-group differences. For women specifically, investigators may emphasize the diversity that characterizes female alcoholics; less often, specific within-sex subtypes are delineated. Studies using this approach, like those that focus on sex differences, are often limited by the number of variables assessed as well as by sample size and selection bias.

Neither of these two approaches alone is sufficient to address the broader theoretical questions and practical clinical issues important for understanding sex and gender in relation to alcohol typologies. Neither considers the similarities and differences between males and females *in the underlying structure of relationships among attribute dimensions.* That is, although mean differences may exist between the sexes, the possibility that patterns of covariation among dimensions may be similar for both sexes is often not explored. In terms of alcoholic subtypes, there may be categories that cross-cut sex as well as subtypes that are sex-specific. In terms of the more general types, the relative proportions of each sex falling into each subtype may differ; the relevance of particular risk factors and etiologic mechanisms may be different for females and males or quite similar regardless of sex.

As just described, the work of Babor and his colleagues[14,26] provides the most comprehensive depiction of sex in relation to alcoholic typologies reported to date. Alcoholic subtypes were empirically derived for both sexes using relatively large and diverse subject samples. Types were identified using a comprehensive set of theoretically meaningful attributes, and similar cluster solutions were obtained for male and female alcoholics. This work provides an important first step in terms of understanding alcoholic subtypes. For both sexes, a powerful risk/severity dimension underlies subtype distinctions. Furthermore, subtypes based on this dimension — Type A alcoholics (Low Risk/Severity) and Type B alcoholics (High Risk/Severity) — have prognostic significance for both sexes. This structure was replicated with two different samples of male alcoholics,[27,28] and treatment-matching effects were demonstrated.[27]

Although this work demonstrates the similarities between the sexes in terms of alcoholic subtypes, differences between the sexes have not been fully explicated. Perhaps most salient are the different proportions of each sex within the two groupings. Whereas males tend to be equally divided into Type A and Type B (47 vs 53%), females are disproportionately represented in the Low Risk/Severity subtype (62 vs 38%). Moreover, this finding is consistent with other attempts to derive female subtypes using cluster analytic techniques.[28]

Inspection of the subtype profiles for each sex reveals other significant differences. Mean values for each of the attributes used in the cluster analysis are listed in TABLE 1, separately by sex of subject. As indicated in the table, Type B males and females tend to have similar profiles. Mean values for the two sexes are significantly different for only 4 of 17 attributes. Consistent with other research on sex differences more generally, Type B males have higher symptom counts for antisocial personality disorder. This difference may explain differences in mean scores on the Michigan

TABLE 1. Profiles of Empirically Derived Alcoholic Subtypes: Type A and Type B Cluster Means

	Type A			Type B		
Defining Characteristic	Males $(n = 107)$	Females $(n = 53)$	t	Males $(n = 121)$	Females $(n = 32)$	t
Premorbid risk factors						
Familial alcoholism	.23	.31	2.22*	.40	.52	2.40*
Childhood disorder	2.12	1.64	1.50	4.08	4.07	.02
Bipolar character disorder	27.12	23.46	5.08***	29.73	29.03	.90
Onset of problem drinking	30.34	34.20	2.30*	21.70	21.46	.22
Alcohol and other substances						
Alcohol use (oz/d)	6.49	4.32	2.55**	9.20	9.56	.25
Relief drinking	14.00	14.66	4.16***	17.74	14.27	58
Dependence severity	13.20	12.88	.52	16.87	15.99	1.24
Benzodiazepine use	.69	1.15	2.80**	1.02	1.08	.28
Polydrug use	.79	1.07	.93	3.48	3.17	.50
Chronicity and consequences						
Physical conditions	12.80	18.58	3.28***	26.06	25.39	.26
Physical consequences	6.84	7.44	2.50*	7.91	8.22	1.00
Social consequences	7.80	7.30	2.17*	9.17	9.09	.22
Lifetime severity (MAST)	35.49	33.29	1.53	42.69	39.52	2.13*
Years of heavy drinking	25.03	18.43	3.88***	18.12	11.18	4.08***
Psychiatric symptoms						
Depression	2.88	5.91	7.39***	5.31	5.31	0.1
Antisocial personality	5.57	2.96	4.74***	10.11	8.13	2.75**
Anxiety	20.41	27.66	4.87***	29.22	29.69	1.14

NOTE: The findings presented are the results of a reanalysis of data presented by Babor et al.[26]

*$p < 0.05$; **$p < 0.01$; ***$p < 0.001$.

Alcoholism Screening Test as well. Items on this instrument, which is considered a good measure of global severity, tend to reflect similar male behavior patterns (e.g., arrests). Although the sexes are similar in terms of age of onset of problem drinking, Type B females have significantly fewer years of heavy drinking, a difference that may reflect the "telescoping" phenomenon already mentioned. In terms of risk factors, females have a significantly greater genetic loading than do their male counterparts in terms of family history of alcoholism. Despite these differences, Type B male and female alcoholics tend to be similar in terms of both risk and severity. Moreover, these similarities are observed in areas in which the sexes are generally found to differ, such as onset age, quantity of alcohol typically consumed, and affective disturbance (depression and anxiety scores).

A very different pattern emerges for Type A alcoholics. Sex differences are evident for 13 of the 17 defining characteristics. Although they are significantly higher in terms of familial alcoholism, Type A females generally score lower on measures of other risk factors. They report fewer (although not statistically significant) childhood conduct problems, and they have a significantly older age of onset for problem drinking. Type A women also score significantly lower on the MacAndrew Alcoholism scale, a putative measure of susceptibility to addiction. In terms of alcohol and other drug use, females consume less alcohol, but report a significantly higher degree of relief drinking. They also report using benzodiapines significantly more often than do their male counterparts. These differential patterns of use are consistent with differences in psychiatric symptoms. Type A males and females differ most markedly in this area, with women scoring significantly higher on measures of depression and anxiety and lower in antisocial personality. Interestingly, although Type A females are clearly less severely affected than Type Bs, depression and anxiety scores are comparable in the two groups. Finally, the sexes differ in terms of the chronicity and consequences of alcohol consumption. Females report more medical and physical consequences of drinking. At the same time, they have shorter drinking histories, again suggesting that the physiologic differences between the sexes that underlie the notion "telescoping" affect women categorized as members of both alcoholic subtypes. Thus, whereas Type B women and men are generally similar in terms of risk/severity profiles, Type A males and females differ significantly across a diverse range of attributes. Furthermore, most of these differences reflect more general sex differences as well as those thought to differentiate male and female alcoholics (e.g., alcohol use, psychiatric symptoms).

The profiles that emerge for both sexes are those of later onset alcoholics; beyond this, marked differences are noted between the sexes. The question arises as to whether these contrasting profiles are simply a reflection of differences that characterize men and women generally or have underlying etiologic significance. Alternatively, some differences may result from differential social responses to men and women with drinking problems. For example, physicians may be less likely to identify female alcoholics, and, consequently, more likely to prescribe benzodiazapines for their presenting problems. Regardless of the source of differences, the profiles for Type A women and men suggest that they may require different treatment approaches.

TOWARD A NEW ANALYTIC STRATEGY

This "second look" at the results reported by Babor and his colleagues suggests that additional analyses may provide further insights on the structural organization of alcoholic subtypes, their defining characteristics, and the relation of these to sex and gender. An analytic strategy designed to explicate the role of sex and gender in alcohol subtyping follows.

The definition of subtypes in any classification scheme requires that investigators first identify relevant attributes for characterizing members of the population under study. In previous studies, alcohol researchers have often relied on a relatively small set of traits. Any alcohol typology, however, must consider risk factors, correlates of use, and alcohol-related consequences as well as drinking styles to adequately characterize particular subtypes. And, as already suggested, cases selected for analysis should be representative of the degree of heterogeneity observed in the population as a whole.

Cluster analysis has been the most popular empirical approach for identifying alcoholic subtypes. However, a variety of techniques are available for representing structure, and alternative methods may be useful in future work. Given the consistent finding that risk/severity constitutes a major dimension along which alcoholics can be ranked, nonmetric multidimensional scaling (MDS) should be considered as a means of uncovering the major dimensions along which alcoholics vary. This approach is particularly well suited for identifying a relatively small number of important dimensions that underlie the relations (similarities and differences) among "objects" (cases or attributes). (Although most approaches to representing structure allow the investigator to treat attributes or cases as objects, and some "two-way" clustering algorithms permit both to be clustered simultaneously, most investigators have applied cluster analytic techniques to cases.) Based on the degree of similarity or dissimilarity between each pair of objects, this procedure produces an array of objects in n-dimensional space. The reference axes in the resulting MDS spatial configuration are arbitrary; however, multiple regression can be used to fit substantive dimensions in the space. The Multiple R and the angle between fitted dimensions allow the investigator to interpret the structural configurations produced by the analysis.

FIGURE 1 depicts a hypothetical MDS configuration of male and female cases. The location of objects in the space corresponds to what might be expected on the basis of the aforementioned Babor *et al.* clustering findings. Several different lines have been placed through the origin of the hypothetical space. As depicted, two slightly correlated major dimensions account for the positions of the objects. The first and primary dimension runs horizontally from left to right and summarizes three vectors that collectively indicate the presence of a risk/severity dimension. Both male and female cases appear at both ends of the risk/severity continuum; Type A alcoholics are primarily located toward the left side (low risk/severity) of the dimension, whereas Type B alcoholics appear on the right (high risk/severity). The second dimension is vertical; properties of the cases in the upper and lower halves of the space suggest that this dimension could be labeled sex of subject or defined in terms of another correlated factor, such as psychiatric symptoms (affective disturbance *vs* antisocial personality). The possible presence of a third, weaker dimension (of unknown meaning) is indicated by the broken arrow that appears to bisect the area between the vertical and horizontal dimensions, but it is actually at approximate right angles to these other vectors. The depictions are hypothetical configurations suggested by the results of previous analyses.

How do the dimensions depicted in the MDS space relate to alcoholic subtypes? FIGURE 2 presents a structural configuration that combines solutions obtained through MDS and clustering. The circles encompassing groups of cases illustrate hypothetical results of clustering analyses of male and female cases. Consistent with the

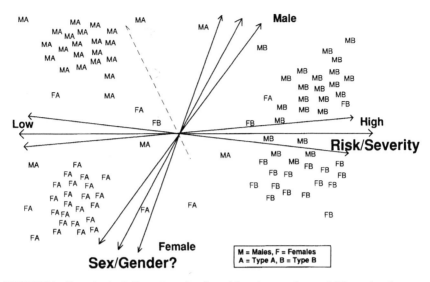

FIGURE 1. Hypothesized dispersion of male and female cases in a multidimensional space.

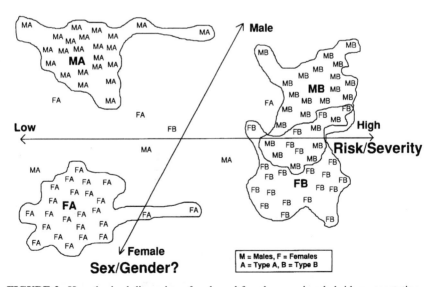

FIGURE 2. Hypothesized dispersion of male and female cases in a hybrid representation.

findings reported above, Type B male and female clusters are highly similar and overlap in the MDS solution. Such relations between groupings can also be demonstrated empirically with the application of clustering algorithms that permit solutions with overlapping clusters. Type A clusters for each sex are similar in that they clearly include cases that are relatively low on the risk/severity continuum. At the same time, these cases differ in terms of a second important dimension in the space, the vector tentatively labeled sex/gender. A third feature of the space portrayed in FIGURE 2 is the depiction in large bold letters of "ideal cases" that provide prototypic profiles of each cluster.

The structural configurations depicted in FIGURES 1 and 2 are hypothetical and should be confirmed by additional analyses. They are presented here to illustrate several hypotheses regarding the role of sex and gender in relation to alcoholic subtypes and to provide concrete examplars of the types of solutions that can be obtained by the application of MDS and clustering analyses. Whereas MDS represents major dimensions well, clustering is better suited to depicting "local structure." Collectively, the two approaches can be combined to develop hybrid models that highlight the importance of powerful dimensions, such as risk/severity, and at the same time permit classification of discrete cases into groups that may have different etiologies or prognoses.[41]

FROM STRUCTURE TO PROCESS:
HIGH PRIORITIES FOR FUTURE RESEARCH

The general findings with respect to sex and alcoholic typologies suggest that biological and social factors are both important causal agents in the development of different forms of alcoholism. This, of course, is true of the origins of sex differences generally. Although there are exceptions, alcohol investigators have often failed to distinguish between biological sex and socially defined gender. The terms sex and gender tend to be used interchangeably, and underlying causes for sex differences are not adequately explored. Sex is a biological fact for each individual. At the same time, sex is a major social category that influences the manner in which individuals are perceived and socialized, the goals, skills, and interests that they develop, the way in which they think about themselves and their behavior, and the situations in which they find themselves. Although there are sex differences that appear to be a direct function of differential physiology (e.g., alcohol metabolism), there are others that are influenced by differential cultural norms that regulate individual behavior and by situational factors that vary as a function of sex. Some sex differences undoubtedly result from biological and physiological differences between men and women, others may be the consequence of distal factors relating to sex-role socialization. Finally, still other differences may occur because of differential social expectations or social status that are functionally related to sex. It is simply not enough to determine that the sexes differ; the source of those differences must be examined to fully understand the role of sex and gender in alcoholic subtypes.

This general point regarding sex and gender points to two important corollaries. First, understanding the role of sex and gender in the etiology of alcoholic subtypes necessitates careful attention to differing levels of analysis (e.g., individual *vs* sociocultural). Second, at the individual level of analysis, *gendered* personality variables (e.g., sex-role orientation) are likely to be important in understanding sex differences in alcohol consumption or problems relating to use.

Despite the identification of risk factors related to the development of alcohol problems in women and men, our understanding of key mechanisms is limited. In particular, findings with regard to the role of stress and negative affect (especially depression) in relation to male and female drinking are equivocal. A variety of methodological approaches are needed to deal with varying levels of analysis and to address the complexity of processes that contribute to alcoholism. The following are priorities for future work:

Genetic and Biological Inputs. The importance of family history as a risk factor for alcoholism is well established. Existing data suggest that genetic risk is particularly important in the etiology of Type B alcoholism; however, the mechanisms involved remain to be explicated. In part because of the different prevalence rates for alcoholism in the general population, it has been assumed that genetic factors are more important for males than for females. This assumption was recently challenged, however. Furthermore, the comparison of Type B profiles for male and female alcoholics summarized above suggests that females do in fact evidence primary alcoholism and that family history may be an important etiologic factor in its development. More studies are needed to determine the degree to which family history influences the development of alcoholism in women and to clarify the mechanisms that underlie family history effects for both sexes.

The comparison of male and female alcoholics just presented suggests that the telescoping phenomenon may be an important contributor to the development of both forms of alcoholism in women. More work is needed to understand how this differential physiological response to alcohol relates to the progression of other problems that characterize alcoholic women.

Prospective Daily Process Studies. Research designed to explicate the temporal relations among events, appraisal, stress, negative affect, and drinking behavior are needed to assess the degree to which these processes contribute to the development of Type A alcoholism. Such research requires an intensive data collection effort, preferably at multiple time points each day over an extended interval of time, and the application of sophisticated time-series analysis techniques. This kind of intensive data collection is essential for determining the significance of depression and anxiety as antecedents to frequent or heavy alcohol use.

Longtitudinal Research. Whereas daily process studies can illuminate the microlevel processes that influence drinking behavior, longtitudinal studies are necessary for investigating the longer-term course of alcoholism in women and men. Ideally, such studies should begin data collection prior to the initiation of regular alcohol use, include a comprehensive assessment of factors potentially related to alcoholism (including gendered psychological variables), and schedule assessment sessions on a relatively frequent basis. The data generated by such research designs can then be analyzed using structural modeling approaches. Different models regarding the etiology of alcoholism can be tested for specific alcoholic subtypes, and the relative importance of sex and gender-related variables can be ascertained.

Cross-National Studies. The present paper suggests that sociocultural factors influence both the distribution of men and women into alcoholic subtypes and the nature of the interrelations among attributes that comprise the subtype profiles. Cross-national studies involving cultures with differing norms or social policies relating to male and female alcohol use, as well as differing gender role expectations generally, are needed to assess the influence of sociocultural factors. Repeated assessments within each site over time may also permit analysis of the effects of social change. Ideally, case level data should be obtained from diverse samples within participating nations; available aggregate statistics may also be used to provide further information about factors that influence changes over time.

REFERENCES

1. ALLAN, C. A. & D. J. COOKE. 1985. Stressful life events and alcohol misuse in women: A critical review. J. Studies on Alcohol **46:** 147–152.

2. JARVIS, T. J. 1992. Implications of gender for alcohol treatment research: A quantitative and qualitative review. Brit. J. Addictions **87:** 1249–1261.

3. LEX, B. W. 1990. Prevention of substance abuse problems in women. *In* Drug and Alcohol Abuse Reviews. R. R. Watson, ed.: 167–221. The Humana Press, Inc. Totawa, NJ.

4. WILSNACK, S. C. & R. W. WILSNACK. 1990. Women and substance abuse: Research directions for the 1990's. Psychol. Addictive Behaviors 46–49.

5. JACOBSON, R. 1986. Female alcoholics: A controlled CT brain scan and clinical study. Brit. J. Addiction **81:** 661–669.

6. GALLANT, D. M. 1987. The female alcoholic: Early onset of brain damage. Alcoholism **11:** 190–191.

7. BECKMAN, L. J. 1984. Treatment needs of women alcoholics. Alcoholism Treatment **1:** 101–114.

8. DUCKERT, F. 1987. Recruitment into treatment and effects of treatment for female problem drinkers. Addictive Behaviors **12:** 137–150.

9. HESSELBROCK, M. N. 1991. Gender comparison of antisocial personality disorder and depression in alcoholism. J. Substance Abuse **3:** 205–219.

10. BECKMAN, L. J. 1978. Self-esteem of women alcoholics. J. Studies on Alcohol **39:** 491–498.

11. VANNICELLI, M. 1984. Barriers to treatment of alcoholic women. Substance Alcohol Actions Misuse **5:** 29–37.

12. INSTITUTE OF MEDICINE. 1990. Broadening the Base of Treatment for Alcohol Problems: Report of a Study by a Committee of the Institute of Medicine: Division of Mental Health and Behavioral Medicine. National Academy Press. Washington, DC.

13. VANNICELLI, M. & L. NASH. 1984. Effects of sex bias on women's studies on alcoholism. Alcoholism: Clinical and Experimental Research **8:** 334–336.

14. BABOR, T. F., Z. DOLINSKI, R. MEYER, M. HESSELBROCK, M. HOFMANN & H. TENNEN. 1992. Brit. J. Addiction **87:** 1415–1431.

15. COOPER, M. L., M. RUSSELL, J. B. SKINNER, M. R. FRONE & P. MUDAR. 1992. Stress and alcohol use: Moderating effects of gender, coping, and alcohol expectancies. J. Abnormal Psychology **101:** 139–152.

16. JELLENIK, E. M. 1960. Alcoholism: A genus and some of its species. Canad. Med. Assoc. J. **83:** 1341–1345.

17. MOREY, L. C. & R. K. BLASHFIELD. 1981. Empirical classifications of alcoholism. J. Studies on Alcohol **42:** 925–937.

18. BABOR, T. F. & R. LAURERMAN. 1986. Classification and forms of inebriety: Historical antecedents of alcoholic typologies. *In* Recent Advances in Alcoholism. M. Galanter, ed. Vol. 5: 145–164. Plenum Press. New York, NY.

19. BABOR, T. B. & Z. DOLINSKY. 1988. Alcoholic typologies: Historical evolution and empirical evaluation of some classification schemes. *In* Alcoholism, Origins, and Outcome. R. M. Rose & J. Barrett, eds.: 245–266. Raven Press. New York, NY.

20. HART, L. S. & D. STUELAND. 1980. Classifying women alcoholics by Cattell's 16PF; a preliminary investigation of an alcoholic typology. J. Studies on Alcohol **41:** 911–921.

21. ESHBAUGH, D. M., D. J. TOSI & C. N. HOYT. 1980. Women alcoholics; a typological description using the MMPI. J. Studies on Alcohol **41:** 310–317.

22. SCHUCKIT, M., F. N. PITTS, T. REICH, L. KING & G. WINOKUR. 1969. Alcoholism: Two types of alcoholism in women. Arch. Gen. Psychiatry **20:** 301–306.

23. SCHUCKIT, M. A. & E. R. MORRISEY. 1976. Alcoholism in women: Some clinical and social perspectives with an emphasis on possible subtypes. *In* Alcoholism Problems in Women and Children. M. Greenblatt & M. A. Schuckit, eds. Grune & Stratton. New York, NY.

24. HELZER, J. E. & T. R. PRYZBECK. 1988. The co-occurrence of alcoholism with other

psychiatric disorders in the general population and its impact on treatment. J. Studies on Alcohol **49:** 219–224.

25. CLONINGER, C. R. 1987. Neurogenetic adaptive mechanisms in alcoholism. Science **236:** 410–416.

26. BABOR, T. F., M. HOFMANN, F. DEL BOCA, V. HESSELBROCK, R. MEYER, Z. DOLINSKY & B. ROUNSAVILLE. 1992. Types of alcoholics, I. Evidence for an empirically-derived typology based on indicators of vulnerability and severity. Arch. Gen. Psychiatry **49:** 599–608.

27. LITT, M. D., T. B. BABOR, F. K. DEL BOCA, R. M. KADDEN & N. L. COONEY. 1992. Types of alcoholics. II. Application of an empirically-derived typology to treatment matching. Arch. Gen. Psychiatry **49:** 609–614.

28. BROWN, J., T. F. BABOR, M. D. LITT & H. R. KRANZLER. The Type A/Type B distinction: Subtyping alcoholics according to indicators of vulnerability and severity. *In* Types of Alcoholics: Evidence from Clinical, Experimental, and Genetic Research. T. F. Babor, V. Hesselbrock, R. E. Meyer & W. J. Shoemaker, eds. Ann. N.Y. Acad. Sci. This volume.

29. FILLMORE, K. M. 1984. "When angels fall": Women's drinking as cultural preoccupation and as reality. *In* Alcohol Problems in Women. S. C. Wilsnak & L. J. Beckman, eds. Guilford Press. New York.

30. FISKE, S. T. & S. E. TAYLOR. 1984. Social Cognition. Addison-Wesley. Reading, MA.

31. TAYLOR, M. E. & S. ST. PIERRE. 1986. A review of current literature. J. Drug Issues **16:** 621–636.

32. GOMBERG, E. S. 1988. Alcoholic women in treatment: The question of stigma and age. Alcohol & Alcoholism **23:** 507–514.

33. YOUNG, G. P., R. RORES, C. MURPHY & R. H. DAILEY. 1987. Intravenous phenobarbital for alcoholic withdrawal and convulsions. Ann. Emergency Med. **16:** 847–850.

34. JOHNSON, P. B. 1982. Sex differences, women's roles, and alcohol use: Preliminary national data. J. Social Issues **38:** 93–116.

35. WILSNACK, S. C. & R. W. WILSNACK. 1992. Women, work, and alcohol: Failures of simple theories. Alcoholism: Clinical and Experimental Research **16:** 172–179.

36. HOROWITZ, A. V. & H. R. WHITE. 1987. Gender role orientations and styles of pathology among adolescents. J. Health & Social Behavior **28:** 158–170.

37. HSER, Y. I., M. D. ANGLIN & W. H. MCGLOTHIN. 1987. Sex differences in addict careers. 1. Initiation of use. Am. J. Drug & Alcohol Abuse **13:** 33–57.

38. ANGLIN, M. D., Y. I. HSER & W. H. MCGLOTHIN. 1987. Sex differences in addict careers. 2. Becoming addicted. Am. J. Drug & Alcohol Abuse **13:** 231–251.

39. KANDEL, D. B. 1984. Marijuana users in young adulthood. Arch. Gen. Psychiatry **41:** 200–209.

40. KANDEL, D. B., M. DAVIES, D. KARUS & K. YAMAGUCHI. 1986. The consequences in young adulthood of adolescent drug involvement. Arch. Gen. Psychiat. **43:** 746–754.

41. SKINNER, H. A. 1979. Dimensions and clusters: A hybrid approach to classification. Applied Psychological Measurement **3:** 327–341.

Components of Antisocial Personality Disorder among Women Convicted for Drunken Driving[a]

BARBARA W. LEX,[b] MARGARET E. GOLDBERG,[b]
JACK H. MENDELSON,[b] NOELLE S. LAWLER,[b]
AND TOMMIE BOWER[c]

[b]Alcohol and Drug Abuse Research Center
Harvard Medical School
Department of Psychiatry
McLean Hospital
115 Mill Street
Belmont, Massachusetts 02178

[c]Longwood Treatment Center
Boston, Massachusetts 02130

A national survey of women's drinking patterns identified driving while intoxicated (DUI) as the most frequently acknowledged alcohol-related problem, followed by fighting with spouses or friends.[1] Drug abuse also is associated with alcohol problems among women.[2] In Massachusetts, increasing numbers of women have come to the attention of the criminal justice system as a result of more stringent enforcement and penalties for drinking and driving.[3] About two thirds of child neglect or abuse cases involve alcohol or polydrug use.[4] There also is growing evidence that a family history of alcoholism has an important impact on the development of alcoholism in women.[5-7]

Using information from public records in Sweden, Cloninger and colleagues conducted adoption studies that identified two types of genetic/familial alcoholism in men.[8-10] Type I was designated "milieu-limited" because expression appeared to be shaped by environmental factors and onset occurred *after* age 25. Type II was designated "male-limited" and generally associated with onset *before* age 25 and with legal infractions.[10] Type II traits are also characteristics of antisocial personality disorder (ASPD).[11,12]

In these Swedish adoption studies, most alcoholic female adoptees had type I alcoholic biological fathers and brothers who needed less alcohol treatment and incurred only minor legal infractions.[10] Adopted alcoholic daughters of Type I fathers also had Type I alcoholism that appeared later, was less severe, and was shaped by environmental conditions. By contrast, daughters of Type II alcoholic fathers showed somatization instead of Type II alcoholism.[8,13-16] Type I and Type II classifications in the Swedish adoption studies emerged from records of work absences, illness treatments, and alcohol problems. No data were obtained directly from these alcohol-dependent men or women whether in treatment or involved in the criminal justice system.

In recent years, several investigators have examined the prevalence of early onset

[a]This research was supported in part by grants DA 00064 and DA 04870 from the National Institute on Drug Abuse and grants AA 06252 and AA 06794 from the National Institute on Alcohol Abuse and Alcoholism.

49

and greater severity of alcohol-related problems in treatment samples including both men and women.[5,12,17,18] Some have argued that Type II alcoholism overlaps with antisocial personality disorder.[12,19] Reanalysis of data from adoption studies in Denmark and Sweden indicates that the cohort effects and small numbers of women are likely to have obscured cases of interest in women.[5] Comparatively smaller numbers of women in treatment populations have early onset (before age 25) and a severe course of illness, but nonetheless their existence has been confirmed.[7,18,20,21]

Furthermore, behaviors of female substance abusers have drawn attention to possible associations between adult antisocial behaviors and antisocial personality disorder concurrent with substance use.[22,23] In men with substance abuse disorders, comorbidity with ASPD may be as high as 50%.[24] The prevalence of ASPD and comorbidity with substance dependence is lower for women. However, some have questioned the relationship between childhood Conduct Disorder and adult personality disorders, especially ASPD.[22,23] In DSM-III and DSM-III-R, antisocial personality disorder criteria require occurrence of at least three childhood antisocial behaviors prior to age 15 and 4 of 10 adult antisocial behaviors after age 18[11,25]; ASPD is *not* to be diagnosed in the absence of Conduct Disorder. Thus, it is possible that comorbidity of substance abuse and ASPD is lower in women than in men in part because the rate of childhood Conduct Disorder is lower among women. In one study of 228 men and 118 women, 53% of the men but 29% of the women were found to have ASPD. However, adult antisocial behaviors also were prevalent in female drug users who did not meet full criteria for ASPD because of the lack of childhood antecedents.[26] Similar observations led Cottler and colleagues[22] to question whether childhood Conduct Disorder should be required to diagnose adult ASPD in women or men.

Women who come to the attention of both civil and criminal courts because of alcohol- or drug-related problems provide an invaluable opportunity for systematic exploration of ASPD behavior and familial/genetic transmission of possible "Type II" characteristics.[7,27] We hypothesized that alcohol-dependent women with Type II alcoholism characteristics or ASPD could be found among a population of women incarcerated for alcohol-related driving offenses.[7] Consequently, we explored the age at onset of alcohol dependence, occurrence of concurrent drug abuse, and ASPD criteria in incarcerated female DUI offenders as a logical component of our continuing studies of alcoholism and family history in women.

METHODS

Subjects. All subjects were residents of the Longwood Treatment Center in Boston, Massachusetts. The Longwood Treatment Center is a 125-bed alcoholism treatment facility within the Massachusetts Department of Correction that receives male and female offenders with DUI related offenses. Average length of stay is about 100 days, which permits development of rapport and extensive data collection.

This sample consists of 33 women admitted between May 1991 and March 1992. These women include all female residents except for four women with medical or psychiatric problems which prevented participation and three who refused. All subjects gave informed consent and were interviewed at a minimum of 3 weeks after admission.

Materials. Subjects were administered Substance Abuse (including Alcohol), Somatoform, Borderline, and Antisocial Personality Disorder sections of the SCID[28] by a trained interviewer from the research team. The personality disorder sections were modified slightly insofar as subjects were asked to answer each question twice, once in terms of their sober behavior and once in terms of their behavior while intoxicated or hung over.

Subjects provided background data, substance use history, including age at first drink, age at onset of regular drinking, age at onset of alcohol dependence, and age at onset of legal troubles, and histories of use of other drugs. A total of 31 women met DSM-III-R[11] criteria for alcohol dependence, and two women met criteria for alcohol abuse (TABLE 1). They also provided sociodemographic information, including marital status, living situation, number of children, and parental alcoholism (TABLE 2).

RESULTS

Characteristics of the 31 subjects who met criteria for alcohol dependence were examined for a pattern of early onset associated with features of ASPD. As currently defined, ASPD consists of two components: (1) Conduct Disorder (a pattern of unruly or unacceptable behavior prior to age 15) and (2) a pattern of unsettled or disruptive behavior as an adult (after age 18).[11] Because it has been suggested that ASPD behavior can occur *after* onset of substance abuse, we designated this pattern "SR ASPD" to indicate the importance of substance abuse-related behavior. To meet criteria for ASPD behavior, individuals must exhibit at least 4 of 12 characteristic behaviors of the adult pattern.[11] Illegal and reckless behaviors are two symptoms prerequisite for sentencing to the Longwood Treatment Center. Consequently, all subjects met at least two ASPD criteria, but in some women the range extended to eight ASPD criteria. The diagnosis of Conduct Disorder requires at least 3 of the 12 characteristic juvenile behaviors.[11] Subjects reported behaviors that ranged from 0 to 6 criteria, but only one woman had Conduct Disorder alone. The most commonly acknowledged behaviors included among the criteria for Conduct Disorder were stealing (32%) and truancy (28%) followed by running away from home (20%), lying (20%), and fighting (8%).

To explore our hypotheses, the sample was divided into three groups according to ASPD status: (1) No ASPD diagnosis (absence of Conduct Disorder *and* SR ASPD behavior), (2) SR ASPD behavior only, and (3) ASPD (with *both* Conduct Disorder *and* ASPD behavior). (The woman diagnosed with Conduct Disorder only was excluded from this analysis.) The relationship between drinking and ASPD behaviors was explored by asking subjects whether each behavior occurred during

TABLE 1. Background Characteristics (33 Patients)

Characteristics	Range	Mean	SD	SE
Age (yr)	21–56	33.7	8.2	1.4
Years of education	8–17	12.7	2.3	0.4
Age at first drink	10–32	15.8	4.8	0.9
Age when regular drinking began	12–35	19.0	5.7	1.0
Age at onset of alcohol dependence[a]	13–45	23.9	6.7	1.2
Age when legal trouble began	14–46	24.5	7.9	1.4
Age at first sexual intercourse	12–22	16.8	2.4	0.4

[a]$N = 31$. Two subjects met criteria for alcohol abuse but not alcohol dependence.

TABLE 2. Behavioral Characteristics

Characteristics	Patients ($n = 33$)	
	Number	Percent
Marital status		
Single	14	42.4
Married	4	12.1
Separated	3	9.1
Divorced	12	36.4
Living situation		
Mate/male friend	10	30.3
Other	23	69.7
Number of children		
None	15	45.5
One	6	18.2
Two	6	18.2
Three	5	15.1
Four	1	3.0
Parental history of alcoholism[a]		
Yes	20	64.5
No	11	35.5

[a]$N = 31$. Two subjects were adopted with no information about family of origin.

abstinence or while drinking. Only 1 of 19 SR ASPD women had four or more symptoms of ASPD (as defined by DSM-II-R) that occurred during intervals of abstinence and drinking. Accordingly, there were 7 women with no ASPD diagnosis (21.9%), 6 with ASPD (18.8%), and 19 with SR ASPD only (59.4%) (F = 5.588, 2 df, $p = 0.0058$). Group characteristics ($n = 32$) are shown in TABLE 3.

The SR ASPD group most resembled the No ASPD diagnosis group in years of education, age of first drink, age of onset of alcohol dependence, age of first legal problems, and history of drug abuse or dependence. Results of one-factor ANOVAs are shown in TABLE 3. Mean age at onset of alcohol dependence was slightly over age 25 for the SR ASPD and No ASPD groups, but 16.8 years for women with ASPD. The SR ASPD group most resembled the ASPD group in rate of parental alcoholism (72.2 and 66.7%) and rate of impulsivity (94.7 and 100%), but it was intermediate in terms of concurrent Borderline Personality Disorder (42.1%). None acknowledged impairment in performance of her parental role.

All but two women, who were adopted, had information about alcoholism in a parent, as ascertained by DSM-III-R criteria.[11] About two thirds of 31 subjects had an alcoholic biological parent (TABLE 2). Subjects with SR ASPD had the greatest percentage of alcoholic parents (about 70%) and women with No ASPD the least (33.3%), but the difference among diagnostic groups was not significant (chi square = 2.97, 2 df, $p = 0.2269$). However, family history of alcoholism had no significant effect on age of onset in any of the diagnostic groups.

Twenty subjects (61%) had a diagnosis of drug abuse or dependence. Marijuana (65%) and cocaine (65%) had been used most frequently, followed by sedatives (50%), hallucinogens (35%), stimulants (30%), and opiates (25%). Drug abuse or dependence was not concentrated in any diagnostic group (chi square = 2.24, 2 df,

TABLE 3. Background Characteristics and Psychiatric Data, by ASPD Status, One-Factor ANOVAs[a]

Characteristics	Subjects (n = 32)			Statistics				
				ANOVA		Fisher's Protected Least Significant Difference**		
	No ASPD (n = 7)	SR ASPD (n = 19)	ASPD (n = 6)	F	P	N/S[d]	S/A[d]	A/N[d]
Age (current)	37.6 ± 4.3	35.0 ± 8.8	26.3 ± 5.0	4.07	0.0277	NS	**	**
Education (yr)	12.7 ± 2.6	12.9 ± 1.9	18.8 ± 3.0	0.48	0.6248	NS	NS	NS
Age at first drink	17.6 ± 3.9	16.6 ± 5.1	12.2 ± 1.3	2.85	0.0741	NS	**	**
Age when regular drinking began	20.3 ± 5.2	20.1 ± 5.8	15.2 ± 2.9	2.15	0.1351	NS	NS	NS
Age at onset of alcohol dependence[b]	26.2 ± 3.4[c]	25.4 ± 7.2	16.8 ± 2.3	4.98	0.0144	NS	**	**
Age when legal problems began	27.7 ± 4.9	25.0 ± 8.6	18.2 ± 3.1	2.90	0.0732	NS	NS	**
Age at first sexual intercourse	17.6 ± 2.2	17.4 ± 2.4	15.2 ± 1.2	2.57	0.0935	NS	**	NS
Number of borderline symptoms	1.8 ± 1.3	3.7 ± 1.9	6.3 ± 1.5	6.18	0.0074	**	**	**
Number of ASPD symptoms	2.7 ± 0.5	5.0 ± 1.2	6.0 ± 1.2	16.14	0.0001	**	NS	**
Number of conduct disorder symptoms	0.3 ± 0.8	0.4 ± 0.8	3.5 ± 1.2	31.88	0.0001	NS	**	**

[a]Results are expressed as means ± SD.
[b]As determined by the SCID.
[c]N = 5; 2 subjects were not alcohol dependent.
[d]Women with No Diagnosis (N); women with SR ASPD (S); women with ASPD (A).
**Indicates significant differences ($p < 0.05$) for all pairwise comparisons with a multiple t-statistic.

$p = 0.3268$). Subjects with a diagnosis of drug abuse or dependence were significantly younger than those in the nondrug-abusing group (mean age 31.0 *versus* 37.9 years). They also were significantly younger than the nondrug-abusing women when they began alcohol use, began marijuana use, and first had sexual intercourse (TABLE 4).

DISCUSSION

In a similar study, Cottler and colleagues[22] analyzed the prevalence of Conduct Disorder and ASPD in a large sample ($n = 605$) comprised of male and female substance abusers. Men and women with "full" ASPD showed different patterns of psychiatric morbidity than did those with "adult only" antisocial personality disorder associated with substance abuse. Adult antisocial behaviors in men and women were associated with later onset of initial use of cannabis, amphetamines, barbiturates, and cocaine, but women generally had about 10% fewer ASPD symptoms and a different pattern of symptoms. Women with the "adult only" ASPD pattern were less likely to have used drugs daily for at least 2 weeks and to stay "high" all day, to report tolerance and trying to "cut down" drug use, and to experience emotional problems. However, about 40% had experienced drug-related health problems, slightly over half of the women in both groups reported caring for children while intoxicated, and about three quarters reported social problems. Men and women with "full" ASPD were more severe drug abusers and had *earlier* onset than "adult only" ASPD patients. When age and onset of substance abuse were controlled, occurrence of any childhood Conduct Disorder symptom predicted "adult only" ASPD behaviors. Inasmuch as men and women in both groups first "got into trouble" by age 10, Cottler suggested that the cut-off age for childhood symptoms (15 years) needs to be reevaluated and that the number of childhood Conduct Disorder criteria required to

TABLE 4. Drug Use Experience

Characteristics	Women with Drug Abuse ($n = 20$)	Women with No Drug Abuse ($n = 13$)	t Value	p Value
Current age (yr)	31.0 ± 5.7	37.9 ± 9.8	2.57	0.02
Formal education (yr)	12.6 ± 2.0	12.7 ± 2.6	0.10	0.92
Age at first drink (yr)	14.1 ± 4.8	18.5 ± 3.6	2.76	0.01
Age began regular drinking (yr)	17.6 ± 5.5	21.2 ± 5.2	1.86	0.07
Age at first marijuana use (yr)	15.3 ± 5.0	21.9 ± 9.2[a]	2.39	0.03
Age regular marijuana use (yr)	17.4 ± 5.8[b]	—[c]	0.56	0.58
Age at first cocaine use (yr)	22.1 ± 7.1[d]	20.8 ± 1.9[e]	0.36	0.72
Age regular cocaine use (yr)	23.6 ± 7.5[f]	—[c]	1.32	0.22
Age legal trouble began (yr)	22.6 ± 7.1	27.5 ± 8.6	1.72	0.10
Age at first sexual intercourse (yr)	16.0 ± 2.3	18.2 ± 2.1	2.72	0.01

[a]Seven subjects.
[b]Fourteen subjects.
[c]One subject.
[d]Nineteen subjects.
[e]Four subjects.
[f]Eleven subjects.

meet the diagnosis of ASPD for women be lowered to one or more. Similarly, Robins and Price[23] suggested that cut-off points for childhood Conduct Disorder criteria required to meet the diagnosis of ASPD be adjusted downward from three childhood Conduct Disorder behaviors to 2+ for males and 1+ for females.

We have evaluated the temporal relationship between ASPD and alcoholism in a special population of female alcohol and drug users incarcerated for DUI offenses. Excluding behaviors that only occurred while drinking, only 1 of the 33 women in our sample met DSM-III-R criteria for ASPD and one met criteria for ASPD with Conduct Disorder. When behaviors while drinking were included, however, 18.2% of this sample met DSM-III-R criteria for ASPD, and an additional 57.6% of the sample met criteria for ASPD except for a history of childhood Conduct Disorder. In contrast to women who had SR ASPD symptoms, the six women with full ASPD (with childhood Conduct Disorder) were younger at onset of alcohol dependence before age 21 (16.8 *vs* 25.6 years), had a higher rate of concurrent borderline personality disorder (85.7% *vs* 42.1%), but the same rate of reported parental alcoholism (71.4% *vs* 72.2%). In this sample, all but one woman who had full ASPD with childhood Conduct Disorder were truant and ran away from home. The means for the number of conduct disorder symptoms for women in the No ASPD and SR ASPD categories were both less than one. Thus, few women in this sample would meet the proposed revised number of Conduct Disorder behavioral criteria for women (1+).[22,23] No women in this sample acknowledged such criteria as acts of vandalism, cruelty to animals, or arson in childhood, all of which are symptoms more typical of men. The pattern for these women suggests that ASPD was a consequence of substance abuse rather than an antecedent, with childhood Conduct Disorder acting only as a limited predictor of ASPD in adulthood. Relations among prodromal behaviors and substance abuse appear more complex for women than we anticipated.

Overlap with Borderline Personality Disorder was not excessive. It is estimated that alcoholism and Borderline Personality Disorder co-occur in about 15%–40% of cases.[29–31] Our data also suggest the need to recognize SR ASPD associated with substance abuse as a legitimate diagnosis that is manifested differently by women and men. We expected a bimodal distribution of subjects with either full ASPD or no diagnosis of Conduct Disorder or Adult ASPD. Instead, a trimodal distribution emerged. The majority of subjects fell between two official DSM-III-R diagnoses by having SR ASPD (without Conduct Disorder). For age of onset, there was a consistent pattern of youngest ages among the ASPD subjects, intermediate ages among the SR ASPD subjects, and oldest ages among the No Conduct Disorder or Adult ASPD subjects.

Results confirmed the initial hypothesis. All women in the sample who met the criteria for ASPD also showed onset of alcohol dependence before age 21. These women appear to fit the description of Type II alcoholism, although the high percentage of parental alcoholism in this sample obscures the effects of heritability in any group. However, behaviors characteristic of ASPD occurred mainly in connection with drinking.

Another interesting finding was that subjects with onset of alcohol problems before age 25 were found in all groups. Age 25 was established as the cut-off for early onset in men.[10] Our findings suggest that age 21 may be more appropriate for *women* with early onset, as indicated by the work of Hesselbrock and colleagues[21] and suggested by Hill and Smith.[5] This downward shift may be an artifact of the legal drinking age or might reflect a secular trend toward earlier use of alcohol and other drugs[32] and concomitant oppositional behavior. Both of these interpretations may also apply to men. Finally, this effect could be influenced by cultural patterns whereby males date younger females and dating is a primary vector for alcohol and drug use in women.

SUMMARY

For women, the temporal relationship between Antisocial Personality Disorder (ASPD) and alcoholism is unclear. Driving while intoxicated is both a symptom of ASPD and the alcohol-related problem most typically reported by women. Accordingly, a period prevalence sample of 33 women incarcerated for drunken driving offenses was assessed with the SCID to identify other symptoms of ASPD. Excluding behaviors that only occurred while drinking, only 1 of the 33 women met DSM-III-R criteria for ASPD. When behaviors while drinking were included, 18.2% ($n = 7$) met criteria for ASPD by having both a history of childhood conduct disorder and characteristic ASPD behaviors as adults. However, 57.6% of the sample displayed the pattern of adult behavioral symptoms without a history of childhood conduct disorder ($n = 19$). Women with a history of conduct disorder and ASPD had a younger mean age of onset of alcohol dependence (16.8 vs 25.6 years) and a higher rate of concurrent borderline personality disorder (85.7 vs 42.1%) than did women who had only adult symptoms of ASPD, but a similar rate of reported parental alcoholism (71.4 vs 72.2%). With one exception, women who were diagnosed with full ASPD with childhood conduct disorder ($n = 6$) had been truant and had run away from home, but none reported cruelty to animals, vandalism, or arson in childhood. Thus, behaviors diagnostic of ASPD were largely consequent to substance abuse, and childhood behaviors were limited predictors of ASPD. Relationships among gender, prodromal behaviors, and substance abuse appear more complex than anticipated, and they indicate the need to recognize adult onset ASPD associated with substance abuse as a legitimate diagnosis manifested differently by women and men.

ACKNOWLEDGMENT

We are grateful to David MacDonald, former Superintendent of the Longwood Treatment Center, a program of the Massachusetts Department of Correction, to Dennis A. Humphrey, Associate Commissioner for Programs and Treatment, and to Michael. W. Forcier, former Deputy Director of Research, Massachusetts Department of Correction, for their cooperation and support of alcoholism treatment and this study.

REFERENCES

1. WILSNACK, R. W., S. C. WILSNACK & A. D. KLASSEN. 1984. Women's drinking and drinking problems: Patterns from a 1981 national survey. Am. J. Public Health **74:** 1231–1238.
2. WILSNACK, S. C., A. D. KLASSEN, B. E. SCHUR & R. W. WILSNACK. 1991. Predicting onset and chronicity of women's problem drinking: A five-year longitudinal analysis. Am. J. Public Health **81:** 305–318.
3. ARGERIOU, M., D. MCCARTY, D. POTTER & L. HOLT. 1986. Characteristics of men and women arrested for driving under the influence of liquor. Alcohol **3:** 127–137.
4. HERSKOWITZ, J., M. SECK, C. FOGG, S. OSGOOD, J. POWERS & D. MAKIN. 1989. Substance Abuse and Family Violence. Part I. Identification of Drug and Alcohol Usage during

Child Abuse Investigations in Boston. Commonwealth of Massachusetts Department of Social Services. Boston, MA.

5. HILL, S. Y. & T. R. SMITH. 1991. Evidence for genetic mediation of alcoholism in women. J. Subs. Abuse **3:** 159–174.

6. KENDLER, K. S., A. C. HEATH, M. C. NEALE, R. C. KESSLER & L. J. EAVES. 1992. A population-based twin study of alcoholism in women. JAMA **268:** 1877–1882.

7. LEX, B. W., J. W. SHOLAR, T. BOWER & J. H. MENDELSON. 1991. Putative type II alcoholism characteristics in female third DUI offenders in Massachusetts: A pilot study. Alcohol **8:** 283–287.

8. BOHMAN, M., C. R. CLONINGER, A. L. VON KNORRING & S. SIGVARDSSON. 1984. An adoption study of somatoform disorder. III. Cross-fostering analysis and genetic relationship to alcoholism and criminality. Arch. Gen. Psychiatry **41:** 872–878.

9. CLONINGER, C. R., B. BOHMAN & S. SIGVARDSSON. 1981. Inheritance of alcohol abuse cross-fostering analysis of adopted men. Arch. Gen. Psychiatry **38:** 861–878.

10. CLONINGER, C. R., M. BOHMAN, S. SIGVARDSSON & A.-L. VON KNORRING. 1985. Psychopathology in adopted-out children of alcoholics: The Stockholm adoption study. *In* Recent Developments in Alcoholism. M. Galanter, ed. Vol. **3:** 37–51. Plenum Press. New York, NY.

11. AMERICAN PSYCHIATRIC ASSOCIATION. 1987. Diagnostic and Statistical Manual of Mental Disorders: DSM-III-R. 3rd Ed. Rev. American Psychiatric Association. Washington, DC.

12. IRWIN, M., M. SCHUCKIT & T. L. SMITH. 1990. Clinical importance of age at onset in Type I and Type 2 primary alcoholics. Arch. Gen. Psychiatry **47:** 320–324.

13. BOHMAN, M., S. SIGVARDSSON & C. R. CLONINGER. 1981. Maternal inheritance of alcohol abuse cross-fostering analysis of adopted women. Arch. Gen. Psychiatry **38:** 965–969.

14. CLONINGER, C. R., S. SIGVARDSSON, A.-L. VON KNORRING & B. BOHMAN. 1984. An adoption study of somatoform disorders. II. Identification of two discrete somatoform disorders. Arch. Gen. Psychiatry **41:** 863–871.

15. CLONINGER, C. R., S. SIGVARDSSON, T. REICH & M. BOHMAN. 1986. Inheritance of risk to develop alcoholism. *In* Genetic and Biological Markers in Drug Abuse and Alcoholism. M. C. Braude & H. M. Chao, eds. NIDA Research Monograph No. 66, DHHS Publication No. (ADM) 86–1444: 86–96. US Government Printing Office. Washington, DC.

16. SIGVARDSSON, S., A.-L. VON KNORRING, M. BOHMAN & C. R. CLONINGER. 1984. An adoption study of somatoform disorders. I. The relationship of somatization to psychiatric disability. Arch. Gen. Psychiatry **41:** 853–859.

17. HESSELBROCK, M. N. 1986. Childhood behavior problems and adult antisocial personality disorder in alcoholism. *In* Psychopathology and Addictive Disorders. R. E. Meyer, ed. : 78–94. The Guilford Press. New York, NY.

18. BABOR, T. F., M. HOFMANN, K. F. DEL BOCA, V. HESSELBROCK, R. E. MEYER, Z. S. DOLINSKY & B. ROUNSAVILLE. 1992. Types of alcoholics. I. Evidence for an empirically derived typology based on indicators of vulnerability and severity. Arch. Gen. Psychiatry **49:** 599–614.

19. SCHUCKIT, M. A. & M. IRWIN. 1989. An analysis of the clinical relevance of Type 1 and Type 2 alcoholics. Br. J. Addict. **84:** 869–876.

20. HESSELBROCK, M. N., R. E. MEYER & J. J. KEENER. 1985. Psychopathology in hospitalized alcoholics. Arch. Gen. Psychiatry **42:** 1050–1055.

21. HESSELBROCK, M. N., V. M. HESSELBROCK, T. F. BABOR, J. R. STABENAU, R. E. MEYER & M. WEIDENMANN. 1983. Antisocial behavior, psychopathology and problem drinking in the natural history of alcoholism. *In* Longitudinal Research in Alcoholism. D. W. Goodwin, K. T. Van Dusen & S. A. Mednick, eds.: 197–214. Kluwer-Nijhoff Publishing. Boston, MA.

22. COTTLER, L. C., R. K. PRICE, W. M. COMPTON & D. MAGER. 1991. Adult antisocial behavior without childhood symptoms among substance abusers. Poster presented at College on Problems of Drug Dependence. Boulder, CO.

23. ROBINS, L. N. & R. K. PRICE. 1991. Adult disorders predicted by childhood conduct

problems: Results from the NIMH epidemiologic catchment area project. Psychiatry **54:** 116–132.

24. GERSTLEY, L. J., A. I. ALTERMAN, A. T. MCLELLAN & G. E. WOODY. 1991. Antisocial personality disorder in patients with substance abuse disorders: A problematic diagnosis? Am. J. Psychiatry **142:** 173–178.

25. AMERICAN PSYCHIATRIC ASSOCIATION. 1980. Diagnostic and Statistical Manual of Mental Disorders. DSM-III. 3rd Ed. American Psychiatric Association. Washington, DC.

26. FELCH, L. J., R. K. BROONER & G. E. BIGELOW. 1991. Gender differences in the antisocial behavior of intravenous drug abusers. Poster presented at College on Problems of Drug Dependence. Palm Beach, FL..

27. LEX, B. W., S. K. TEOH, I. LAGOMASINO, N. K. MELLO & J. H. MENDELSON. 1990. Characteristics of women receiving mandated treatment for alcohol or polysubstance dependence in Massachusetts. Drug Alc. Depend. **25:** 13–20.

28. SPITZER, R. L., J. B. W. WILLIAMS, M. GIBBON & M. B. FIRST. 1990. Structured Clinical Interview for DSM-III-R. American Psychiatric Press. Washington, DC.

29. NACE, E., J. SAXON & N. SHORE. 1986. Borderline personality disorder and alcoholism treatment. J. Stud. Alcohol **47:** 196–200.

30. NACE, E., J. SAXON & N. SHORE. 1983. A comparison of borderline and nonborderline alcoholic patients. Arch. Gen. Psychiatry **40:** 54–56.

31. GUNDERSON, J. G., M. C. ZANARINI & C. L. KISIEL. 1991. Borderline personality disorder: A review of data on DSM-III-R descriptions. J. Pers. Disorder **5:** 340–352.

32. REICH, T., C. R. CLONINGER, P. VAN EERDEWEGH, J. P. RICE & J. MULLANEY. 1988. Secular trends in familial transmission of alcoholism. Alcohol: Clin. Exper. **12:** 458–464.

Evidence of Heterogeneity of Genetic Effect in Iowa Adoption Studies[a]

REMI CADORET,[b] ED TROUGHTON,[b]
AND GEORGE WOODWORTH[c,d]

Departments of Psychiatry,[b] Statistics,[c] and Preventive Medicine[d]
University of Iowa
200 Hawkins Drive, #2887 JPP
Iowa City, Iowa 52242

The adoption paradigm is one of the most powerful study designs available to separate out genetic/environmental effects and to determine gene environment interaction.[1] Starting with the senior author's move to Iowa in 1973, a series of studies were initiated using both private and public adoption agencies that operate throughout the state of Iowa. These studies resulted in a number of publications dealing with genetic effects in alcoholism, antisocial personality, major depression, childhood attention deficit hyperactivity disorders, and obesity.[2–9] This paper concentrates on the evidence for a genetic factor in alcoholism and shows how the manifestation of this factor has changed with samples from different adoption agencies.

METHOD

General Procedures

The adoption paradigm used in all studies to be reported here is shown in FIGURE 1. Index adoptees are determined from agency records for the earlier studies, whereas in the most recent study hospital and prison records were used in addition to the agency records. Index adoptees are selected on the basis of having a biologic parent with a psychiatric problem such as alcoholism. Index adoptees are matched with a control adoptee for age and sex and age of biologic mother at the time of the child's birth. Control adoptees are determined, as far as possible, to be free of any psychiatric or behavioral problem in biologic parents as determined from records available to investigators. All adoptees are separated at birth from their biologic parents and have been successfully placed in a permanent adoptive home. This procedure eliminates children who were thought or found to be defective at birth and who were placed not in permanent homes but in institutions or permanent foster care. It also eliminates adoptees who for some reason failed to adjust to an adoptive home with the result that the adoption was not finalized. All adoptees were required to be placed with nonrelatives. Following the determination of index and control adoptees from the adoption agency records, contact was made with adoptees and their adoptive parents. At the time of each study, adoptees were included in the study if they ranged in age from 18 to 45 years. Older adoptees' adoptive parents were not available for interview usually because they had died, and older adoptees are more difficult to trace. Information about the adoptee and his or her upbringing as well as

[a] This work was supported by grants NIDA R01 DA05821, NIAAA R01 AA06159, and NIAA R01 AA04750.

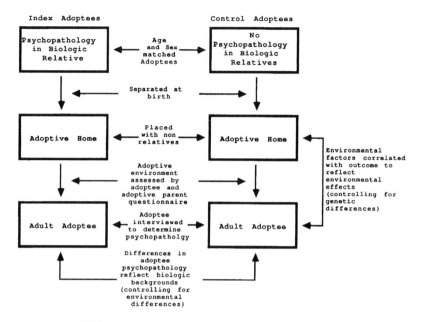

FIGURE 1. Adoption study paradigm used in Iowa studies.

adolescent and adult psychopathology was determined from an adoptive parent questionnaire and an adoptee interview.

Brief Description of Specific Studies

In the first study to be reported, adoptees and their parents were interviewed by telephone. In subsequent studies, face-to-face personal interviews were carried out by research assistants blind to the biologic diagnosis of the adoptee's biologic family history. The first study was done with Iowa Children and Family Services (ICFS) adoptees, and data were collected from 1974 to 1977. These interviews were done on the phone, and for the adoptee interview an early experimental version of the Diagnostic Interview Schedule (DIS)[11] was used.

The second study was done with Lutheran Social Services (LSS) during the years 1979 to 1982. Lutheran Social Services study differed from the Iowa Family and Children Services in that index subjects were recruited to come from only antisocial and alcoholic backgrounds; in contrast, in the Iowa Children and Family Services study all types of psychopathology in the biologic background were selected for index subjects. The other change that was made in the Lutheran Social Services study was the introduction of personally administered interviews. In half of the study the *SADS-L* was used[10] and in the second half the DIS.[11]

The third study was done with a sample from Catholic adoption agencies (CAA) across the state. The Dubuque diocese contributed by far the largest number of adoptees to this study. Index adoptees were selected with a biologic background of alcoholism or antisocial problems. This determination was made from adoption agency records. The material given to assess the environment was enlarged for this interview over that of the previous two studies and consisted of two questionnaires designed to assess the environmental factors: the Family Environment Scale[12] and the Home Environment Questionnaire.[13] Additional information was also collected to determine availability and ease of obtaining alcohol and illegal drugs. The DIS was also used for the adult adoptee interview.

Toward the end of the data collection period in the third study, arrangements were made with the Department of Human Services in the state of Iowa and the Department of Corrections to search their institutional records for the names of biologic parents who had given up a child for adoption at birth. It proved possible to get both state psychiatric hospital and prison records of parents who had given up a child at birth for adoption and to maintain confidentiality of biologic parent names as required by Iowa state law. These records were evaluated independently by two psychiatrists and DSM-III-R diagnoses were made of antisocial personality, alcohol abuse/dependency, and drug abuse/dependency. These diagnoses in biologic parents were then used to determine index adoptees. This improvement in data collection was added to the fourth and final study which was designed to test the hypothesis that drug abuse/dependency involved genetic factors of alcoholism and antisocial personality. Data were collected from four independent adoption agencies with two of them representing agencies that had participated in the past, the Dubuque diocese and Lutheran Social Services, and two new agencies, the Department of Human Services and Hillcrest, the latter a private agency. Because of the involvement of four different adoption agencies, we call this the Four Agency Study (4AS). Similar interviews were used to assess the environment in the adoptive home as in the previous study, and the DIS was again used for the assessment of the adult adoptee. In all cases interviews were administered in person.

ANALYSIS OF RESULTS

In earlier years, *t* tests and simple chi-square analyses were used to demonstrate differences between control and index outcomes. However, since the early 1980s, log-linear modeling has been used to explore the data.[14] Log-linear modeling acknowledges that alcohol abuse/dependency, in addition to the genetic element, is a product of many etiologic factors such as the environmental and gene-environmental interactions. It is also possible to do causal modeling with dichotomous variables.

RESULTS OF STUDIES

Only the findings of the male sample and the results of the ICFS study are shown in TABLE 1. Seventy males from a control biologic background showed 18.6% (n = 13) with alcohol abuse and four of six from an alcoholic biologic background or 66.7% alcoholic. This contrast was significant with an odds ratio of 8.8. These results along with other odds ratios found in the entire dataset are shown in the interaction diagram in FIGURE 2. A significant relation was found between alcohol biologic problem and adoptee alcohol abuse. A significant relation between adoptee outcome of alcohol abuse and adoptee antisocial personality is present as well as a relation between a biologic background of antisocial problems and adoptee antisocial personality. In addition, the model also showed a suggestive but not significant environ-

TABLE 1. Iowa Children and Family Service (ICFS) Study Data Collected from 1974–1977

	Control Biologic Background (n = 70)	Alcoholic Biologic Background (n = 6)	Odds Ratio	Confidence Limits	
				Lower	Upper
Males	13 (18.6%)[a]	4 (66.7%)	8.8	1.4	53.1

[a]Number of adoptees with alcohol abuse/dependence in sample, with percentage in sample in parentheses.

mental effect, that is, the presence of an alcohol problem in an adoptive family member. This was the best fitting model and seemed to agree with a specificity of inheritance of antisocial behavior as well as alcoholic behavior. When this model was first developed, results were not yet available from the Environmental Catchment Area (ECA) study which showed indeed a significant association in the general population between alcohol abuse and antisocial personality.[15]

The second series of adoptees was obtained from Lutheran Social Services for the purpose of confirming a genetic effect in alcoholism. After the data had been collected and the model just shown was discovered in ICFS data, the Lutheran Social Service data provided an excellent independent opportunity to replicate that model. The results for males in the Lutheran Social Service study are shown in TABLE 2, where 18% (9/50) from a control background were alcoholic in contrast to 62.5% (10/16) from an alcoholic biologic background. This results in a raw odds ratio of 7.6 and is clearly significantly different from 1. The best fitting model (essentially the same as that reported for the ICFS data) is displayed in FIGURE 3. In this study the relation between antisocial biologic background and adoptee antisocial personality was positive but not significant, whereas the adoptive family alcohol problem-

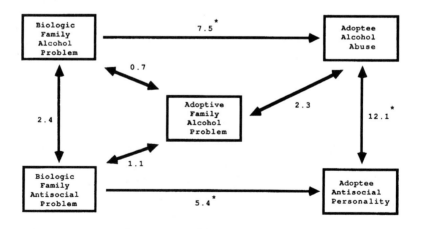

* = significant difference

FIGURE 2. Interaction diagram for Iowa Children and Family Services male adoptees.

TABLE 2. Lutheran Social Service (LSS) Study Data Collected from 1979–1982

	Control Biologic Background ($n = 50$)	Alcoholic Biologic Background ($n = 16$)	Odds Ratio	Confidence Limits	
				Lower	Upper
Males	9 (18.0%)[a]	10 (62.5%)	7.6	2.2	26.3

[a]Number of adoptees with alcohol abuse/dependence in sample, with percentage in sample in parentheses.

adoptee alcohol abuse relationship was significant. The LSS dataset was presented by Cadoret *et al.* in 1985, and the ICFS material by Cadoret *et al.*[16] in 1987.

At this point it was felt that this adoption paradigm was robust and could be used to gather more information about genetic effects, environmental effects, and possible gene environment interactions that might be involved in the etiology of alcohol abuse. By this time (mid-1980s) there was additional interest in the possibility that the log-linear model was compatible with the Bohman-Cloninger hypothesis of two types of alcoholism.[17,18] However, more data were needed to further delineate the environmental factors and to determine the presence of gene-environment interaction as suggested by the Bohman-Cloninger model. Accordingly, a large third study was launched and data were collected from 1986 to 1989. Data collection was done at three of the four Iowa dioceses; the Dubuque, the Des Moines, and the Sioux City dioceses. Most cases came from the Dubuque diocese (TABLE 3). The data showed no genetic effect for alcoholism. In fact, the results appeared to go in the opposite direction with an odds ratio of less than 1. The data from each of the three dioceses are shown in TABLE 3. The overall results were negative for demonstrating an alcohol genetic factor. The log-linear model that was fitted to the data appears in FIGURE 4. The only significant relation in the model was found in the positive correlation

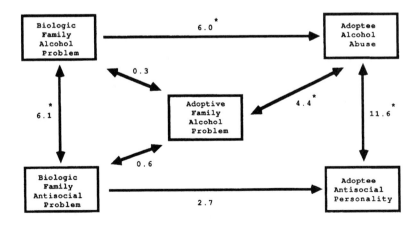

* = significant difference

FIGURE 3. Interaction diagram for Lutheran Social Services male adoptees.

TABLE 3. Catholic Adoption Agency Study (CAA) Data Collected from 1986–1989

Agency	Control Biologic Background	Alcoholic Biologic Background	Odds Ratio	Confidence Limits	
				Lower	Upper
Dubuque diocese	31/53 (58.5%)[a]	19/38 (50%)	0.7	0.3	1.6
Des Moines diocese	5/10 (50%)	2/7 (28.6%)	0.4	0.05	3.1
Sioux City diocese	4/6 (66.7%)	3/4 (75.0%)	1.5	0.09	25.4
Total	40/69 (58.0%)	24/49 (49.0%)	0.7	0.3	1.5

[a]Numerator of fraction is number with alcohol abuse/dependence; denominator is sample size. Percentage alcoholic in sample is in parentheses.

between the adoptee outcome of alcohol abuse and antisocial personality, a finding compatible with the two previous studies. The lack of genetic association was entirely unexpected and led the investigators to consider and explore a number of possibilities that will be reported here.

One concern that has been frequently expressed about the Iowa adoption studies is that diagnoses have been made on biologic parents from material contained within adoption agency records. Obviously these records vary considerably in quality and quantity between agencies. One possible explanation of the lack of a genetic effect in study three is that a number of alcoholic parents were overlooked in the control group. To investigate this possibility, it was planned to examine state mental hospital and prison records for biologic parents who had given up children for adoption. However, the only agency that agreed to allow data to be released for this institutional record search was the Dubuque diocese. The remaining two dioceses were

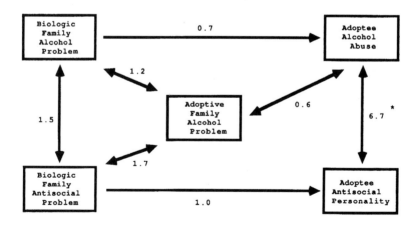

* = significant difference

FIGURE 4. Interaction diagram for Catholic adoption agencies male adoptees.

concerned about confidentiality and decided against releasing the biologic parents' names for this search. However, the search confirmed that the original determination of alcohol abuser and control from the agency records was quite valid. In the index group, 50 male adoptees were diagnosed from agency information as having a biologic parent with an alcohol problem. Of these 50 males, 10 had a biologic parent with a hospital or prison record of treatment for alcohol abuse. An additional 4 of these 50 males had a biologic parent with antisocial personality. *None* of the control male adoptees had a biologic parent with a hospital or prison record. These findings suggest that mistaken categorization of biologic parents as alcoholic from adoption agency records was probably not a determining factor in explaining the negative results in this dataset.

Another hypothesis involved the possibility of a bias being introduced by refusals. These adoption studies had always been characterized by rather high refusal rates (41–54%) over the years, with a slight increase in refusal rates with more recently collected data. However, the rates in the present study were not different from those in previous studies. There has been no tendency across studies for more index subjects to refuse than control subjects.

The third hypothesis focused on the possibility of a coding error, in which the coding of the records of the control and experimental subjects was mixed. However, careful review of the coding failed to find any errors in this area. It was possible to check for coding errors by tracing individual adoptees through the process of coding, because the birth date was unique for each individual and interviewed adoptees' birth dates always checked with the birth date listed in the agency for that specific adoptee. A further check was available to determine a possible mixup of index and controls. An analysis inheritance of body mass index had been done with the ICFS and LSS datasets,[19] and analyses of the CAA data resulted in correlations similar to those found in the ICFS and LSS data. This "biologic marker" suggests that correct identification of adoptees occurred; otherwise, the correlations would have been different.

A fourth hypothesis proposed heterogeneity of expression of the alcoholism diathesis. Not all adoption studies for alcoholism have been positive for a genetic factor. The first adoption study done by Ann Roe in the 1940s failed to show a drinking difference between adoptees from an alcoholic background and control adoptees, despite a fairly sizable sample size.[20] This suggested the present hypothesis, namely, that the alcoholism genetic factor (or whatever genetic mechanism is responsible for alcoholism) may manifest differently in different populations. This hypothesis does not propose a simple sampling error, but it actually predicts that the manifestation of the genetic factor may vary considerably between populations as a function of some population characteristic. Other differences in rates of alcoholism reported in adoption studies from Denmark[21,22] and Sweden[18,17] could well reflect differences in diagnostic practice rather than real population differences in the manifestation of a genetic factor. The present studies would allow us to determine if such a population variation existed because we could keep constant across agencies, the general methodology, and biologic parent and adoptee diagnostic practices. We proposed to test this hypothesis by building into our next study as many comparisons or repetitions as could be made with previously used adoption agencies.

The fourth and final study had as its original purpose the testing of the hypothesis that certain biologic parent backgrounds (antisocial personality and alcohol or drug abuse) are factors in drug abuse in adopted away offspring. In setting up this study four different agencies were used, hence the title Four Agency Study (4AS). The Dubuque diocese and Lutheran Social Services represented agencies that had produced diametrically opposite results in expression of a genetic factor for alcoholism. Two new agencies were added, the Iowa Department of Human Services and the Hillcrest Family Services. Unfortunately we had had no previous samples from the latter two. Iowa Children and Family Services and the remaining Catholic dioceses in the state were asked to participate but refused because of concerns about confidentiality in releasing names of biologic parents for the record search to determine a hospital or prison diagnosis for biologic parents. Results for the current study are shown in TABLE 4 and are separated by adoption agency. It can be seen that the Lutheran Social Services and Hillcrest showed results more compatible with the genetic factor, whereas the Department of Human Services showed little evidence of a genetic factor. The Dubuque diocese showed similar differences to those found in the third study, that is, a tendency for a control nonalcoholic background to have more alcoholic adoptees. FIGURE 5 shows an interaction diagram for all males from these four agencies. One of the odds ratios in this log-linear model, the positive relationship between biologic parent with antisocial personality and adoptee antisocial personality, is significant. Most relationships are in the direction predicted by the original model developed from the ICFS and LSS datasets but are not significant.

TABLE 5 compares the two independent studies with the Dubuque diocese adoptees. The first is the study of 1986–1989 (line 1), and the second is the more recent study of 1990–1992 (line 3). It can be seen that the odds ratio for the biologic parent–alcoholism–adoptee alcoholism correlates in both studies is similar and indicates a negative association. It is possible to subdivide the 1986–1989 study using the experimentals determined from hospital and prison records, because names from the Dubuque agency, at least, were run through the hospital and prison record search. This modified sample in the second line indicates an odds ratio showing a positive relationship. Totals for the two samples using hospitalized or incarcerated parents to determine index cases are shown in the bottom two lines, the upper of the two showing the difference using all the data and the bottom line showing the difference between control and index cases determined from hospital/prison records. All of the odds ratios are indistinguishable from one. In contrast to this finding, TABLE 6 shows the results with the Lutheran Social Service Agency. The first line of TABLE 6 shows the odds ratio for the 1979–1982 Lutheran Social Services Study and the second line the study using Lutheran Social Services adoptees from 1990–1992. The number of

TABLE 4. Four Agency Study (4AS) Data Collected from 1990–1992

Agency	Control Biologic Background	Alcoholic Biologic Background	Odds Ratio	Confidence Limits	
				Lower	Upper
Dubuque diocese	11/14 (78.6%)[a]	5/8 (62.5%)	0.5	0.1	3.1
Iowa Dept. of Human Services	7/14 (50.0%)	4/8 (50.0%)	1.0	0.2	5.7
Lutheran Social Services	4/10 (40.0%)	11/13 (84.6%)	8.3	1.2	59
Hillcrest	5/11 (45.5%)	4/5 (80.0%)	4.8	0.4	58.0
Total	27/49 (55.1%)	24/34 (70.6%)	2	0.8	4.9

[a]Numerator of fraction is number with alcohol abuse/dependence; denominator is sample size. Percentage alcoholic in sample is in parentheses.

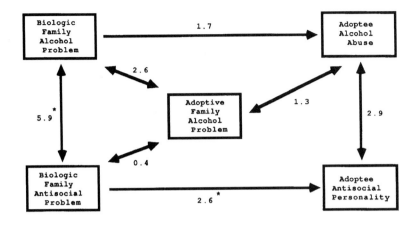

* = significant difference

FIGURE 5. Interaction diagram for the Four Agency Study male adoptees.

adoptees diagnosed as abusing alcohol is probably increased compared to that of earlier studies. In part this may represent changes in diagnostic practice over the 10-year period, but it could also reflect different cohorts of individuals (average age for males in both studies is similar). The 1990–1992 study shows an odds ratio similar to the earlier one and is by itself significantly different from one. To quantitatively assess the apparent heterogeneity of expression of the genetic factor, a hierarchical random effects meta-analysis was done.[23] The odds ratios for the nine samples presented in this report (ICFS, LSS2, three CAA samples, and four samples from 4AS) are shown in FIGURE 6. This figure shows the odds ratios as natural logarithms of the unadjusted odds ratios from each study. A comparison of the crude odds ratios with available adjusted odds ratios from the log-linear models indicated little difference between crude and adjusted odds ratios. Accordingly, the crude odds ratios were used in the heterogeneity analyses that follow.

TABLE 5. Comparison of Dubuque Studies

Study	Control Biologic Background	Alcoholic Biologic Background	Odds Ratio	Confidence Limits	
				Lower	Upper
(1) 1986–1989	31/53 (58.5%)[a]	19/38 (50.0%)	0.7	0.3	1.6
(2) Hospitalized experimentals	31/53 (58.5%)	7/10 (70.0%)	1.7	0.4	7.1
(3) 1990–1992	11/14 (78.6%)	5/8 (62.5%)	0.5	0.1	3.1
Total of (1) & (3)	42/67 (62.7%)	24/46 (52.2%)	0.6	0.3	1.4
Total of (2) & (3)	42/67 (62.7%)	12/18 (66.7%)	1.2	0.4	3.6

[a]Numerator of fraction is number with alcohol abuse/dependence; denominator is sample size. Percentage alcoholic in sample is in parentheses.

TABLE 6. Comparison of Lutheran Social Services Studies

	Control Biologic Background	Alcoholic Biologic Background	Odds Ratio	Confidence Limits	
				Lower	Upper
First study (1979–1982)	9/50 (18.0%)[a]	10/16 (62.5%)	7.6	2.1	26.3
Second study (1990–1992)	4/10 (40.0%)	11/13 (84.6%)	8.3	1.2	59.0

[a]Numerator of fraction is number with alcohol abuse/dependence; denominator is sample size. Percentage alcoholic in sample is in parentheses.

As a group, the nine odds ratios are significantly heterogeneous ($\chi^2 = 20.4$, $df = 8$, $p = 0.01$). Heterogeneity was tested with the Fast*Pro computer package for meta-analysis.[24] The principal source of heterogeneity is the difference between LSS1 and CAA combined ($\chi^2 \leq 9.4$, $df = 1$, $p = 0.002$) and ICFS and CAA combined ($\chi^2 = 6.7$, $df = 1$, $p = 0.01$). These differences could reflect methodologic variation from one study to the other as well as heterogeneity from agency to agency. The heterogeneity analyses do show that results from the Catholic adoption agencies are homogeneous and are significantly different from the results from either the ICFS study or the first LSS study. This leads to a test of the heterogeneity hypothesis with the last study (4AS). For this study, the heterogeneity χ^2 is 5.3 with 3 df and a probability of 0.15. The previous experience with Catholic agencies *versus* LSS agencies suggests that evidence of heterogeneity would be concentrated in the contrast between the Catholic agency (Dubuque) and LSS. Breaking down the 3 df in the heterogeneity χ^2, we find that the 1 df for the contrast between Dubuque and LSS is 4.28 with a probability of 0.039. This supports the hypothesis of heterogeneity of expression of a genetic diathesis.

All nine Iowa studies show a combined posterior mean odds ratio (Hasselblad's confidence profile) of 1.9 with confidence limits of .94 and 3.80, and a 95% chance

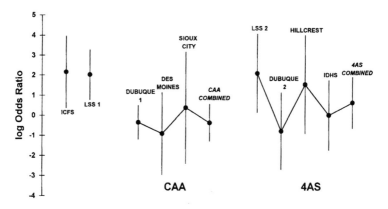

FIGURE 6. Heterogeneity of Iowa crude odds ratios.

that the true odds ratio is greater than one. This statistical estimate of the true odds ratio is very close to the combined odds ratio of the Swedish and Danish adoption studies in which the posterior mean odds ratio is 2.3 with confidence limits of 0.8 to 6.7. These limits are larger than those reported for the nine Iowa samples, but they reflect both within-study variability and especially between-study variance. For the Danish and Swedish studies there is a 94% likelihood that the true odds ratio is greater than one.

DISCUSSION

The results suggest that the manifestation of a genetic factor may vary between different groups of adoptees. In this study different groups were defined in terms of the adoption agency that handled their adoption. There are many factors confounded with the adoption agency that must be explored to determine a likely explanation. Agencies in the Iowa studies obviously differed in their religious orientation as well as ethnicity, and socioeconomic status of both biologic and adoptive parent clients. Data on some of these factors, such as SES and ethnicity, are available and are being explored. The differential effect of genetic factors in different populations could be due to a variety of causes: (1) genes that cause alcoholism could differ from one group to another; (2) the same genes shared by different groups could interact with different environmental factors; and (3) different environmental factors, independent of genes, could increase or decrease the rate of alcoholism in a sample so as to obscure or magnify differences.

Whatever the reason, heterogeneity of expression of genotype could be a significant problem in genetic behavioral studies. One wonders if such an effect could account for some of the difficulty reported in replicating linkage studies with conditions such as affective disorder and schizophrenia.

The finding of heterogeneity of genetic expression raises the possibility that other factors important in alcoholic etiology, such as the environment, could also show different effects depending on the subpopulation. Obviously other genetic factors such as those leading to criminality or antisocial personality could also show heterogeneity of expression which could well be independent of the expression of alcohol genotypes. The possible heterogeneity of subpopulations of adoptees raises a number of technical problems. If genetic and/or environmental factors manifest differently depending on the subpopulation, then studies may require sampling specific subpopulations rather than a large number of individuals from one population. Models will also have to be constructed by subpopulation, and better, more meaningful ways to identify subpopulations (other than by adoption agency) must be found. The present finding of heterogeneity of expression in males might also apply to female samples. Studies with the Iowa data are underway to explore these possibilities. Heterogeneity also affects the conclusions made from models fit to such data. In the present example, interest in various genetic and environmental factors correlating with alcoholism lies in their relevance to hypotheses such as that of Bohman and Cloninger[17,18] that there are two genetic types of alcoholism. The model found in the LSS and ICSF studies is compatible with that hypothesis, with the pathway to alcohol abuse/dependency through antisocial personality representing Type II alcoholism. The model pathway direct from alcoholic biologic parent to alcohol abuse/dependency would represent Type I alcoholism which might be more influenced by environmental factors (such as adoptive family alcoholism). The heterogeneity discovered here complicates how persons become alcoholic.

SUMMARY

This paper describes male adoption studies at the University of Iowa using private and public adoption agencies within the state of Iowa from 1974 to the present time. This research involves four large studies, the first two of which demonstrated significant genetic as well as environmental effects in the etiology of alcoholism as well as significant correlations between adoptee conditions of antisocial personality and alcohol abuse. Findings in the first two studies were similar. However, the third study, using a sample from Catholic-sponsored adoption agencies across the state, failed to show a genetic effect. The final and fourth study was designed in part to look for heterogeneity in the manifestation of a genetic factor from one sample to another in the Iowa studies. This was done by sampling from two agencies, one of which had shown a genetic effect and the other that had not. In this fourth study, the agency that had not shown a genetic effect in the previous study also failed to show an effect, but the agency that had shown a significant genetic effect in the past did demonstrate a significant genetic effect. There were two remaining agencies in the fourth study for which no comparison with past samples could be made. One of these agencies showed a marginally significant genetic effect and the other showed no effect. Statistical analysis of this last study suggested that observed variability in the odds ratio from sample to sample was due to differences in manifestation of the genetic effect.

REFERENCES

1. CADORET, R. J. 1986. Adoption studies: Historical and methodological critique. Psych. Dev. **1:** 45–64.
2. CADORET, R. J., T. W. O'GORMAN, E. TROUGHTON *et al.* 1985. Alcoholism and antisocial personality: Inter-relationships, genetic, and environmental factors. Arch. Gen. Psychiatry **42:** 161–167.
3. CADORET, R. J. 1985. Genes, environment and their interaction in the development of psychopathology. *In* Genetic Aspects of Human Behavior. T. Sakai & Tsuboi, eds. Igaku-Shoin Ltd. Tokyo.
4. CADORET, R. J., T. W. O'GORMAN, E. HEYWOOD & E. TROUGHTON. 1985. Genetic and environmental factors in major depression. J. Affect. Dis. **9:** 155–164.
5. CADORET, R. J., E. TROUGHTON & T. W. O'GORMAN. 1987. Genetic and environmental factors in alcohol abuse and antisocial personality. J. Stud. Alcohol **48:** 1–8.
6. CADORET, R. J., E. TROUGHTON, T. W. O'GORMAN & E. HEYWOOD. 1986. An adoption study of genetic and environmental factors in drug abuse. Arch. Gen. Psychiatry **43:** 1131–1136.
7. PRICE, R. A., R. J. CADORET, A. J. STUNKARD & E. TROUGHTON. 1986. Genetic contributions to human fatness: An adoption study. Am. J. Psychiatry **144:** 1003–1008.
8. CADORET, R. J., E. TROUGHTON, J. BAGFORD & G. WOODWORTH. 1990. Genetic and environmental factors in adoptee antisocial personality. Arch. Eur. Psychiatry & Neuro. Sci. **239:** 231–240.
9. CADORET, R. J. & M. A. STEWART. 1991. An adoption study of attention deficit/hyperactivity/aggression and their relationship to adult antisocial personality. Comp. Psychiatry **32:** 73–82.
10. SPITZER, R. L. & J. ENDICOTT. 1979. Schedule for Affective Disorders and Schizophrenia—Lifetime Version (SADS-L), 3rd Ed. New York State Psychiatric Institute. New York, NY.
11. ROBINS, L. N., J. E. HELZER, J. L. CROUGHAN *et al.* 1981. NIMH Diagnostic Interview Schedule, Version III. Washington University. St. Louis, MO.

12. Moos, R. H. & B. S. Moos. 1981. Family Environmental Scale Manual. Consulting Psychologists Press. Palo Alto, CA.
13. Sines, J. O., W. M. Clarke & R. M. Lauer. 1984. Home Environment Questionnaire. J. Abn. Child Psychol. **12:** 519–429.
14. Fienberg, S. E. 1980. The Analysis of Cross-Classified Data. MIT Press. Cambridge.
15. Regier, D. & L. N. Robins. 1991. Psychiatric Disorders in America: The Epidemiologic Catchment Area Study. Free Press. New York.
16. Cadoret, R. J., E. Troughton & T. O'Gorman. 1987. Genetic and environmental factors in alcohol abuse and antisocial personality. J. Stud. Alcohol **48:** 1–8.
17. Bohman, M., C. R. Cloninger, S. Sigvardsson & A. L. von Knorring. 1982. Predisposition to petty criminality in Swedish adoptees. I. Genetic and environmental heterogeneity. Arch. Gen. Psychiatry **39:** 1233–1241.
18. Cloninger, C. K., M. Bohman & S. Sigvardsson. 1981. Inheritance of alcohol abuse: Cross-fostering analysis of adopted men. Arch. Gen. Psychiatry **38:** 861–868.
19. Price, R. A., R. J. Cadoret, A. J. Stunkard & E. Troughton. 1987. Genetic contributions to human fatness: An adoption study. Am. J. Psychiatry **144:** 1003–1008.
20. Roe, A. 1944. Adult adjustment of children of alcoholic parents raised in foster homes. Q. J. Stud. Alcohol **5:** 378–393.
21. Goodwin, D. W., R. Schulsinger & L. Hermansen. 1973. Alcohol problems in adoptees raised apart from alcoholic biological parents. Arch. Gen. Psychiatry **28:** 238–243.
22. Goodwin, D. W., F. Schulsinger & N. Moller. 1974. Drinking problems in adopted and nonadopted sons of alcoholics. Arch. Gen. Psychiatry **31:** 164–171.
23. Hasselblad, V., D. M. Eddy & R. Schachter. 1992. Meta-Analysis by the Confidence Profile Method. Academic Press. Boston.
24. Eddy, D. M. & V. Hasselblad. 1992. Fast*Pro Software for Meta-Analysis by the Confidence Profile Method. Academic Press. Boston.

Genetic Influences on Alcohol Consumption Patterns and Problem Drinking: Results from the Australian NH&MRC Twin Panel Follow-up Survey[a]

A. C. HEATH[b] AND N. G. MARTIN[c]

bDepartments of Psychiatry, Psychology and Genetics
Washington University School of Medicine
4940 Children's Place
St. Louis, Missouri 63130

cQueensland Institute for Medical Research
Brisbane, Australia

The role of genetic factors in the etiology of alcoholism, particularly alcoholism in women, remains controversial. Findings consistent with an important genetic influence in males on alcoholism risk, or other measures of alcohol problems, have come from studies of adoptees,[1-4] half-siblings,[5] and monozygotic and dizygotic twin pairs.[6-13] Negative findings or findings suggesting at most a modest genetic contribution to male alcoholism risk have emerged from three adoption samples[14] (see also Cadoret, this volume) and three twin samples.[15-17] Whilst the majority of these studies supports a genetic contribution to male alcoholism risk, the extent to which inconsistencies between studies yielding positive and negative results can be explained by differences in sociocultural environment, differences in assessment procedure (e.g., reliance on official records of hospitalization or temperance board contact *versus* medical records *versus* self-report questionnaire *versus* structured psychiatric interviews), differences in sample ascertainment procedure (epidemiologic sampling *versus* ascertainment through probands identified through treatment facilities or official records), or merely differences in sample size and consequent statistical power remains uncertain.

For alcoholism in females, findings have been even less consistent than those in males. An early consensus emerged that genetic factors were less important in females than in males, supported by the relatively weak evidence for a genetic influence on female alcoholism risk both from adoption studies[18,19] and from twin studies.[9-11] A recent interview study of an epidemiologic sample of young adult female twin pairs, however, yielded heritability estimates, varying as a function of narrowness or breadth of diagnostic criteria, in the range of 50–60%.[20] A mailed questionnaire assessment of problem drinking by an older, predominantly female, twin panel, likewise found evidence for substantial genetic influence in females.[13,21] Once again, there are many possible reasons for these inconsistencies; however, findings indicating a substantial heritability in females have been restricted to those studies which used an epidemiologic sampling strategy, rather than identifying alcoholic probands through treatment facilities or official records.

aThis research was supported by ADAMHA grants AA03539, AA07535, and AA07728 and a grant from the Australian NH&MRC.

In contrast to these findings for alcoholism risk, surveys of drinking practices in general population twin samples, unselected for alcoholism risk, have consistently found evidence for a genetic contribution to individual differences in drinking pattern (average quantity consumed per drinking occasion) in both males and females.[22] Self-report drinking practices assessed in general community (rather than treatment) samples have good reliability and validity and, at least beyond age 25, considerable longitudinal stability.[22,23] Twin studies of alcohol consumption patterns in the United States,[21,22,24] Scandinavia (Finland[8,15,25]; Sweden[26]), Australia,[27,28] and the United Kingdom[29] have all yielded results consistent with a substantial genetic influence on alcohol consumption patterns in both sexes, with estimates of heritability (i.e., proportion of variation in alcohol consumption levels attributable to genetic factors) in most studies falling in the range of 30–60%.[22] Prospective studies indicate that a substantial proportion of the longitudinally stable component of variation in drinking practices may be determined by genetic factors.[23,30–32] In contrast to findings for alcoholism, studies from Finland, Sweden, and Australia, which used sample sizes sufficiently large to give good statistical power for resolving sex differences in the genetic and environmental determinants of drinking practices, yielded heritability estimates that were at least as high in females as in males.[25–27]

Given the robust evidence for genetic influences on alcohol consumption patterns, it is natural to question whether these genetic influences on consumption can explain at least part of the genetic variation in alcoholism risk. If genetic factors play a major role in determining an individual's "alcohol consumption set-point," the level of consumption at which an individual's drinking practices will settle by age 25, it seems likely that because of genetic influences, some individuals will maintain a level of consumption too low to put them at risk of drinking problems,[33] whilst other individuals will maintain consumption levels that may put them at much higher risk. High risk studies comparing adult offspring of alcoholics and controls in an alcohol challenge paradigm have found differences in such measures as subjective reactions to alcohol and objective measures of increase in body sway, after a standard body-weight adjusted dose of alcohol.[34] Data from the Australian alcohol challenge twin study confirm a significant genetic contribution to differences in performance after alcohol challenge,[35] but they also show a significant and substantial correlation between genetic effects on alcohol challenge performance and genetic effects on normal variation in alcohol consumption patterns.[36–38] It would therefore be surprising if genetic influences on consumption were not also having an influence on risk of alcohol problems. In the present paper, we attempt to address more directly the question of whether genetic influences on alcohol consumption pattern can explain in part genetic influences on alcoholism risk, using self-report data on consumption and history of problem drinking obtained in the 1988–1989 follow-up survey of the Australian National Health and Medical Research Council twin panel.

METHOD

Sample

The Australian National Health and Medical Research Council (NH&MRC) twin panel is a volunteer twin panel, recruited through the media, schools, and a variety of other sources.[27,39] In 1980–1981, a mailed questionnaire survey of adult twin pairs enrolled on the panel was conducted, questionnaires being mailed to 5,967 twin pairs aged 18 or older. Completed questionnaires were received from both members of 3,808 twin pairs (64% pairwise response rate) and from one twin only from an additional 576 pairs (69% individual response rate). The median age of respondents was 30. The questionnaire contained questions about personality, health, and lifestyle, including self-report drinking patterns, but no questions about alcohol-related problems. Although a volunteer sample, the "1981 survey" sample consists of individuals from a broad range of socioeconomic backgrounds. Among twins who returned both questionnaires, 33% of the female respondents and 19% of the male respondents had completed 10 or fewer years of education, and 27.1% and 23.7%, respectively, reported unskilled occupations. Eight percent of female respondents and 20% of male respondents were university graduates. Self-report drinking practices of this sample were broadly comparable to figures reported for the general population of Australia by the Australian Bureau of Statistics, modestly lower for male respondents and higher for female respondents.[27,40] As is commonly found in twin studies,[41] however, female twins and monozygotic (Mz) twins were over-represented in the sample; 1,232 MZ female, 567 MZ male, 747 DZ female, 350 DZ male, and 912 unlike-sex twin pairs returned questionnaires.

In 1988–1989, a three-phase follow-up survey of the 1981 survey sample was begun, to explore the changes in drinking practices and history of alcohol problems in this sample. A follow-up questionnaire was mailed to all 3,808 complete twin pairs participating in the 1981 survey who were still living and for whom a current address could be traced. Twins who did not return a mailed questionnaire were given the option of an abbreviated telephone interview. In subsequent phases of the follow-up (still in progress), twins receive a telephone screening interview to identify pairs in which at least one twin has a history of alcohol problems, for more intensive follow-up by in-person interview. In the first phase, mailed questionnaire or abbreviated telephone interview data were obtained from both members of 2,997 twin pairs (82% pairwise response rate if we exclude from the target sample those 139 twin pairs in which one or both twins were decreased by the time of follow-up). The same self-report questionnaire was remailed in 1990 to 500 male and 500 female twin individuals who had returned follow-up questionnaires, to provide 2-year test-retest data, with completed returns being returned by 427 male and 442 female respondents. Data on self-report drinking practices and drinking problems from this first questionnaire phase are analyzed in this paper. Inasmuch as data from lifetime abstainers were excluded, as were data from twins whose cotwin either had not responded or had omitted to answer either problem-drinking or alcohol consumption questions, final sample sizes for the analyses presented in this paper were: 726 MZ female pairs, 406 DZ female pairs, 321 MZ male pairs, 178 DZ male pairs, and 454 unlike-sex pairs.

Measures

Drinking problems were assessed using items selected from a self-report questionnaire based on Feighner criteria for alcoholism,[42] which had been used for a national survey in the United States,[43] with the original hope of improving discrimination between type I and type II alcoholics.[44] Questions about alcohol withdrawal symptoms and alcoholic blackouts were not used because of concern about whether respondents would understand these items when given in a self-report questionnaire. Three items relating to self-perceived excessive consumption of alcohol, objections about one's drinking from others, and guilty feelings about one's drinking appeared to have low specificity when used to assess problem drinking in this sample, particularly in female respondents, and these items are not used in the analyses presented here. TABLE 1 (in Results section) summarizes the items used to assess alcohol problems and the corresponding Feighner group for each item. Respondents were asked to indicate whether they had ever experienced any of these problems, and if so, whether the problem had occurred during each of the following time periods: (1) last 12 months; (2) 1–3 years ago; (3) 4–6 years ago; and (4) 7 or more years ago (which would overlap with the time of the 1981 mailing). For each time period the number of reported problems was summed, and a lifetime problem-drinking score was computed as the maximum score from any time period. Whilst a self-report questionnaire format does not allow detailed probing to permit accurate identification of symptoms that occurred together in time, the scoring procedure used at least provides some limited information about clustering of symptoms in the same period of the respondent's life.[13] Self-report average weekly alcohol consumption was assessed by a single question which asked respondents to rate their own typical weekly consumption on a 9-point scale, with response categories (1) none, (2) 1–3 drinks, (3) 4–6 drinks, (4) 7–12 drinks, (5) 13–18 drinks, (6) 19–24 drinks, (7) 25–42 drinks, (8) 43–70 drinks, and (9) 70+ drinks per week, with one drink defined as "a can/stubby of beer, a glass of wine or nip of spirits."

Data Analyses

Initial exploratory analyses of the psychometric properties of the problem-drinking scale (computation of Cronbach's alpha[45]; principal factor analysis[46]) ignored the twin structure of our data, that is, the nonindependence of observations on twin pairs. Under random sampling (which we assume is at least approximately achieved here), this will still lead to unbiased estimates of population parameters, although the sampling variance of those parameters will be underestimated. Exploratory analyses also used product-moment correlations between items, ignoring the complication that these were dichotomous and their joint distribution therefore highly non-normal. As a consequence, estimates of factor loadings and Cronbach's alpha will be lower than would have been the case had we used tetrachoric correlations, but they will be more comparable to publications on other scales in which a similar approach was typically used. For these exploratory analyses, alcohol problem items were recoded as lifetime presence or absence of symptoms, without regard for clustering of symptoms in the same period of the respondent's life. In computing 2-year test-retest correlations for average weekly consumption and maximum lifetime problem score, polychoric correlations were used.[47,48]

For genetic analyses, we computed 4×4 matrices of polychoric correlations between problem-drinking and average weekly alcohol consumption scores of first and second twins, separately for each zygosity group. Genetic models were fitted by asymptotic weighted least squares using LISREL.[49–53] We fitted a general bivariate genetic model,[53,54] allowing for sex-dependent common factor genetic, shared environmental and within-family environmental effects on both weekly consumption and history of problems as well as residual or specific-factor genetic and environmental effects on problem drinking. Parameters of the full bivariate genetic model[53,54] are summarized in FIGURE 1. In the most general model, separate parameters were estimated for males and females (distinguished by subscripts m and f). We also fitted submodels of the general bivariate genetic model, deleting one or more parameters of the full model (e.g., fixing the genetic common factor effect on problems to

"COMMON-FACTOR" EFFECTS   "SPECIFIC FACTOR" EFFECTS

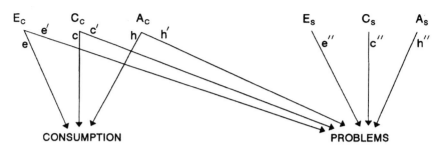

FIGURE 1. Bivariate genetic path model[53,54] for genetic and environmental contributions to the variation and covariation of alcohol consumption levels and problem drinking. E_c, C_c, and A_c denote within-family environmental, shared environmental, and additive genetic common factor effects, which influence both consumption and problems (paths e, c, and h for consumption, e', c', and h' for problems); and E_S, C_S, and A_S denote residual or specific-factor within-family environmental, shared environmental, and additive genetic effects on problems (paths e'', c'', and h'', respectively).

zero [h'=0], to test the hypothesis that there is no correlation between genetic effects on consumption and genetic effects on problem drinking, or fixing the genetic specific factor effect on problems to zero [h'' = 0], to test the hypothesis that the same genes that determine level of consumption also determine risk of problem drinking). For each model we obtained weighted least squares estimates of model parameters as well as an overall chi-square test of the goodness-of-fit of the model.[50,51,53] The overall goodness-of-fit of nested models was compared by likelihood-ratio chi-square ("chi-square difference") test, with number of degrees of freedom equal to the number of free parameters of the more general model fixed to zero in the full model.

Fitting a bivariate genetic model allowed us to estimate the correlation between genetic effects on problem drinking and genetic effects on alcohol consumption level, estimated as

$$r_G = h' \, h'' \, (h''[h^2 + h'^2]^{0.5})^{-1},$$

and to estimate in similar fashion correlations between shared environmental effects on consumption level and on problem drinking and correlations between within-family environmental effects on consumption level and on problem drinking. However, since we were analyzing current drinking patterns, but lifetime history of problem drinking, significant and substantial genetic and environmental correlations might arise because of a causal effect of problem drinking on alcohol consumption pattern. To test this hypothesis, we also fitted strong causal models (FIG. 2) which assumed that a positive genetic correlation between consumption and problem drinking arises because of the causal influence of problem drinking on consumption pattern ($i_a>0$) or because of the causal influence of consumption pattern on problem drinking ($i_p>0$). Parameters of the most general reciprocal causation submodel are summarized in FIGURE 2. We have shown elsewhere[53,54] that the strong unidirectional causation and reciprocal causation models are a submodel of the general bivariate genetic model of FIGURE 1.

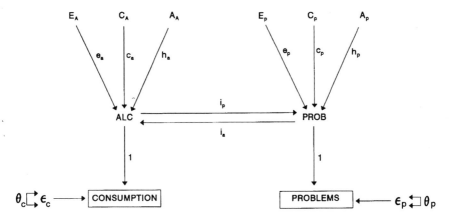

FIGURE 2. Reciprocal causation path model allowing for the reciprocal causal influence of alcohol consumption level on problem drinking and of problem drinking on consumption pattern. ALC and PROB denote true phenotypes for consumption level and for problem drinking, with reciprocal causal paths i_p and i_a denoting the causal influence of consumption on problems, and of problems on consumption, respectively; and CONSUMPTION and PROBLEMS denote the corresponding observed variables, with error variances θ_c and θ_P. Variables E_A, C_A, and A_A denote direct within-family environmental, shared environmental, and additive genetic effects on consumption (i.e., excluding indirect effects mediated through the causal influence of problems on consumption), with corresponding paths e_a, c_a, and h_a; and E_P, C_P, and A_P denote direct within-family environmental, shared environmental, and additive genetic effects on alcohol problems, excluding those effects mediated through the causal influence of consumption on problems.

RESULTS

TABLE 1 summarizes factor loadings under a one-factor model, estimated separately for males and for females, for items of the problem-drinking scale. In females ($n = 3,426$), the item "Drinks in morning to get rid of hangover" had only a modest factor loading (0.21), but loadings of all other items were substantial, ranging from a high of 0.75 ("treated for a drinking problem") to a low of 0.53 (adverse effects of drinking on health). In males ($n = 1,819$), several items had modest loadings of about 0.4 ("drunk driving"; "treated for a drinking problem"; "drinks in the morning to get rid of a hangover"), but the remaining items had substantial loadings in the range of 0.5–0.6. In two-factor solutions, the second factor in females had moderately high loadings only on the two items relating to cutting down on drinking, and the second factor in males had no clear interpretation, implying that the scale used is assessing a single global factor of alcohol-related problems. Internal consistency was high in both sexes: Cronbach's alpha for lifetime problems was 0.87 in females and 0.83 in males; and for current problems, that is, problems during the preceding 12 months, was 0.92 in females and 0.84 in males. In the reliability substudy, 2-year test-retest correlations for lifetime total problems were also high, with polychoric correlations of 0.72 in females ($n = 373$) and 0.85 in males ($n = 350$) being obtained. Thus, psychometric properties of the problem-drinking scale appear satisfactory. The test-retest correlation for average weekly consumption was 0.83 in females ($n = 414$) and 0.84 in males ($n = 408$).

TABLE 1. Factor Loadings and Endorsement Frequencies of Items of the Problem-Drinking Scale: Items were Recorded as Lifetime Presence or Absence of Problems

Feighner Group	Item Content	Factor Loadings Females	Factor Loadings Males	Endorsement Frequency (%) Females	Endorsement Frequency (%) Males
A	Adverse effects on health	0.53	0.54	4.8	11.5
A	Binge drinking	0.70	0.57	1.0	7.2
A	Neglected responsibilities while drinking	0.68	0.61	1.9	9.0
B	Inability to cut down on drinking	0.59	0.56	2.2	6.0
B	Failure to stick to plan to stop drinking	0.57	0.57	3.2	6.5
B	Drinks in morning to get rid of hangover	0.21	0.39	1.5	8.8
C	Adverse effects on work/employment opportunities	0.73	0.63	0.9	4.9
C	Physical fights while drinking	0.61	0.50	1.3	8.4
C	Drunk driving	0.54	0.38	1.5	11.0
D	Adverse effects on friendships/social life	0.65	0.65	2.5	11.1
D	Adverse effects on marriage/home life	0.65	0.60	2.4	11.4
—	Treatment for drinking problem	0.75	0.41	0.7	1.7

Also shown in TABLE 1 are endorsement frequencies for the individual items of the problem-drinking scale. These emphasize that the Australian twin sample is a general community sample, in which only a small minority of individuals have received treatment for alcohol problems (0.7% of women, 1.7% in men) and in which the most common alcohol problems relate to adverse social and health consequences of the respondent's drinking and (particularly in men) drunk driving.

TABLE 2 gives the 4 × 4 matrices of polychoric correlations between lifetime problem-drinking score and current consumption pattern of first and second twins from each zygosity group. In female like-sex twin pairs, twin pair correlations for average weekly alcohol consumption and for reported history of drinking problems are approximately twice as high in monozygotic as in dizygotic twin pairs (0.59 vs. 0.28 and 0.43 vs. 0.24), consistent with important additive genetic influences on both consumption and problem drinking. Furthermore, cross-correlations between problem drinking by one twin and alcohol consumption patterns of the cotwin are on average substantially higher in monozygotic than dizygotic pairs, implying a genetic contribution to the covariance of consumption pattern and problem drinking. In male like-sex pairs, for whom sample sizes are much smaller, monozygotic and dizygotic correlations for problem drinking are approximately equal in magnitude. However, before we interpret this as evidence against a genetic influence on problem drinking in Australian males, two complications must be noted. A significant correlation is observed between opposite-sex pairs for problem drinking (0.31), instead of the zero correlation which would be predicted if there was a genetic but no shared environmental influence on female problem drinking, and a shared environmental but no genetic influence on male problem drinking. Furthermore, a higher MZ than DZ male like-sex correlation is observed for current consumption, and substantially higher MZ than DZ cross-correlations between current consumption by one twin and problem drinking by the cotwin, suggesting that there may in fact be genetic influences on problem drinking by males, and that our failure to observe higher MZ than DZ male like-sex correlations for problem drinking may be a consequence of sampling variability, given the rather small sample sizes for the male like-sex groups.

TABLE 2. Twin Pair Correlations for Current Average Weekly Alcohol Consumption and Lifetime Problem Drinking Score

	MZ Female Pairs (n = 726)				DZ Female Pairs (n = 406)			
	I	II	III	IV	I	II	III	IV
I. Problem drinking—twin A	1.00				1.00			
II. Current consumption—twin A	0.43	1.00			0.22	1.00		
III. Problem drinking—twin B	0.43	0.19	1.00		0.24	0.17	1.00	
IV. Current consumption—twin B	0.25	0.59	0.18	1.00	-0.02	0.28	0.32	1.00

	MZ Male Pairs (n = 321)				DZ Male Pairs (n = 178)			
I. Problem drinking—twin A	1.00				1.00			
II. Current consumption—twin A	0.47	1.00			0.31	1.00		
III. Problem drinking—twin B	0.44	0.20	1.00		0.43	0.12	1.00	
IV. Current consumption—twin B	0.27	0.43	0.33	1.00	0.03	0.33	0.42	1.00

	DZ Unlike-Sex Pairs (n = 454)			
I. Problem drinking—female twin	1.00			
II. Current consumption—female twin	0.40	1.00		
III. Problem drinking—male twin	0.31	0.03	1.00	
IV. Current consumption—male twin	0.09	0.20	0.40	1.00

Results of fitting bivariate genetic models to the alcohol consumption and problem-drinking data are summarized in TABLE 3. The full model allowing for sex-dependent parameters gives an acceptable fit to the data ($p = 0.25$). Models which specify no genetic influences on consumption (model 2) or no common factor genetic influences on problems (model 4) are rejected by chi-square test of goodness-of-fit. Models which specify no shared environmental effects on consumption (model 3), no specific factor shared environmental effects on problems (model 6), no specific factor genetic effects on problems (model 5), or neither specific factor genetic nor shared environmental effects (model 7) all give acceptable fits to the data, and fits which are not significantly worse than that of the full model, by likelihood-ratio chi-squares test ($\chi^2 < 2.5$ in all cases). However, a model which specifies no shared environmental effects on problems or consumption and no specific factor genetic effects on problems is rejected by chi-square test of goodness of fit (model 8: p <0.001). We must retain in the model either specific factor genetic effects (model 9) or specific factor shared environmental effects (model 10), with the latter model giving a slightly (though not significantly) better fit. Constraining the parameters of model 9 or model 10 to be the same in both sexes in each case leads to a significant worsening of fit, by likelihood-ratio chi-square test (model 9 vs 11: $\chi^2 = 9.47$, d.f. = 4, $p = 0.05$; model 10 vs 12: $\chi^2 = 9.79$, d.f. = 4, $p < 0.05$).

FIGURE 3 summarizes parameter estimates under models 9 and 10, the two best-fitting models. Model 9 yields a total heritability for problem drinking of 44%

TABLE 3. Results of Fitting Bivariate Genetic Models to Twin Pair Correlation Matrices for Problem Drinking and Alcohol Consumption Patterns

	Goodness-of-Fit		
	d.f.	χ^2	p
1. Full sex-dependent bivariate genetic model	16	19.37	0.25
2. No genetic effects on consumption (hm=hm'=hf=hf'=0)	20	62.36	<0.001
3. No shared environmental effects on consumption (cm=cm'=cf=cf'=0)	20	21.57	0.37
4. No common factor genetic effects on problems (hm'=hf'=0)	18	29.52	0.04
5. No specific factor genetic effects on problems (hm''=hf''=0)	18	20.44	0.31
6. No specific factor shared environmental effects on problems (cm''=cf''=0)	18	20.49	0.31
7. No specific factor genetic or shared environmental effects (hm''=hf''=cm''=cf''=0)	20	21.00	0.40
8. Model 1 with hm''=hf''=cm=cm'=cf=cf'=cf''=0	24	64.18	<0.001
9. No shared environmental effects (cm=cm'=cm''=cf=cf'=cf''=0)	22	25.90	0.26
10. Model 1 with hm''=hf''=cm=cm'=cf=cf'=0	22	22.66	0.42
11. Model 9 with no sex-dependent parameters	26	35.37	0.10
12. Model 10 with no sex-dependent parameters	26	32.45	0.18

in females, 50% in males; a heritability for average weekly alcohol consumption of 58% in females, 45% in males; and a genetic correlation (i.e., correlation between genetic effects on risk of problem drinking and genetic effects on alcohol consumption) of 0.42 in females, 0.45 in males. Under model 10, heritability estimates for average weekly alcohol consumption remain the same, but the heritability of risk of problem drinking is reduced to 8% in females, 10% in males, with shared environmental effects accounting for an additional 29% and 35% of the variance, respectively. Under model 10, there is a perfect correlation between genes which affect consumption and genes which determine risk of problem drinking, because there are no specific factor genetic effects on consumption.

When we fitted unidirectional causation models (FIG. 2), to test whether the significant genetic correlations observed between problem-drinking history and current consumption levels could be explained by the effects of problem drinking on consumption pattern rather than of consumption pattern on problem drinking, results were rather mixed. When we fixed $h_{pm} = h_{pf} = 0$ (for comparison with model 10), and instead estimated causal parameters i_{pm} and i_{pf} representing the causal influence of problem drinking on consumption patterns, this model gave a much worse fit to the data ($\chi^2 = 31.89$, d.f. = 24, $p = 0.13$), and this fit was little improved by estimating male and female error variances for problem drinking as free parameters ($\chi^2 = 31.36$, d.f. = 22, $p = 0.09$). However, if instead we allowed for a causal influence of consumption pattern on risk of problem drinking (i.e., $i_{pm} = i_{pf} = 0$ and $i_{am}, i_{af} > 0$), this model gave an excellent fit to the data ($\chi^2 = 24.34$, d.f. = 24, $p = 0.44$), consistent with the hypothesis that genetically determined differences in drinking style are contributing to risk of problem drinking. However, when we tested direction-of-causation models using the assumptions about mode of inheritance of model 9 (i.e., $h_{pm}, h_{pf} > 0$), there was no power to resolve alternative causal hypotheses. For

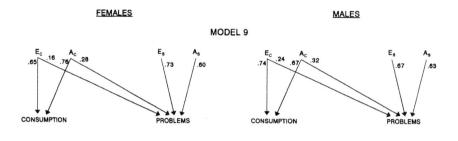

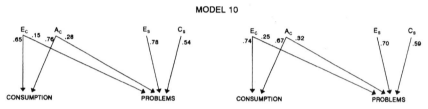

FIGURE 3. Parameter estimates under best-fitting models.

example, under this model, the hypothesis that problem-drinking history is affecting current consumption pattern yielded a chi-square of 27.67, and the alternative hypothesis that consumption pattern is influencing risk of problem drinking yielded an almost identical chi-square of 27.58.

DISCUSSION

We have analyzed self-report questionnaire data on history of alcohol problems and current alcohol consumption patterns obtained in a follow-up survey of the Australian NH&MRC twin register, to explore the relation between genetic influences on consumption and genetic influences on risk of problems. Our measure of problem drinking, based on Feighner criteria, had good psychometric properties, though the much higher loadings of items such as "treatment for a drinking problem" in females than in males is consistent with the possibility that the scale is a better measure of alcohol problems in women (the primary focus of our study) than in men. Model-fitting confirmed significant genetic influences on both consumption and problem drinking. However, whilst estimates of the heritability of average weekly consumption were consistent across the two best-fitting models (58% in females, 45% in males), estimates of the heritability of alcohol problems, though significantly different from zero, differed markedly according to whether we allowed for specific-factor genetic effects on problems (model 9: heritability estimates of 44% in females, 50% in males) or specific-factor shared environmental effects on problems (model 10: 8% in females, 10% in males). The imprecision of our heritability estimates for alcohol problems is undoubtedly a consequence of the reduced power for resolving genetic and shared environmental effects on dichotomous variables,[53] particularly when the degree of familial resemblance is relatively modest.[55] As interview data become available, allowing for assessment of tolerance and withdrawal, we anticipate that higher twin-pair correlations will be observed,[20] allowing for a much more powerful resolution of genetic and nongenetic effects.

Interpretation of our results is highly model dependent. Under model 10, which yielded a very modest heritability for alcohol problems in both sexes, all of the genetic variance in risk for alcohol problems could be explained by genetic factors that also influenced alcohol consumption patterns. Under model 9, which yielded more substantial heritability estimates for alcohol problems, genetic correlations were more modest (0.42 in females, 0.45 in males), though significantly different from zero. Furthermore, under model 9 we could not exclude the possibility that these genetic correlations were the consequences of the influence of history of drinking problems on current consumption patterns. Inclusion of baseline data on alcohol consumption patterns from the 1981 survey, and interview data on history of alcoholism from phases two and three of the follow-up survey, may allow these ambiguities to be resolved with considerably greater statistical power.

SUMMARY

Self-report questionnaire data from 3,000 adult twin pairs participating in the 1988–1989 follow-up survey of the Australian NH&MRC twin panel were analyzed to determine (1) the contribution of genetic factors to risk of problem drinking in males and females; and (2) the magnitude of the correlation between genetic effects on problem drinking and genetic effects on alcohol consumption level. Significant genetic contributions were found both for average weekly consumption of alcohol and for problem-drinking history. For level of consumption, genetic factors accounted for approximately 58% of the variation in females and 45% of the variation in males. Heritability estimates for problem drinking, though significantly greater than zero, were variable in magnitude, ranging (under different models) from 8–44% in females and 10–50% in males. Likewise, estimates of the magnitude of the genetic correlation, whilst in all cases significantly greater than zero, ranged from 0.42–1.00 in females and 0.45–1.00 in males under different models.

ACKNOWLEDGMENTS

We are grateful to Ann Eldredge, Sue Mason, and Olivia Zhang for data collection, to Ulrich Kehren (deceased) and John Pearson for data management, and to Michael Hodge for assistance with programming and data analyses.

REFERENCES

1. GOODWIN, D. W., F. SCHULSINGER, L. HERMANSEN, S. B. GUZE & G. WINOKUR. 1973. Alcohol problems in adoptees raised apart from alcoholic biological parents. Arch. Gen. Psychiatry 28: 238–243.
2. CLONINGER, C. R., M. BOHMAN & S. SIGVARDSSON. 1981. Inheritance of alcohol abuse: Cross-fostering analysis of adopted men. Arch. Gen. Psychiatry 38: 861–868.
3. CADORET, R. J., C. A. CAIN, E. TROUGHTON & E. HEYWOOD. 1985. Alcoholism and antisocial personality. Interrelationships, genetic and environmental factors. Arch. Gen. Psychiatry 42: 161–167.

4. CADORET, R. J., E. TROUGHTON & T. W. O'GORMAN. 1987. Genetic and environmental factors in alcohol abuse and antisocial personality. J. Stud. Alcohol **48**: 1.
5. SCHUCKIT, M. A., D. W. GOODWIN & G. WINOKUR. 1972. A study of alcoholism in half siblings. Am. J. Psychiatry **128**: 1132–1136.
6. KAIJ, L. 1960. Alcoholism in Twins: Studies on the Etiology and Sequels of Abuse of Alcohol. Almqvist & Wiksell International. Stockholm.
7. HRUBEC, Z. & G. S. OMENN. 1981. Evidence of genetic predisposition to alcoholic cirrhosis and psychosis: Twin concordances for alcoholism and its biological points by zygosity among male veterans. Alc. Clin. Exper. Res. **5**: 207–215.
8. KAPRIO, J., M. D. KOSKENVUO, H. LANGINVAINIO, K. ROMANOV, S. SARNA & R. J. ROSE. 1987. Genetic influences on use and abuse of alcohol: A study of 5638 adult Finnish brothers. Alc. Clin. Exper. Res. **11**: 349–356.
9. PICKENS, R. W., D. S. SVIKIS, M. McGUE, D. T. LYKKEN, L. L. HESTON & P. J. CLAYTON. 1991. Heterogeneity in the inheritance of alcoholism: A study of male and female twins. Arch. Gen. Psychiatry **48**: 19–28.
10. CALDWELL, C. B. & I. I. GOTTESMAN. 1991. Sex differences in the risk for alcoholism: A twin study. Behav. Genet. **21**: 563.
11. McGUE, M., R. W. PICKENS & D. S. SVIKIS. 1992. Sex and age effects on the inheritance of alcohol problems: A twin study. J. Abnorm. Psychol. **101**: 3–17.
12. ROMANOV, K., J. KAPRIO, R. J. ROSE & M. KOSKENVUO. 1992. Genetics of alcoholism: Effects of migration on concordance rates among male twins. Alcohol Alcohol. Suppl. **1**: 137–140.
13. PRESCOTT, C. A., J. K. HEWITT, K. R. TRUETT, A. C. HEATH, M. C. NEALE & L. J. EAVES. 1993. Alcohol use and abuse in a community sample of older twins. J. Stud. Alcohol, in press.
14. ROE, A. & B. BURKS. 1945. Adult Adjustment of Foster Children of Alcoholic and Psychotic Parentage and the Influence of the Foster Home. No. 3. Memoirs of the Section on Alcohol Studies. Yale University Press. New Haven, CT.
15. PARTANEN, J., K. BRUUN & T. MARKKANEN. 1966. Inheritance of drinking behavior: A study on intelligence, personality, and use of alcohol of adult twins. *In* Alcohol Research in the Northern Countries. The Finnish Foundation for Alcohol Studies, Vol. 14. Amqvist & Wiksell. Stockholm.
16. GURLING, H. M. D., S. GRANT & J. DANGL. 1985. The genetic and cultural transmission of alcohol use, alcoholism, cigarette smoking and coffee drinking: A review and an example using a log linear cultural transmission model. Br. J. Addict. **80**: 269–279.
17. ALLGULANDER, C., J. NOWACK & J. P. RICE. 1991. Psychopathology and treatment of 30,344 twins in Sweden. II. Heritability estimates of psychiatric diagnosis and treatment in 12,884 twin pairs. Acta Psychiatr. Scand. **83**: 12–15.
18. BOHMAN, M., S. SIGVARDSSON & C. R. CLONINGER. 1981. Maternal inheritance of alcohol abuse: Cross fostering analysis of adopted women. Arch. Gen. Psychiatry **38**: 965.
19. GOODWIN, D. W., F. SCHULSINGER, J. KNOP, S. MEDNICK & S. GUZE. 1977. Psychopathology in adopted and nonadopted daughters of alcoholics. Arch. Gen. Psychiatry **34**: 1005.
20. KENDLER, K. S., A. C. HEATH, M. C. NEALE, R. C. KESSLER & L. J. EAVES. 1992. A population-based twin study of alcoholism in women. JAMA **268**: 1877–1882.
21. PRESCOTT, C. A., J. K. HEWITT, A. C. HEATH, K. R. TRUETT, M. C. NEALE & L. J. EAVES. 1993. Genetic and environmental contributions to alcohol use in a community sample of older twins. J. Stud. Alcohol, in press.
22. HEATH, A. C. 1993. Genetic influences on drinking behavior in humans. *In* Alcohol and Alcoholism, Vol. 1, Genetic Factors and Alcoholism. H. Begleiter & B. Kissin, eds. Oxford University Press. Oxford, in press.
23. FILLMORE, K. M., E. HARTKA, B. M. JOHNSTONE, E. V. LEINO, M. MOTOYOSHI & M. T. TEMPLE. 1991. A meta-analysis of life course variation in drinking. Br. J. Addict. **86**: 1221–1268.
24. CARMELLI, D., G. E. SWAN, D. ROBINETTE & R. R. FABSITZ. 1990. Heritability of substance use in the NAS-NRC twin registry. Acta Genet. Med. Gemellol. **39**: 91–98.
25. KAPRIO, J., R. J. ROSE, K. ROMANOV & M. KOSKENVUO. 1991. Genetic and environmental

determinants of use and abuse of alcohol: The Finnish twin cohort studies. Alcohol Alcohol. Suppl. **1**: 131–136.

26. MEDLUND, P., R. CEDERLOF, B. FLODERUS-MYRHED, L. FRIBERG & S. SORENSEN. 1977. A new Swedish twin registry. Acta Med. Scand. (Suppl. 600).

27. JARDINE, R. & N. G. MARTIN. 1984. Causes of variation in drinking habits in a large twin sample. Acta Genet. Med. Gemellol. **33**: 435–450.

28. HEATH, A. C., J. MEYER, R. JARDINE & N. G. MARTIN. 1991. The inheritance of alcohol consumption patterns in a general population twin sample. II. Determinants of consumption frequency and quantity consumed. J. Stud. Alcohol **52**: 425–433.

29. CLIFFORD, C. A., J. L. HOPPER, D. W. FULKER & R. M. MURRAY. 1984. A genetic and environmental analysis of a twin family study of alcohol use, anxiety, and depression. Genet. Epidemiol. **1**: 63–79.

30. KAPRIO, J., R. VIKEN, M. KOSKENVUO, K. ROMANOV & R. J. ROSE. 1992. Consistency and change in patterns of social drinking: A 6-year follow-up of the Finnish twin cohort. Alc. Clin. Exper. Res. **16**: 234–240.

31. CARMELLI, D., A. C. HEATH & D. ROBINETTE. 1993. A genetic analysis of drinking behavior in World War II veteran twins. Genet. Epidemiol. **10**: 201–213.

32. HEATH, A. C. & N. G. MARTIN. 1991. Persistence and change in drinking habits. Paper presented at the Research Society on Alcoholism, Marco Island, Florida.

33. HEATH, A. C., J. M. MEYER & N. G. MARTIN. 1990. The inheritance of alcohol consumption patterns in the Australian twin survey, 1981. *In* Banbury Report 33: Genetics and Biology of Alcoholism. H. Begleiter & C. R. Cloninger, eds. Cold Spring Harbor Laboratory Press. Plainview, NY.

34. SCHUCKIT, M. A. & E. O. GOLD. 1988. A simultaneous evaluation of multiple markers of ethanol/placebo challenges in sons of alcoholics and controls. Arch. Gen. Psychiatry **45**: 211–216.

35. MARTIN, N. G., J. G. OAKESHOTT, J. B. GIBSON, G. A. STARMER, J. PERL & A. V. WILKS. 1985. A twin study of psychomotor and physiological response to an acute dose of alcohol. Behav. Genet. **15**: 305–347.

36. HEATH, A. C. & N. G. MARTIN. 1991. The inheritance of alcohol sensitivity and of patterns of alcohol use. Alcohol Alcohol. Suppl. **1**: 141–145.

37. HEATH, A. C. & N. G. MARTIN. 1991. Intoxication after an acute dose of alcohol. An assessment of its association with alcohol consumption patterns by using twin data. Alc. Clin. Exper. Res. **15**: 122–128.

38. HEATH, A. C. & N. G. MARTIN. 1992. Genetic differences in psychomotor performance decrement after alcohol: A multivariate analysis. J. Stud. Alcohol **53**: 262–271.

39. JARDINE, R. 1985. A twin study of personality, social attitudes, and drinking behavior. Ph.D. thesis, Australian National University.

40. Australian Bureau of Statistics. 1978. Alcohol and tobacco consumption patterns. *In* Catalogue No. 4312.0.

41. LYKKEN, D. T., A. TELLEGEN & R. DeRUBIES. 1978. Volunteer bias in twin research: The rules of two thirds. Soc. Biol. **25**: 1–9.

42. FEIGHNER, J. P., E. ROBINS, S. B. GUZE, R. A. WOODRUFF, G. WINOKUR & R. MUNOZ. 1972. Diagnostic criteria for use in psychiatric research. Arch. Gen. Psychiatry **26**: 57–63.

43. CLONINGER, C. R., T. R. PRZYBECK & D. M. SVRAKIC. 1991. The tridimensional personality questionnaire: U.S. normative data. Psychol. Rep. **69**: 1047–1057.

44. CLONINGER, C. R. 1987. Neurogenetic adaptive mechanisms in alcoholism. Science **236**: 410–416.

45. CRONBACH, L. J. 1951. Coefficient alpha and the internal structure of tests. Psychometrika **16**: 297–334.

46. HARMAN, H. H. 1976. Modern Factor Analysis. University of Chicago Press. Chicago.

47. OLSSON, U. 1979. Maximum-likelihood estimation of the polychoric coefficient. Psychometrika **44**: 443–460.

48. JORESKOG, K. & D. SORBOM. 1988. PRELIS: A preprocessor for LISREL. Scientific Software. Mooresville, IN.

49. BROWNE, M. W. 1984. Asymptotically distribution-free methods for the analysis of covariance structures. Br. J. Math. Stat. Psychol. **37**: 1–21.
50. JORESKOG, K. & D. SORBOM. 1988. LISREL VII. Analysis of Linear Structural Relationships. Scientific Software. Mooresville, IN.
51. HEATH, A. C., M. C. NEALE, J. K. HEWITT, L. J. EAVES & D. W. FULKER. 1989. Testing structural equation models for twin data using LISREL. Behav. Genet. **19**: 9–35.
52. NEALE, M. C. A. C. HEATH, J. K. HEWITT, L. J. EAVES & D. W. FULKER. 1989. Fitting genetic models with LISREL: Hypothesis testing. Behav. Genet. **19**: 37–49.
53. NEALE, M. C. & L. CARDON. 1992. Methodology for Genetic Studies of Twins and Families, NATO ASI Series. Kluwer Academic Publishers. Dordrecht, the Netherlands.
54. HEATH, A. C., R. C. KESSLER, M. C. NEALE, J. K. HEWITT, L. J. EAVES & K. S. KENDLER. 1993. Testing hypotheses about direction of causation using cross-sectional family data. Behav. Genet. **23**: 29–50.
55. MARTIN, N. G., L. J. EAVES, M. J. KEARSEY & P. DAVIES. 1978. The power of the classical twin study. Heredity **40**: 97–116.

Variants of Alcoholism: Patterns in Development, Course, and Prognosis[a]

ANDREY G. VRUBLEVSKY

Department of Polysubstance Dependence
State Research Center on Addictions
Russian Federation, Ministry of Health
Moscow, Russia

Numerous factors are thought to play a role in the development of alcoholism. Clinical investigations conducted in the last 10–15 years, which include twin studies and experimental investigations, indicate that there are genetic determinants in a number of behavioral, physiologic, and biochemical responses to ethanol. Several investigators have proposed the presence of a biological predisposition for the development of alcohol dependence in a certain percentage of the population.[1-5] Nevertheless, it is evident that environmental factors and psychological mechanisms also play a major role in the development of alcoholism. Some investigators even attach primary importance to these.[6]

Thus, existing evidence is in good agreement with the theory of polygenic diseases which specifies that an individual's susceptibility is determined by a genetically determined predisposition and the detrimental effect of exogenous environmental factors.[7] However, major individual differences exist in genetic predisposition, other biological features, personality traits, reactions to external stimuli, and various exogenous factors, including craniocerebral trauma and somatic pathology. All of these factors produce variations in the development of alcoholism, course and clinical signs, response to treatment, appearance of cross-dependence to various psychoactive substances, and markedness of social and medical consequences. This can also explain the existence of controversial and at times directly contradictory points of view on the nature of addictive disorders.

Virtually no data exist on the formation, course, and clinical signs of alcoholism in patients with a marked genetic predisposition for the disease in relation to those in whom genetic factors are much less significant; however, there are isolated reports on the effect of hereditary factors on the clinical picture of alcoholism.[8-10] The problem of predicting the clinical course of the disease and the effectiveness of treatment methods is particularly difficult.

A limitation of most studies is the absence of a complex multifactorial approach. Moreover, even in those studies in which efforts are made to evaluate the role of multiple parameters (diagnostic, prognostic, therapeutic, etc.), the scope essentially does not proceed beyond the establishment of typical "average" patterns. The problem with this approach is that the statistical average model of a patient does not exist in reality. Nevertheless, the selection of an appropriate approach to preventive, clinical, and therapeutic objectives depends on a vast number of interrelated variables. These are related to biological characteristics, clinical signs of the disease, and numerous environmental factors. At present, primary research questions are generally approached from the viewpoint of group prognosis. Group prognosis permits the prediction, for example, of an average percentage of certain features of a disease, or average effectiveness of treatment, for an entire group of patients. But the wide variation in parameters, as is generally observed in group prognosis, does not permit objective evaluation of the development of a disease or the effectiveness of a given treatment method in each specific case. To resolve these questions, current research sought to evaluate individual diagnostic and prognostic methods, which allow characterization of each specific patient and the prediction of the course of the disease in the individual. This can then be used as a basis for instituting optimal therapeutic and preventive measures.

Individual prognosis is a highly complicated undertaking because it takes into account the mutual effect of clinical, biological, social, and other variables. Previously, statistical methods did not provide an easy solution to this problem because of the need for a cumbersome mathematical system. Similar individual prognostic systems in medicine became possible only in recent years as a result of the development of special algorithms and computer programs based on multifactorial mathematical analysis.[11] The use of multifactorial mathematical analysis in medicine makes it possible to solve isolated problems of individual prognosis in various pathological conditions, such as cancer.[12] Similarly, the use in this study of not only the statistical but also the analytical capabilities of computer technology made it possible to consider a multitude of interrelations among different variables—clinical, biological, social, and therapeutic—that affect clinical signs and the effectiveness of treatment.

The goal of this research was to study patterns in the formation and course of alcoholism as a function of the biological predisposition for its development, in order to identify variants of the disorder and to develop methods for individual clinical and therapeutic prognosis. The objectives were:

1. To isolate a group of patients with an endogenous (biological) predisposition for the development of alcoholism on the basis of a variety of clinical and time-based indicators.

2. To compare defined groups of alcoholic patients with a biological predisposition (endoform variant) and without it (exoform variant) and to produce criteria for a differential diagnosis.

3. To conduct a comparative analysis of biological and social indicators in the two groups in the premorbid period and to identify indicators that point to an unfavorably progressing course of the disease.

4. To evaluate the effectiveness of various methods of treatment for endoform and exoform alcoholism.

5. To study the clinical features of psychoactive drug dependence in patients with endoform and exoform variants of alcoholism.

METHODS

A sample of 352 patients with alcoholism were studied with the use of psychopathologic evaluations, prospective follow-ups, and retrospective methods. Substance dependence secondary to the alcoholism developed in 121 of these patients. The primary patient group, for which clinical patterns of alcoholism and features of its course and therapy were identified, consisted of 318 patients (231 with alcoholism and 87 with substance dependence secondary to alcoholism). We could not detail the development of clinical symptoms of alcoholism in 34 patients with substance dependence arising secondary to alcoholism. The development and course of substance dependence secondary to alcoholism were studied in all patients in this group (121 patients). All patients were men. The overwhelming majority of patients ranged from 21–50 years of age. The study includes patients with all stages of alcoholism and different lengths of illness. The majority of patients had been ill for 6–15 years.

Most had completed secondary school or specialized education, but some individuals had partial secondary education and some had completed higher education. The majority of patients were employed or students. Manual laborers predominated; white collar workers made up a smaller portion, and students and retired individuals the smallest portion. Thus, the group can be regarded as representative of the general population.

Univariate and multivariate analyses revealed the most informative parameters in terms of diagnosis and prognosis, which were evaluated by an informativeness coefficient (IC).

The distribution of indicators in the groups with different types of alcoholism, different treatment methods, and different lengths of therapeutic remission made it possible to establish internal associations among them and to express these in mathematical form, as differential diagnostic and prognostic algorithms. An algorithm is the numerical expression of the gradations for each indicator, which is a sum used to determine the individual prognosis.

The algorithms were developed for the diagnosis of alcoholism types, the prediction of clinical course, and the length of therapeutic remission as a whole. We took into account treatment methods (the effectiveness of radical alcohol deterrent therapy specifically) and the occurrence of substance dependence in the endoform variant of alcoholism and alcoholism in general. The algorithms were developed by analyzing 135 biological, clinical, social, therapeutic, and other indicators for 318 patients.

RESULTS

The first goal was to identify two groups of patients, one with a marked predisposition for alcoholism, the other in whom this predisposition was minor and alcoholism developed because of the much more pronounced effect of exogenous factors. Patients without a hereditary load for alcoholism or any other psychiatric disorder ($n = 160$) and patients with a hereditary load of alcoholism ($n = 121$) were selected from the study group. In some cases, either a hereditary load for other diseases was noted or information on heredity was incomplete.

TABLE 1 shows that the clinical signs of alcoholism, psychopathology, premorbid characteristics, and reactions to initial alcohol use correlated with the hereditary load of alcoholism. Nevertheless, the highest level of correlation with the specific heredity factor was the rate of alcoholism development, that is, the length of the period from the start of systematic alcohol abuse to the appearance of clinical signs of psychological and/or physical dependence.

Differences in the rate of alcoholism development (period from the start of systematic alcohol abuse to the appearance of signs of psychological and/or physical dependence) in the groups of patients with and without the hereditary alcoholism load are clearly evident in TABLE 2. In patients with a family history of alcoholism, the development of alcohol dependence during the first 3 years after the start of systematic alcohol abuse was observed in 78% of the cases compared with 35% for such cases in the other group. On the other hand, instances of alcoholism development after 4 or more years were encountered much more frequently in patients without the hereditary load.

The relationship between the type of alcoholism and the time of alcoholism development is shown in TABLE 3. After 3 years a sharp decline is noted in the frequency of hereditary alcoholism. In general, a family history of alcoholism was noted in 56.3% of families with patients with an alcoholism development period of

TABLE 1. Indicators in Order of Significance of Their Association with the Hereditary Alcoholism Load

No.	IC[a]	Indicators
1	0.0753	Length of alcoholism development period
2	0.0495	Quantitative control
3	0.0494	Paroxysmal dysphoric state
4	0.0478	Initial tolerance to alcohol
5	0.0462	Length of alcohol abuse
6	0.0444	Depressive states
7	0.0439	Emotional reaction to initial alcohol use
8	0.0426	Age at start of alcohol abuse
9	0.0371	Suicidal tendencies
10	0.0340	Age at time of examination
11	0.0314	Structure of AAS
12	0.0306	Markedness of pathologic attraction
13	0.0301	Hyperactivity, minimal brain dysfunction in childhood
14	0.0291	Evaluation of feelings after initial alcohol use
15	0.0264	Psychopathic-like syndromes
16	0.0261	Personality deviations in premorbid period
17	0.0259	Spontaneous remission
18	0.0245	Dynamics of the form of alcohol abuse
19	0.0190	Pregnancy- and parturition-related problems
20	0.0180	Neurotic episodes in childhood
21	0.0173	Constitutional-personality type
22	0.0165	Pathological adolescent crisis
23	0.0164	Psychological infantilism
24	0.015?	Social adjustment in premorbid period
25	0.0145	Alcoholic psychosis
26	0.0136	Pharyngeal reflex

[a]Informativeness coefficient = R^2.

up to 3 years and in 17.9% of patients with a later development period, that is, more than 3 times less frequently.

Thus, the difference between the familial and the nonfamilial types was especially great in the two groups in whom alcoholism developed either in the first 3 years after the start of abuse or later. On this basis, cases of alcoholism in the entire patient group in which dependence developed in the first 3 years after the start of alcohol abuse were classified as the variant with an endogenous predisposition, that is, endoform, and the others in which dependence developed after 4 or more years were classified as the exoform variant.

Clinical Features of Endoform and Exoform Types of Alcoholism

Comparative analysis of the development, course, and clinical signs of endoform and exoform variants of alcoholism was conducted. These results, presented elsewhere in greater detail,[13] revealed that patients with endoform alcoholism were younger and had an earlier onset of alcohol abuse, absence of clear staging in the development of the disease (episodic alcohol use and "heavy social drinking"), and a lower frequency of a detrimental alcohol environment fostering alcohol abuse. In

TABLE 2. Length of Alcoholism Development As a Function of Hereditary Load Factor

| Length of Alcoholism Development (yr) | Heredity Relative to Alcoholism | | | | |
| | Negative Family History | | Positive Family History | | |
	n	%	n	%	IC[a]
Less than 1	40	25.0	73	60.3	
1–3	16	10.0	21	17.5	
4–6	42	26.3	15	12.4	
7–9	34	21.2	8	6.6	
10–12	13	8.2	2	1.6	0.0753
13–15	11	6.9	2	1.6	
16–18	2	1.2	—	—	
19 or more	2	1.2	—	—	
Total	160	100.0	121	100.0	

[a]Informativeness coefficient = R^2.

endoform alcoholism, the alcohol abstinence syndrome developed much more rapidly, and marked progression of the illness was typical.

Major differences were found between the two variants in the pathologic desire for alcohol. Endoform alcoholism was characterized by an earlier onset of the behavioral component of this desire—goal-directed alcohol "seeking" behavior—and the predominance of a spontaneous desire for alcohol attaining the degree of compulsiveness. Desire was virtually constant. Exoform alcoholism, in contrast, was characterized by the periodic occurrence of a situationally determined desire for alcohol. A more pronounced disruption of control over drinking was typical for the endoform variant. Instances of impaired control from the onset of drinking occurred much more frequently. Psychological disorders were more frequent in the alcohol abstinence syndrome than in exoform alcoholism. A typical feature was the occurrence of psychological disorders within the structure of abstinence even in early stages of the syndrome.

Differences were found between the two alcoholism types in the form of alcohol abuse. In endoform alcoholism, true inebriation or constant inebriation was observed

TABLE 3. Relationship between Type of Alcoholism and the Alcoholism Development Period

| Hereditary Load | Length of Alcoholism Development | | | | | | | | | |
| | <1 Year | | 1–3 Years | | 4–6 Years | | >6 Years | | Total | |
	n	%	n	%	n	%	n	%	n	%
Absent	40	31.3	16	41.0	42	61.8	62	74.7	160	50.3
Alcoholism	73	57.0	21	53.9	15	22.1	12	14.5	121	38.1
Other or not known	15	11.7	2	5.1	11	16.1	9	10.8	37	11.6
Total	128	100.0	39	100.0	68	100.0	83	100.9	318	100.0

much more frequently when the full clinical picture was present, and pseudoinebriation less often. Early signs of alcohol dependence appeared much more frequently during the period when alcohol abuse was characterized by single alcohol excesses. In a percentage of cases, true drunken states appeared even in the early stages. This was not encountered in the exoform type. For patients with exoform alcoholism, the periodic form of alcohol abuse was more typical throughout the illness, whereas in the endoform variant replacement of the constant form of abuse by periodic drinking was encountered more frequently.

In endoform alcoholism, cases of increased tolerance as well as a higher percentage of cases with total alcohol amnesia associated with inebriation were observed more frequently.

Major differences were found between types relative to psychopathologic disorders. Paroxysmal dysphoric states, depressive, psychopathic-like, and neurosis-like disorders were generally noted much more frequently in endoform alcoholism. The difference was evident outside of inebriation periods. Signs of organic CNS damage, predominantly without clear definition, were noted with a higher frequency in exoform alcoholism. This fact probably can be explained by the effect of chronic alcohol intoxication, because with an essentially identical length of illness, the total period of alcohol abuse was substantially longer in the exoform variant of alcoholism.

The risk of suicide was much higher in endoform alcoholism. Suicidal tendencies were found to be almost 3.5 times greater and suicidal attempts 4.5 times more frequent than in patients with exoform alcoholism.

Clinical analysis of the two types of alcoholism revealed indicators with a high differential diagnostic significance and produced an algorithm with the use of multifactorial mathematical analysis using the generalized portrait method. The algorithm made it possible to determine the alcoholism variant for each specific patient with an accuracy of 96%.

Prognosis of the Alcoholism Types

The next task was to compare the two types according to 28 prognostic indicators, describing the biological and social parameters of patients with endoform and exoform alcoholism. The indicators included hereditary load in first- and second-degree relatives, personality traits, characteristics of development in the premorbid period, somatic and organic pathology, family background, reactions to initial alcohol use, and several others.

The hereditary alcoholism load in patients with the endoform variant was present most frequently in the father, then in other blood relatives, and least often in the mother. In comparison with the exoform type, alcoholism in fathers was observed 3 times as frequently in the endoform variant, in other relatives 6 times as frequently, and in mothers 16 times as frequently. Other psychological disorders, primarily endogenous psychoses, were noted more frequently in mothers. In addition, alcoholism was found in two or more relatives in 22.7% of families of alcoholism patients with endoform. This was not at all typical for exoform alcoholism. If all relatives are analyzed together, then the hereditary alcoholism load in the endoform variant was three times higher. Moreover, cases of the concurrent presence of alcoholism and other psychological disorders, primarily endogenous psychoses, in families of patients were observed more frequently.

Comparison of premorbid personality traits in the two groups showed the absence of any constitutional personality type in either variant. However, personality deviations, such as accentuation and psychopathies, were encountered more frequently in endoform alcoholism, but in general the difference was not very great. In addition, signs of psychologic infantilism (2.5 times more frequently), signs of hyperactivity and attention-deficit disorder, and neurotic episodes were more typical of the patients with endoform alcoholism. Here signs of pathologically progressing adolescent crises were observed much more frequently. Cases of pregnancy- and parturition-related pathology were encountered more frequently in endoform alcoholism.

No pronounced differences were noted between the two groups relative to age of initial alcohol use, but first use generally occurred somewhat earlier in the endoform variant (15.9 vs 16.5 years). In endoform cases, a marked positive emotional reaction to initial alcohol use (7 times more often) and a high level of initial tolerance to the effect of alcohol were much more typical. The initial use of alcohol in doses producing marked alcohol intoxication was noted very often here. In general, cases of initial loss of control in a state of inebriation (14 times more often) and the absence of pharyngeal reflex during overdosing (twice as often) were observed much more often in endoform alcoholism. It should be emphasized that the presence of marked alcoholic inebriation during initial alcohol use did not necessarily indicate the initial lack of quantitative control. The latter indicator was established only when the state of marked intoxication was reached systematically, almost from the very start of alcohol use, despite efforts to stay at a level of moderate inebriation. In the other cases, this referred to episodes of marked inebriation with initial use, which was reached consciously and alternated with moderate episodic use of alcohol, depending on ambient conditions.

The differences determined in heredity, features of the premorbid period, and reactions to initial alcohol use permitted the development of an algorithm for individual prognosis of alcoholism variants with the use of multifactorial mathematical analysis. It permitted prediction of the endoform variant of alcoholism with a 75% probability. Thus, identification of the group at higher risk for an unfavorable course of the disease may be possible even at the stage of initial alcohol use, before the start of systematic abuse.

Comparative Effectiveness of Pharmacologic Therapy

The effectiveness of various pharmacologic treatment methods was also analyzed. The results showed that traditional nonradical conditioned reflex therapy (using short-acting sensitizing preparations) was ineffective in patients with endoform alcoholism. The probability of prolonged remission was close to zero in these patients. Radical deterrent methods (Disulfiram) produced better results, but here as well the length of remission depended greatly on the type of alcoholism.

Considerable attention has been focused on the evaluation of psychotropic therapy for alcoholics to suppress the pathologic desire for alcohol, to normalize the psychological state of patients, and to increase the length of remissions. It appeared that the psychotropic agents (lithium salts, sustained-release neuroleptics, etc.) permitted major prolongation of the length of remission, particularly in patients with the endoform variant of alcoholism.

Predictors of Therapeutic Remissions

In the next analysis, patients with brief remissions (less than 1 year) and long remissions (more than 1 year) were compared, and the significant predictors of outcome were determined. Factors affecting the length of remission were rather numerous. In addition to the alcoholism subtype, age-related parameters, progression of alcoholism, desire for alcohol, psychopathologic disorders, various pharmacologic treatment methods, the socio-occupational and family status of the patients, and their attitude toward therapy were related to length of remission.

Clinical Features of Substance Dependence According to Subtype

Analysis of other drug dependence in patients with endoform and exoform variants of alcoholism revealed major differences. TABLE 4 demonstrates that the probability of substance dependence is much higher in the endoform subtype. A much earlier age for the start of narcotics use, a broader range of psychoactive agents, and a higher frequency of substance dependence are typical of endoform alcoholism. In contrast, the use of drugs readily accessible during this time was encountered predominantly in exoform alcoholism (TABLE 5). The differences also relate to initial reactions to narcotics use and the rate of dependence formation. The differences again point to the unfavorable prognosis for endoform alcoholism.

DISCUSSION

The existence of individuals with a marked biological predisposition to alcoholism was confirmed by this study. Analysis of heredity showed the presence of an association between clinical and premorbid parameters in patients with alcoholism and the hereditary alcoholism load. The highest correlation was found between hereditary factors and the rate of alcoholism development.

Two clinical variants of alcoholism were identified, which differ in terms of the significance of biological and external factors on their development. Patterns in their course and clinical signs were described in comparative terms, and a differential diagnostic algorithm was developed on the basis of multifactorial mathematical analysis. This enabled separation of patients with the two variants and determination of the presence of a biological predisposition to alcoholism in a specific patient with a high probability from clinical indicators.

A set of prognostic indicators was identified on the basis of the premorbid characteristics of patients and their reactions to initial alcohol use. The comparative effectiveness of various methods of pharmacologic therapy in patients with the two

TABLE 4. Occurrence of Substance Dependence in Patients with Alcoholism As a Function of Disease Variant

Diagnosis	Endoform Variant		Exoform Variant		Total		
	n	%	n	%	n	%	IC[a]
Alcoholism	98	58.7	133	88.1	231	72.6	
Alcoholism complicated by substance dependence	69	41.3	18	11.9	87	27.4	0.0797
Total	167	100.0	151	100.0	318	100.0	

[a]Informativeness coefficient = R^2.

TABLE 5. Type of Alcoholism According to the First Dependence-Inducing Narcotic

Type of Basic Introductory Narcotic	Endoform Variant		Exoform Variant		Total	
	n	%	n	%	n	%
Hashish	10	10.6	—	—	10	8.3
Opiates	18	19.1	2	7.4	20	16.5
Ephedrine	13	13.8	1	3.7	14	11.6
Barbiturates	23	24.5	15	55.5	38	31.4
Other sedative-hypnotic substances	20	21.3	8	29.7	28	23.1
Inhalational substances	4	4.3	—	—	4	3.3
Others	6	6.4	1	3.7	7	5.8
Total	94	100.0	27	100.0	121	100.0

alcoholism variants was also studied. The hypothesis of unique pathogenetic mechanisms in alcoholism and substance dependence, particularly in endoform variants, was confirmed.

The use of mathematical programs makes it possible to conduct research studies on the pathogenesis, prevention, clinical picture, and treatment of alcoholism and substance dependence in a more uniform group. The principles for developing the procedure of individual prognosis, presented in this work, can be used not only in psychiatry and the study of substance dependence but also in other fields of medicine as well. The algorithm for a differential diagnosis of variants of alcoholism will enable practicing physicians to separate alcoholics into groups with an unfavorable and relatively favorable course and to carry out in a more differential manner specific therapeutic and rehabilitative measures.

The use of an algorithm enables the identification of adolescents who are subject to an unfavorable disease course, that is, the group with a high "biological risk," and the direction of preventive measures at them. Moreover, the identification among the young of those with a marked biological predisposition for alcoholism, in all probability, will at the same time identify the group at "high risk" for the development of substance dependence. Consequently, this study has practical significance not only within the scope of primary prevention of alcoholism but also for dependence disorders in general.

The following points summarize the main conclusions of the study:

1. A clear link exists between the hereditary load, premorbid factors, clinical features of patients, and psychopathologic syndromes; in this regard, the significance of this association varies as a function of the indicator and is highest relative to the rate of alcohol dependence formation.

2. The course, clinical signs, and psychopathology of alcoholism differs greatly as a function of biological predisposition for its development; this can be used as the basis for defining the alcoholism variant, endoform or exoform, in patients by clinical indicators.

3. There is a variety of interrelated indicators, including features of the premorbid period and responses to initial alcohol use, that have a high prognostic value and enable individual prognosis of the course of alcoholism.

4. The effectiveness of pharmacologic therapy differs in endoform and exoform variants of alcoholism and, in addition, depends on many other clinical and social parameters, which may be used in the prediction of therapeutic response. A differ-

ential course of psychotropic treatment, particularly in the endoform variant, enables the achievement of substantially greater therapeutic results.

5. The probability of developing substance dependence in alcoholics is much higher in the endoform variant and depends on a set of premorbid and clinical factors, particularly pathologic desire and psychopathologic disorders, which are interrelated, have a prognostic value, and can be used for individual prediction of the appearance of cross-dependence on other substances.

6. Clinical features of substance dependence developing in patients with endoform and exoform alcoholism were identified. It was established that more severe forms of substance dependence are typical of the endoform variant. In the exoform type, the development of drug dependence was secondary to signs of alcoholic deterioration.

7. The results suggest the possibility of separating alcoholic patients into more uniform groups according to the presence or absence of biological predisposition, which may improve the prognosis associated with specific treatments. Thus, the typological approach can be conducive to a more optimal solution of both scientific and applied problems of alcohol and drug dependence.

REFERENCES

1. GOODWIN, D. N., F. SCHULSINGER, L. HERMANSEN et al. 1973. Alcohol problems in adoptees raised apart from alcoholic biological parents. Arch. Gen. Psychiatry **28**: 238–243.
2. BOHMAN, M. 1978. Some genetic aspects of alcoholism criminality. Arch. Gen. Psychiatry **35**: 269–276.
3. CADORET, R. J., C. A. CAIN & W. M. GROVE. 1980. Development of alcoholism in adoptees raised apart from alcoholic biological relatives. Arch. Gen. Psychiatry **37**: 561–563.
4. CLONINGER, D. E., M. BOHMAN & S. SIGVARDSSON. 1981. Inheritance of alcohol abuse. Arch. Gen. Psychiatry **38**: 861–868.
5. LI, T. K. 1984. Genetic variability in responses to ethanol in humans and experimental animals. *In* Proceedings of NIAAA-WHO Collaborating Center Designation Meeting and Alcohol Research Seminar. L. H. Towle, ed: 50–62. Rockville, MD.
6. PATTISON, E. M., M. B. SOBELL & L. C. SOBELL, eds. 1977. Emerging Concepts of Alcohol Dependence. Springer. New York, NY.
7. GINDILIS, V. M. 1979. Genetics of schizophrenic psychoses. Doctoral dissertation. Moscow, Russia.
8. KAIJ, L. 1960. Studies on the etiology and sequels of abuse of alcohol. Department of Psychiatry, University of Lund, Sweden.
9. PARTANEN, J. K., K. BRUNN & T. MARKKANEN. 1966. Inheritance of drinking behavior: A Study on Intelligence, Personality and Use of Alcohol in Adult Twins. Vol. 14. Finnish Foundation of Alcohol Studies. Helsinki, Finland.
10. HRUBEC, Z. & G. S. OMENN. 1981. Evidence of genetic predisposition to alcoholic cirrhosis and psychosis: Twin concordances for alcoholism and its biological endpoints by zigosity among male veterans. Alcoholism: Clinical and Experimental Research **5**: 207–212.
11. ANOKHINA, I. P. 1985. Dopamine and opiate system peculiarities as a basis of development of alcohol dependence: Role of genetic factors. *In* Proceedings of the 34th International Congress on Alcoholism and Drug Dependence. Calgary, Alberta, Canada.
12. MARDALEISHVILI, K. M. 1984. Clinical and morphological basis for predicting and modeling of individual treatment of mouth cavity cancer. Doctoral dissertation. Moscow, Russia.
13. VRUBLEVSKY, A. G. 1989. Variants of alcoholism: Patterns of formation, course and

prognosis. Dissertation for a Doctorate in Medical Sciences, USSR Ministry of Public Health, Moscow, Russia.

14. IVANETS, N. N. 1985. Individual sensitivity to alcohol as a basis to predict general clinical regularities of alcoholism. *In* Proceedings of the 34th International Congress on Alcoholism and Drug Dependence: 270. Calgary, Alberta, Canada.

15. ANOKHINA, I. P. 1968. About the common pathogenesis of alcoholism and drug addiction. *In* The Materials of VIII All Union Conference on Neurology and Psychiatry/Narcology. 1:307–370. Moscow, Russia.

16. IVANETS, N. N. 1988. Actual clinical problems of alcoholism. *In* Pathogenesis, Clinical Features and Treatment of Alcoholism. 1: 357–360. Moscow, Russia.

17. ANOKHINA, I. P. 1976. Neurochemical aspects of pathologenesis of chronic alcoholism. *In* Pathogenesis, Clinical Features and Treatment of Alcoholism. 15–19. Moscow, Russia.

Thirty-Year Follow-Up of Men at High Risk for Alcoholism

DONALD W. GOODWIN,[a] JOACHIM KNOP,[b] PER JENSEN,[b]
WILLIAM F. GABRIELLI, JR.,[c] FINI SCHULSINGER,[b]
AND ELIZABETH C. PENICK[a]

[a]University of Kansas Medical Center
Kansas City, Kansas 66160

[b]The Institute for Preventive Medicine
Copenhagen, Denmark

[c]Kansas City VA Medical Center and
University of Kansas Medical Center
Kansas City, Kansas 66160

Alcoholism runs in families. This has been recognized for centuries. During much of this century it has been generally believed that "familial alcoholism" resulted from children modeling their behavior after their parents. Recent reviews of twin and adoption studies conducted in Europe and the United States indicate that alcoholism, to some extent, is influenced by heredity.[1,2] Whether alcoholism is partly inherited or not, children of alcoholics are at "high risk" for alcoholism. Perhaps 20–30% of sons of alcoholics and 5–10% of daughters become alcoholic, a rate three to five times greater than that of the general population.[1] The term "high risk study" usually refers to a study of children of afflicted families, in this case alcoholics, before they begin heavy drinking. Heavy use of alcohol may produce physical and psychological problems that can be mistaken for causal factors. The present study continues the longitudinal examination of a cohort of Danish men whose biological fathers were identified as alcoholic in a psychiatric registry or community alcohol treatment program and therefore were considered at high risk for alcoholism. The goal is to identify early predictors of future alcoholism. Predictors are not always, but sometimes, related to causation. Exploring causal factors of alcoholism is a goal of the study.

METHOD

In the late 1950s and early 1960s a group of Danish physicians joined forces to search for causes of birth anomalies. The physicians were obstetricians, pediatricians, neonatologists, and neurologists. They collected data about hundreds of perinatal variables, including the health of the parents, the circumstances of pregnancy and delivery, and the condition of the children 1 year after birth. The cohort consisted of 9,125 consecutive deliveries from pregnancies that lasted at least 20 weeks' gestation and that occurred between December 1959 and January 1961 at an inner city Copenhagen hospital. The findings were published in a two-volume monograph.[3] Of the more than 1,000 variables studied, only maternal venereal disease was associated with birth anomalies.

For many years, the perinatal data were stored and largely ignored. Later Fini Schulsinger, a psychiatrist and co-director of what was then the Psychological

Institute in Copenhagen, obtained access to the information. With authorities in Copenhagen assured that total confidentiality would be maintained, the names of the Danish children and their biological parents were processed through local and central registers in Denmark, identifying those children with a father treated in a Danish psychiatric facility or community alcohol treatment clinic who received the diagnosis of "spiritus abusus" (alcoholism). By this method 220 high risk sons were identified and 110 controls matched for age and sex were selected. Only males were studied because men are more often alcoholic than women and a smaller number of subjects is needed to achieve a large enough number of alcoholics for study. The controls were considered at "low risk" for alcoholism because there was no mention in psychiatric registries or local alcoholism clinics that either parent had been treated for alcoholism. It was appreciated that some parents of the control subjects might have had alcoholism, but if so, they had not received treatment by the time the records were searched in 1978.

In the early 1980s, high- and low- risk individuals were studied by a multidisciplinary team consisting of psychiatrists, psychologists, neurologists, social workers, and electrophysiologists. The subjects were in the 19–21–year age range at the time they were studied. Some were brought into the laboratory and challenged with alcohol; electrophysiological measures were obtained. They were interviewed by psychiatrists and social workers and received neuropsychological testing.[4–6]

In the 20-year followup, differences found between the high- and low-risk groups included the following:

1. Challenged with alcohol, the high risk group generated more slow wave alpha activity on the EEG than did the low risk group.[7]

2. The high risk group made more errors on the Categories Test of the Halstead Battery.[8] Interestingly, male alcoholics often perform poorly on neuropsychological tests, a finding usually attributed to the harmful effects of alcohol on the brain. The fact that their sons also made excessive cognitive testing errors, before they engaged in long-term heavy drinking, suggested that a relationship between poor cognitive performance and alcoholism may not be related to the effects of alcohol, but perhaps may have causal implications for the development of alcoholism. Future data analysis will focus on findings at the 20-year followup and their relationship to subsequent alcohol problems. Again, the goal is to identify predictors of future alcoholism in *individuals* and not in groups. The data reported here, as with the 20-year followup, still reflect group findings.

Recently, a 30-year followup on 241 (71%) of the original 330 subjects was completed; 67% of those followed up were at high risk and 33% at low risk. As part of this followup effort, a structured comprehensive diagnostic interview was administered to them by a psychiatrist (P. J.). The interview permitted diagnoses by DSM-III-R criteria[9] as well as the diagnosis of alcoholism according to Feighner criteria and the Michigan Alcoholism Screening Test (MAST).[10–12] The interviewer was blind to the risk status of the subjects.

RESULTS

TABLE 1 summarizes the clinical diagnostic findings of the 30-year followup. About half the patients in both the high- and low-risk groups were assigned at least one psychiatric diagnosis according to DSM-III-R criteria; the difference was not significant. The high-risk group had significantly more substance use disorders, including alcoholism, than did the low-risk group (48% *vs* 32%, chi square = 5.49,

TABLE 1. Prevalence of Single and Combined Categorical Diagnoses in Danish Men at High Risk (n = 162) or Low Risk (n = 79) for Alcoholism Based on Different Diagnostic Systems

	High		Low		
Diagnosis	n	%	n	%	p^a
DSM-III-R[9]					
Any DSM-III-R diagnosis	92	56.8	36	45.6	0.10
Any nonsubstance use DSM-III-R diagnosis	52	32.1	27	34.2	0.75
Any substance abuse/dependence	77	47.5	25	31.6	0.02
Any substance abuse	38	23.5	14	17.5	0.31
Any substance dependence	47	29.0	12	15.2	0.02
Any alcohol abuse/dependence	66	40.7	22	27.8	0.05
Any alcohol abuse	36	22.2	13	16.5	0.30
Any alcohol dependence	30	18.5	9	11.4	0.16
Any drug abuse/dependence	26	16.0	6	7.6	0.07
Any drug abuse	6	3.7	1	1.3	0.29
Any drug dependence	20	12.3	5	6.3	0.15
Antisocial personality	17	10.5	8	10.1	0.93
Any anxiety disorder	16	9.9	5	6.3	0.36
Depression/dysthymia/cyclothymia	9	5.6	5	6.3	0.81
Schizoid/schizophrenia schizophreniform	2	1.2	1	1.3	0.98
PDI-R (Feighner)[10,12]					
Definite alcoholism	30	18.5	9	11.4	0.16
MAST[b] (>5 or more)[11]					
Definite alcoholism	48	44.4	13	26.5	0.03

[a]The p values are based on chi square.
[b]Scores on the MAST were available for only the first 157 subjects.

p <0.02) The difference in substance use disorders between the two groups was largely explained by a higher rate of substance dependence among the high-risk subjects. Alcohol dependence/abuse also occurred significantly more often in the high-risk group (41% *vs* 28%, chi square = 3.81, p <0.05) Again, the diagnosis of alcohol dependency made a larger contribution to the difference than did that of alcohol abuse, but alcohol dependence by itself did not significantly discriminate these relatively small groups.

No significant differences for drug abuse/dependence (excluding alcohol dependence) were noted between the two groups despite the trend for greater drug use disorders in the high-risk group (chi square = 3.30, p <0.07). There were no significant group differences for the diagnoses of antisocial personality, anxiety disorder, depressive disorder, or schizophrenia. These findings demonstrated that a history of parental alcoholism, determined three decades ago, was mainly associated with increased rates of substance dependence in the offspring.

The criteria proposed by Feighner *et al.*[9] have been fully operationalized by the revised Psychiatric Diagnostic Interview (PDI-R).[2] This structured diagnostic interview uses 20 questions, divided into four symptom groups, to reflect the original Feighner criteria for alcoholism. "Definite alcoholism" was diagnosed by Feighner *et al.*[9] when one or more positive symptoms were found in three of the four criteria groups. When the Feighner *et al.* categorical diagnosis of "definite alcoholism" was applied, the difference between the two risk groups was not significant; however, more of the high-risk subjects (18%) than low-risk subjects (11%) satisfied the Feighner criteria (chi square = 1.99, p <16). Nevertheless, when the 20 alcoholism items of the PDI-R were employed as a continuous measure, the difference between the high-risk and low-risk group was statistically significant [F(1,239) = 5.13, p <0.02], with the high-risk groups reporting significantly more symptoms of alcoholism (HR = 3.2 and LR = 2.0 symptoms out of 20). PDI-R alcoholism items found significantly more often in the high-risk group included: (1) family concern about the subject's drinking, (2) treatment or hospitalization for drinking, (3) binge drinking, and (4) loss of control.

On the Michigan Alcoholism Screening Test (MAST), a cutoff score of five is recommended to categorize an individual as alcoholic. With this MAST cutoff score, 44% of the high-risk subjects were positive for alcoholism compared to 27% of the low-risk group (chi square = 4.55, p <0.03). The total MAST score also significantly discriminated the two groups. The mean MAST scores were 8.2 and 3.8 for the high- and low-risk subjects, respectively [F(1,155) = 7.2, p <0.008]. On the MAST, high-risk subjects more often endorsed items indicating: (1) trouble at work because of drinking, (2) requests for help from others to stop drinking, and (3) legal problems associated with their drinking.

High- and low-risk groups did not differ in the prevalence of other, DSM-III-R assigned, *non*substance use disorders (TABLE 1). Analyses were performed to determine if psychiatric comorbidity differed according to risk status of those subjects who were assigned a diagnosis of either alcohol abuse or alcohol dependence. Among the alcohol-dependent high-risk subjects, 43% were also given the diagnosis of antisocial dependent personality disorder (ASP), while among the alcohol-dependent low-risk subjects, 67% were labeled ASP. In contrast, among the alcohol abuse high-risk subjects, only 6% were also assigned the diagnosis of ASP, while 8% of the low-risk subjects with alcohol abuse were assigned the ASP diagnosis. Thus, in both high- and low-risk groups, antisocial personality occurred significantly more often among those subjects diagnosed as alcohol dependent than those diagnosed as alcohol abusers. The powerful relationship between ASP and alcohol dependence but not alcohol abuse was independent of risk status (HR:chi square = 13.3, p <0.001; LR:chi square = 8.53, p <0.004). No significant differences were noted for anxiety disorder, schizophrenia, or depression across groups of high- and low-risk subjects who received the diagnosis of either alcohol dependence or alcohol abuse.

DISCUSSION

As has been reported numerous times, sons of alcoholics in this study were found to be drinking alcoholically more often than were sons of nonalcoholics. More of the high-risk subjects received a DSM-III-R diagnosis of alcohol dependence or abuse than did low-risk subjects. More of the high-risk subjects satisfied the MAST cutoff for definite alcoholism. On two continuous measures of drinking severity, high-risk subjects reported significantly more lifetime problems with alcohol than did low-risk subjects.

As groups, the high- and low-risk subjects did not differ in lifetime prevalence of DSM-III-R diagnosed depression, antisocial personality disorder, anxiety disorder, or schizophrenia. Nevertheless, alcohol-dependent subjects in both risk groups were more likely to receive the comorbid diagnosis of antisocial personality disorder than were subjects in both risk groups who received the diagnosis of alcohol abuse or no substance abuse diagnosis. Neither risk status nor the alcohol-related diagnoses of dependence or abuse were associated with differential rates of other, *non*substance use comorbid diagnoses.

These data suggest that alcohol dependence is more strongly associated with high-risk status and antisocial personality disorder than alcohol abuse, using DSM-III-R criteria. Unlike twin and adoption studies, high-risk studies do not separate genetic from environmental influences. However, the elevated rate of antisocial personality disorder among alcohol-dependent men in *both* risk groups suggests that the association between antisocial personality disorder and alcoholism may be based on environmental rather than genetic factors.

It has been suggested that there are two types of alcoholism, the type that runs in families and the type that does not.[1] So-called "familial alcoholism" is sometimes characterized as being a more severe form of the illness than is nonfamilial alcoholism. The results of this study seem to support that idea. By combining the DSM-III-R diagnosis of alcohol abuse with alcohol dependence, 41% of the high-risk group compared to 28% of the low-risk group satisfied criteria for a diagnosis of alcoholism. However, the prevalence of DSM-III-R alcohol abuse did not differentiate the two groups, whereas a DSM-III-R diagnosis of alcohol dependence tended to discriminate the high- from the low-risk group. Inasmuch as alcohol dependence generally reflects greater severity than does alcohol abuse, this finding is consistent with the notion that familial alcoholism involves a particularly virulent form of alcoholism.

REFERENCES

1. GOODWIN, D. W. 1991. Genetic determinants of alcoholism. *In* Diagnosis and Treatment of Alcoholism. 3rd Ed. J. Mendelson & N. Mello, eds. McGraw-Hill. New York.
2. GOODWIN, D. W. 1990. The genetics of alcoholism. *In* Genes, Brain and Behavior. P. R. McHugh & V. A. McKusick, eds. Raven Press. New York.
3. ZACHAU-CHRISTIANSEN & E. M. ROSS. 1975. Human Development during the First Year. Wiley & Sons. New York.
4. KNOP, J. 1980. Selection of variables in a prospective study of young men at high risk for alcoholism. Acta. Psychiat. Scand. **285** (Suppl.): 347–352.
5. KNOP, J., T. W. TEASDALE, F. SCHULSINGER & D. W. GOODWIN. 1985. A prospective study of young men at high risk for alcoholism: School behavior and achievement. J. Stud. Alcohol **46**: 273–278.
6. POLLOCK, V. E., T. W. TEASDALE, W. F. GABRIELLI & J. KNOP. 1986. Subjective and objective measures of response to alcohol in young men at high risk for alcoholism. J. Stud. Alcohol **47**: 297–304.
7. POLLOCK, V. E., J. VOLAVKA, S. A. MEDNICK, D. W. GOODWIN, J. KNOP & F. SCHULSINGER. 1984. Electroencephalagraphic findings. *In* Longitudinal Research in Alcoholism. D. W. Goodwin, K. T. Van Dusen & S. A. Mednick, Eds. 125–145. The Hague Kluwer-Nijhoff Publishing. Boston, MA.
8. DREJER, K., A. THEILGAARD, T. W. TEASDALE, D. W. GOODWIN & F. SCHULSINGER. 1985. A prospective study of young men at high risk for alcoholism: Neuropsychological assessment. Alcohol. Clin. Exp. Res. **9**: 298–302.
9. American Psychiatric Association. 1987. Diagnostic and Statistical Manual of Mental Disorders. 3rd Ed. Rev. Washington, D.C.
10. FEIGHNER, J. P., E. ROBINS, S. B. GUZE, R. A. WOODRUFF, G. WINOKUR & R. MUNOZ. 1972. Diagnostic criteria for use in psychiatric research. Arch. Gen. Psychiatry **26**: 57–63.
11. SELZER, M. L. 1971. The Michigan alcoholism screening test: The quest for a new diagnostic instrument. Am. J. Psychiatry **126**: 1653–1658.
12. OTHMER, E., E. C. PENICK, B. J. POWELL, M. R. READ & S. OTHMER. 1989. The Psychiatric Diagnostic Interview (Revised). Western Psychological Services. Los Angeles, CA.

Assortment for the Liability to Substance Abuse and Personality Traits[a]

MICHAEL M. VANYUKOV, HOWARD B. MOSS,
AND RALPH E. TARTER

Center for Education and Drug Abuse Research (CEDAR)
Department of Psychiatry
University of Pittsburgh School of Medicine
Pittsburgh, Pennsylvania 15213

Assortative mating, that is, nonrandom selection of a mating partner with regard to a trait or traits, can exert considerable influence on the genetic structure of the population. Although it generally does not change the frequencies of alleles, positive assortative mating (positive correlation between the phenotypic values of the partners) induces genetic correlations between unrelated spouses. This in turn leads to an increase in the additive genetic variance of the trait in offspring and, consequently, in the heritability of the trait and in the frequency of genotypes associated with extreme phenotypes. For the multifactorially inherited liability to a disorder, the extreme, suprathreshold phenotype is the disorder itself. Hence, the genetic events associated with positive assortment for the liability may result in an increment in the frequency and severity of the disorder. In proportion to the heritability of the trait, the genetic resemblance between the parents and the offspring increases. It is important, however, that assortative mating increases parent-offspring phenotypic correlations regardless of whether they are mostly genetic or mostly environmental in origin.

There is, however, a dearth of information on the assortativeness of mating for the liability to substance abuse in males and females and its influence on the offspring's phenotypes. Moreover, it is not known if marital assortment for personality traits is associated with that liability. This preliminary study is an initial step in addressing this deficiency.

METHODS

Sample Selection and Patients' Characteristics

Subjects were males (probands) who had either a diagnosis of Psychoactive Substance Use Disorder (PSUD) (Group 1: drug abuse, $n = 66$; Group 2: alcoholism only, $n = 22$) or no psychiatric disorder after age 18 (Group 3: control, $n = 76$) according to DSM-III-R diagnostic criteria, their wives, and their 10–12-year-old sons. The subjects were recruited from the community using advertisement and public announcements as well as from clinical facilities as part of an ongoing longitudinal multidisciplinary study of the factors contributing to the risk for the

[a]This work was supported in part by the National Institute on Drug Abuse (P50 DA 05605) and the National Institute of Mental Health (MH 16804–11)

development of PSUD (Center for Education and Drug Abuse Research, CEDAR). The demographic characteristics of the sample are presented in TABLE 1.

Procedure and Analysis

To evaluate marital assortment on psychological traits, correlations between probands and their wives' phenotypes were estimated for the traits measured by the Dimensions of Temperament Survey—Revised (DOTS-R), Multidimensional Personality Questionnaire (MPQ), IQ (Wechsler Adult Intelligence Scale—Revised, WAIS-R), and childhood conduct disorder and antisocial personality disorder symptom counts (expanded version of the SCID[1]). Also, spousal correlations for education level and age were estimated. Correlation coefficients were computed using the DOS-PRELIS (version 1.20) software (preprocessor for LISREL) as PM matrix (product moment correlations for continuous variables, polychoric correlations for ordinal variables, and polyserial correlations for pairs of ordinal and continuous variables).[2] As this preliminary analysis was directed specifically to selecting indicators for further study, low significance thresholds were preferred to procedures such as Bonferroni correction which could lead to the deletion of potentially valuable variables at this early stage of the analysis.

RESULTS

Among the wives of 164 probands, 47 (29%) had a diagnosis of substance abuse/dependence/polydrug abuse. Only 8 of them were spouses of the control probands. This corresponded to the tetrachoric correlation of 0.61 (p <0.001) between the substance use phenotype (i.e., having or not having a diagnosis of abuse/dependence) of the proband and that of his wife. Interestingly, not only quantitative, but also qualitative assortment was observed, that is, the wife's substance use phenotype was specific for the group of abusers her husband belonged to; although spouses of the group 1 probands (defined by a drug abuse disorder) abused or were dependent on a variety of drugs, including alcohol (25%), opioids (17%), marijuana (13%), sedatives (6%), hallucinogens (4%), amphetamines (3%), and

TABLE 1. Demographic Characteristics of the Sample

	Group			
	Substance/Alcohol Abuse (n = 88)		Control (n = 76)	
Characteristic	Mean	SD	Mean	SD
Father's age	40.3	5.54	42.0	5.84
Mother's age	36.8	4.91	37.0	4.16
Child's age	10.8	0.93	11.0	0.88
Father's education	15.0	1.65	13.4	2.00
Child's grade	4.4	1.09	4.7	1.16
Father's SES	38.3	12.12	49.2	13.07
Race (% white)	88.0		96.0	

cocaine (2%), only alcohol dependence or abuse was diagnosed among the wives of the Group 2 (alcohol abuse only) probands (22%).

Significant correlations between probands and their wives' phenotypes were found for personality and temperament measures of stress reactivity (0.18, p <0.05), alienation (0.45, p <0.001), traditionalism (0.38, p <0.001), absorption (0.19, p <0.05), and negative affectivity (0.35, p <0.001) (MPQ); and mood quality (0.14, p <0.05) (DOTS). Significant correlations were also found for verbal IQ (0.46, p <0.001), performance IQ (0.22, p <0.01), and full scale IQ (0.42, p <0.001) (WAIS-R) and for age (0.66, p <0.001) and education level (0.56, p <0.001). Positive assortment was also demonstrated for antisocial propensity in probands and their wives. This dimension was assessed using lifetime ASP symptom count (r = 0.31, p <0.001) and retrospective childhood conduct disorder (CD) symptom count (r = 0.35, p <0.001). These findings confirm the results of a previous study obtained on a smaller, but similar sample.[3]

Given the significant heritabilities of the traits discussed, the correlations found may imply a high probability of the further increase in the additive genetic component of the variance of these traits in offspring. In turn, this may lead to an increase in the risk and severity of substance abuse, including alcoholism, in the affected parents' children, inasmuch as those traits are associated with the liability to this disorder. The strength of these associations is addressed below.

Data on the relationship of the traits on which the marital assortment was observed with probands and their wives' addiction status, presented in TABLE 2, show a relationship between the traits for which spousal phenotypes correlate and the individual's dichotomously defined liability to substance abuse (having/not having the disorder). It is noteworthy that this relationship is more pronounced in females, the probands' wives.

To test for the influence of parental assortment on the child's personality traits, correlations were computed between parental traits for which the assortment was observed and the following child's traits: measures of *delinquency* and *internalizing* and *externalizing behavior* based on ratings from the Child Behavior Checklist (Mother's version) and estimates of the child's (one 10–12-year-old son per nuclear family) behavioral pathology—*conduct disorder symptom count*, the *presence/absence of a diagnosed internalizing disorder* (mood and anxiety disorders), and the *presence/absence of an externalizing disorder* (attention deficit/hyperactivity, oppositional-defiant disorder, conduct disorder, and adjustment disorders). These data are presented in TABLE 3.

As can be seen, there is a significant association of the child's traits (particularly those that can be considered as indices of disruptive behavior) with the parental characteristics related to marital assortment. Moreover, this association with the

TABLE 2. Point-Biserial Correlations between Assortment Traits and Substance Use Phenotype

Spouses	Stress Reactivity	Alienation	Traditionalism	Absorption	Negative Affectivity	Mood Quality	IQ (Full Scale)	Education
Probands	0.10	0.06	−0.15*	−0.07	0.08	−0.14*	−0.28***	−0.49***
Wives	0.29***	0.40***	−0.05	0.00	0.31***	−0.22**	−0.45***	−0.43***

NOTES: *p <0.05; **p <0.01; ***p <0.001.

TABLE 3. Association of the Child's Characteristics with the Parental Assortment Traits

Trait/Parent	Delinquent	Externalizing (CBCL)	Internalizing (CBCL)	CD Count	Internalizing Disorders	Externalizing Disorders
Stress reactivity:						
Mother	0.30***	0.36***	0.40***	0.11	0.29***	0.11
Father	0.27***	0.28***	0.37***	0.17*	0.29***	0.31***
Alienation:						
Mother	0.33***	0.39***	0.46***	0.14*	0.34***	0.21**
Father	0.17*	0.19*	0.30***	0.08	0.15*	0.13
Tradition:						
Mother	0.03	−0.03	−0.04	−0.18*	0.18*	−0.15*
Father	−0.01	−0.07	−0.08	−0.22**	0.14*	−0.21**
Absorption:						
Mother	−0.04	0.08	0.27***	−0.06	0.09	0.06
Father	0.04	0.01	0.19*	0.01	0.01	0.07
Mood quality:						
Mother	−0.26**	−0.30***	−0.22**	−0.18*	−0.21**	−0.21
Father	−0.13	−0.17*	−0.30***	−0.12	−0.19*	−0.12
Negative affectivity:						
Mother	0.29***	0.37***	0.45***	0.11	0.33***	0.15*
Father	0.19*	0.24**	0.39***	0.10	0.20*	0.21**
ASP count:						
Mother	0.24**	0.33***	0.31***	0.22*	0.10	0.39***
Father	0.26***	0.21*	0.19*	0.30***	0.26***	0.33***
CD count:						
Mother	0.33***	0.32***	0.21*	0.32***	0.29***	0.34***
Father	0.23**	0.18*	0.13	0.22**	0.14*	0.19*
Addiction:						
Mother	0.24**	0.28***	0.17*	0.31***	0.12	0.43***
Father	0.31***	0.31***	0.32***	0.39***	0.34***	0.47***

NOTES: *p <0.05; **p <0.01; ***p <0.001.

mother's traits in most cases, including her addiction status, is at least as strong as that with the father's.

DISCUSSION

The findings indicate a high phenotypic similarity between husbands and wives for the presence/absence of a diagnosis of substance abuse. This is consistent with the data on marital assortment for the liability to alcoholism.[4-8] The assortativeness of mating for the latter depended on the family history and sex of the proband,[4] namely, assortment was expressed in female probands, but not in male probands. Moreover, the probability of marrying an alcoholic correlated with the number of affected relatives of the female proband, which indicates the presence of true assortment as opposed to the possible effect of spousal influence ("contagion"). The existence of positive assortative mating and the suggestion of gender differences in assortment may have important ramifications. For example, the low heritability values previously reported for the liability to substance (other than alcohol) abuse/dependence in females were based on the assumption of the absence of assortment.[9] Consequently, these values may underestimate the genetic component of the variation in this trait. In this regard, it is salient that the data from genetic studies[9–18] suggest that interindividual differences in the liability to alcoholism in both males and females are to a large degree determined by the variation in the individual genotypes. Analogous results have been obtained with regard to drug abuse.[19]

Importantly, phenotypic similarity of parents for the liability to alcoholism correlated not only with the risk of alcoholism in offspring, but also with the risk of antisocial personality/conduct disorder and attention deficit disorder (ADHD).[4,7] In turn, it was found that both CD and ADHD are associated with an increased risk for the development of substance abuse and with the risk of psychopathology in general.[20] Taking into account the significant heritabilities of personality and temperament traits[21–23] and the liabilities to alcoholism and substance abuse, the spousal correlations found for these traits may imply a high probability of the further increase in the additive genetic component of their variance. In turn, associations of the liability to substance abuse with these traits may result in an increase in the risk and severity of substance abuse in the affected parents' children. Considering that the direction of the associations observed corresponds to a higher behavioral deviation in the groups of abusers than in the control probands and their wives, this is consistent with the hypothesis that affected females have, on the average, a higher liability and/or genetic predisposition to substance abuse than do affected males.[4] Marital assortment on these traits can therefore lead to the aggregation of both genetic and environmental factors underlying the high-risk–associated values of these characteristics in the affecteds' offspring. This further supports the hypothesis about an increase in the liability to a substance abuse disorder in substance abusers' children beyond their already increased risk directly due to having an affected parent. The relationship between childhood conduct deviations and the risk for substance abuse and the associations revealed in this study between the indices of disruptive behavior in children and the parental characteristics related to marital assortment may be indicative of the trend in that untoward direction.

CONCLUSION

The data presented herein suggest the existence of positive assortative mating for the liability to substance abuse that may be secondary to the assortment for other personality characteristics. The magnitude of this assortment is greater in females than in males. The presence of the association between the liability to substance abuse and personality traits of the parents and their children's behavior phenotypes indicates the possibility of the significant influence of parental assortment on the probability of the development of alcoholism and other substance abuse in offspring.

REFERENCES

1. SPITZER, R., J. WILLIAMS & G. MIRIAM. 1987. Instruction Manual for the Structural Clinical Interview for DSM-III-R. New State Psychiatric Institute. New York.
2. JÖRESKOG, K. G. & D. SÖRBOM. 1988. PRELIS: A Program for Multivariate Data Screening and Data Summarization. *A preprocessor for LISREL*. 2nd Ed. Scientific Software. Chicago.

3. VANYUKOV, M. M., H. B. MOSS, J. A. PLAIL, T. BLACKSON, A. C. MEZZICH & R. E. TARTER. 1993. Antisocial symptoms in preadolescent boys and in their parents: Associations with cortisol. *Psychiatry Res.*, **46**: 9–17.
4. MOSKALENKO, V. D., M. M. VANYUKOV, Z. V. SOLOVYOVA, T. V. RAKHMANOVA & M. M. VLADIMIRSKY. 1992. A genetic study of alcoholism in the Moscow population: Preliminary findings. *J. Stud. Alcohol* **53**: 218–224.
5. HALL, R. L., V. M. HESSELBROCK & J. R. STABENAU. 1983. Familial distribution of alcohol use: I. Assortative mating in the parents of alcoholics. *Behav. Genet.* **13**: 361–372.
6. HALL, R. L., V. M. HESSELBROCK & J. R. STABENAU. 1983. Familial distribution of alcohol use. II. Assortative mating of alcoholic probands. *Behav. Genet.* **13**: 373–382.
7. PENICK, E. C., B. J. POWELL, S. F. BINGHAM, B. I. LISKOW, N. S. MILLER & M. R. READ. 1987. A comparative study of familial alcoholism. *J. Stud. Alcohol* **48**: 136–146.
8. JACOB, T. & D. A. BREMER. 1986. Assortative mating among men and women alcoholics. *J. Stud. Alcohol* **47**: 219–222.
9. PICKENS, R. W., D. S. SVIKIS, M. MCGUE, D. T. LYKKEN, L. L. HESTON & P. J. CLAYTON. 1991. Heterogeneity in the inheritance of alcoholism: A study of male and female twins. *Arch. Gen. Psychiatry* **48**: 19–28.
10. KAIJ, L. 1960. Studies on the Etiology and Sequels of Abuse of Alcohol. Hakan Ohlssons Boktryckery. Lund.
11. MCGUE, M., R. W. PICKENS & D. S. SVIKIS. 1992. Sex and age effects on the inheritance of alcohol problems: A twin study. *J. Abnorm. Psychol.* **101**: 3–17.
12. KENDLER, K. S., A. C. HEATH, M. C. NEALE, R. C. KESSLER & L. J. EAVES. 1992. A population-based twin study of alcoholism in women. *J. Am. Med. Assoc.* **268**: 1877–1882.
13. BOHMAN, M., S. SIGVARDSSON & C. R. CLONINGER. 1981. Maternal inheritance of alcohol abuse. *Arch. Gen. Psychiatry* **38**: 965–968.
14. BOHMAN, M., C. R. CLONINGER, S. SIGVARDSSON & A.-L. VON KNORRING. 1983. Gene-environment interaction in the psychopathology of Swedish adoptees: Studies of the origins of alcoholism and criminality. *In* S. B. Guze, F. J. Earls & J. E. Barrett, eds.: 265–278. Childhood Psychopathology and Development. Raven Press. New York.
15. CADORET, R. J., T. O'GORMAN, E. TROUGHTON & E. HEYWOOD. 1985. Alcoholism and antisocial personality: Interrelationships, genetic and environmental factors. *Arch. Gen. Psychiatry* **42**: 161–167.
16. CLONINGER, C. R., M. BOHMAN & S. SIGVARDSSON. 1981. Inheritance of alcohol abuse: Cross-fostering analysis of adopted men. *Arch. Gen. Psychiatry* **38**: 861–868.
17. GOODWIN, D. W., F. SCHULSINGER, L. HERMANSEN, S. B. GUZE & G. WINOKUR. 1973. Alcohol problems in adoptees raised apart from alcoholic parents. *Arch. Gen. Psychiatry* **28**: 238–243.
18. GOODWIN, D. W., F. SCHULSINGER, N. MOLLER, L. HERMANSEN, G. WINOKUR & S. B. GUZE. 1974. Drinking problems in adopted and nonadopted sons of alcoholics. *Arch. Gen. Psychiatry* **31**: 164–169.
19. CADORET, R. J., E. TROUGHTON, T. W. O'GORMAN & E. HEYWOOD. 1986. An adoption study of genetic and environmental factors in drug abuse. *Arch. Gen. Psychiatry* **43**: 1131–1136.
20. MERIKANGAS, K. 1982. Assortative mating for psychiatric disorders and psychological traits. *Arch. Gen. Psychiatry* **39**: 1173–1180.
21. BUSS, D. M. 1984. Marital assortment for personality dispositions: Assessment with three different data sources. *Behav. Genet.* **14**: 111–123.
22. PLOMIN, R., N. L. PEDERSEN, G. E. MCCLEARN, J. R. NESSELROADE & C. S. BERGEMAN. 1988. EAS temperaments during the last half of the life span: Twins reared apart and twins reared together. *Psychol. & Aging* **3**: 43–50.
23. BOUCHARD, T. J., D. T. LYKKEN, M. MCGUE, N. L. SEGAL & A. TELLEGEN. 1990. Sources of human psychological differences: The Minnesota study of twins reared apart. *Science* **250**: 223–228.

Second Messenger and Protein Phosphorylation Mechanisms Underlying Possible Genetic Vulnerability to Alcoholism[a]

ERIC J. NESTLER, XAVIER GUITART, JORDI ORTIZ,
AND LOUIS TREVISAN

Laboratory of Molecular Psychiatry
Departments of Psychiatry and Pharmacology
Yale University School of Medicine
and Connecticut Mental Health Center
34 Park Street
New Haven, Connecticut 06508

Studies of the basic neurobiology of drug addiction, including alcoholism, should be given a high priority for two reasons. First, from a clinical perspective, drug abuse continues to exact enormous human and financial costs on society, yet all currently available treatments for drug addiction are notoriously ineffective. The search for a better understanding of the neurobiological mechanisms underlying the addictive actions of drugs of abuse and of the genetic factors that contribute to addiction will result in crucial advances in our ability to treat and prevent drug addiction.

Second, from a basic neuroscience perspective, study of the neurobiology of drug addiction offers a unique opportunity to establish the biological basis of a complex and clinically relevant behavioral abnormality. This is based largely on the availability of good animal models of drug addiction, which make it possible to study the detailed underlying mechanisms involved. Advances made in the field of drug addiction should provide important insights into mechanisms involved in other neuropsychiatric disorders, for which animal models are much less straightforward and much more difficult to interpret.

CENTRAL IMPORTANCE OF INTRACELLULAR MESSENGER PATHWAYS IN NEURONAL REGULATION

It is now well established that all types of neurotransmitters and hormones produce most of their diverse effects on target neuron functioning through the regulation of cascades of intracellular messengers. These intracellular messengers, illustrated in FIGURE 1, include G-proteins (which couple plasma membrane receptors to intracellular effector systems) and the effector systems themselves, which include second messengers (such as cyclic AMP, calcium, and phosphatidylinositol), protein kinases and protein phosphatases, and phosphoproteins.[1–3] Among the effects mediated through intracellular messenger pathways is the regulation of gene expression. Inasmuch as many important aspects of drug addiction develop gradually

[a]This work was supported by the VA-Yale Alcoholism Research Center, U.S. Department of Veterans Affairs, and by the Abraham Ribicoff Research Facilities, Connecticut Mental Health Center, State of Connecticut Department of Mental Health.

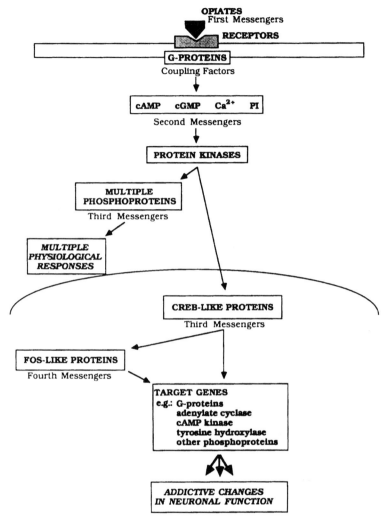

FIGURE 1. *Schematic illustration of intracellular messenger pathways through which diverse extracellular signals produce multiple types of physiological responses, including the regulation of gene expression, in target neurons.* The figure illustrates the likely central role played by these intracellular messengers in mediating the addictive actions of opiates and other drugs of abuse. CREB-like transcription factors refer to those that are expressed constitutively and regulated by extracellular agents primarily through changes in their degree of phosphorylation. Fos-like transcription factors refer to those that are expressed at very low levels under basal conditions and are regulated by extracellular agents primarily through induction of their expression (presumably by CREB-like proteins). Both types of mechanisms could contribute to the addictive actions of opiates and other drugs of abuse. Modified from ref. 1.

and progressively in response to continued drug exposure and can persist for a long time after discontinuation of drug exposure, it seems likely that mechanisms involving the regulation of intracellular messenger pathways and neuronal gene expression are of particular relevance to drug addiction phenomena.[4] This information has provided a conceptual framework within which our laboratory has investigated the molecular mechanisms of drug addiction. We have focused on changes in specific intracellular signal transduction pathways that drugs of abuse induce in specific target brain regions after chronic administration.

STUDIES OF OPIATE ACTION IN THE LOCUS COERULEUS

An early focus was on opiate action in the locus coeruleus (LC). The LC is the major noradrenergic nucleus in brain that plays an important role in mediating physical aspects of opiate addiction, namely, physical dependence and physical withdrawal.[4,5] We found that whereas acute opiate treatment inhibits the cyclic AMP signal transduction pathway in the LC, chronic opiate exposure up-regulates the cyclic AMP pathway at every major step between receptor and physiological response, with increases seen in levels of the G-protein subunits $Gi\alpha$ and $Go\alpha$, adenylate cyclase, cyclic AMP-dependent protein kinase, and specific phosphoprotein substrates for the protein kinase.[6–10] In collaboration with Dr. George K. Aghajanian, we have provided direct electrophysiological evidence that such up-regulation of the cyclic AMP system is one mechanism by which opiates induce tolerance, dependence, and withdrawal in this brain region.[11] Although there are likely to be many mechanisms underlying opiate addiction in the nervous system, this work represents the first such mechanism to be established at the molecular, biochemical, electrophysiological, and behavioral level.

Efforts are now aimed at delineating the mechanisms by which chronic opiate exposure up-regulates the cyclic AMP pathway in the LC. Because the up-regulation involves altered levels of specific proteins and their mRNAs, it is possible that regulation of neuronal gene expression is involved.[4] The most important mechanism by which gene expression is regulated in the nervous system involves transcription factors, diverse types of DNA-binding proteins that bind to the regulatory regions of specific genes and thereby increase or decrease the rate at which those genes are transcribed (FIG. 1).[1] To date, chronic morphine has been shown to regulate two major families of transcription factors in the LC: c-fos and related immediate early genes and CREB (cyclic AMP response element binding) protein.[12,13] Studies are underway to relate regulation of these specific transcription factors to the up-regulated expression of the cyclic AMP system in the LC. These studies will provide details concerning the precise mechanism, at the molecular level, by which opiates alter the expression of particular target genes in LC neurons and thereby lead to tolerance, dependence, and withdrawal.

STUDIES OF OPIATE AND COCAINE ACTION IN THE MESOLIMBIC DOPAMINE SYSTEM

The ability to identify intracellular mechanisms of opiate addiction in the LC led us to study whether similar mechanisms may be involved in the mesolimbic dopamine system, which has been implicated in psychological dependence (i.e., drug reinforcement and craving) not only to opiates, but also to many other types of drugs of abuse including cocaine and alcohol.[14–16] The mesolimbic dopamine system consists of dopaminergic neurons in the ventral tegmental area (VTA) and their various projection areas, notably the nucleus accumbens (NAc).

To date, we have identified a series of common chronic actions that opiates and cocaine exert in the mesolimbic dopamine system. Chronic, but not acute, administration of morphine or cocaine was shown to increase levels of tyrosine hydroxylase[17] and decrease levels of neurofilament proteins[18] in the VTA and to decrease levels of the G-protein subunit $Gi\alpha$ and increase levels of adenylate cyclase and cyclic AMP-dependent protein kinase, in the NAc.[19,20] These intracellular adaptations are specific to the mesolimbic dopamine system; they were not observed in several other brain regions studied, including the substantia nigra and caudate putamen, components of the nigrostriatal dopamine system generally not implicated in drug reward mechanisms. Moreover, chronic treatment of rats with other classes of psychotropic drugs that lack reinforcing properties, for example, the antipsychotic drug haloperidol and the antidepressant drugs imipramine and fluoxetine, failed to produce these changes in the VTA and NAc. These results supported the hypothesis that drug-induced changes in these intracellular messengers may represent part of a common biochemical mechanism of drug addiction.[4] Further support for this view is provided by preliminary observations that chronic self-administration of alcohol also increases levels of tyrosine hydroxylase in the VTA and of cyclic AMP-dependent protein kinase in the NAc.[21]

INHERENT DIFFERENCES IN INTRACELLULAR MESSENGER PROTEINS IN LEWIS AND FISCHER RATS

To study further the relevance of morphine, cocaine, and alcohol regulation of intracellular messenger proteins in the mesolimbic dopamine system to drug reward mechanisms, these intracellular messengers were studied in the VTA and NAc of Lewis and Fischer 344 rats, inbred strains known to differ inherently in drug preference. Lewis rats have been shown to self-administer opiates, cocaine, and alcohol at higher rates compared to Fischer rats[22,23] and to develop greater degrees of conditioned place preference to opiates and cocaine.[24] In addition, cannabinoids facilitate self-stimulation of the VTA in Lewis rats, but not in Fischer rats.[25]

We have shown that the VTA of drug-naive Lewis rats contains higher levels of tyrosine hydroxylase[26] and lower levels of neurofilament proteins[24] compared to the VTA of drug-naive Fischer rats. Moreover, the NAc of Lewis rats contains lower levels of $Gi\alpha$ and higher levels of adenylate cyclase and cyclic AMP-dependent protein kinase compared to the NAc of Fischer rats.[27] These Lewis-Fischer strain differences were not observed in several other regions of brain and spinal cord examined. Thus, the drug-preferring Lewis rat (as compared to the relatively non-preferring Fischer rat) resembles outbred Sprague-Dawley rats treated chronically with several types of drugs of abuse.

BIOCHEMICAL MODEL OF THE DRUG-ADDICTED/DRUG-PREFERRING STATE

The common actions of morphine and cocaine (and possibly alcohol) on specific intracellular messenger proteins in the mesolimbic dopamine system, and the identification of different inherent levels of these same intracellular messengers specifically in this neural pathway between Lewis and Fischer rats, led to the hypothesis that the drug-addicted and drug-preferring state is associated with higher levels of tyrosine hydroxylase and lower levels of neurofilament proteins in the VTA and with lower levels of Giα and higher levels of adenylate cyclase and cyclic AMP-dependent protein kinase in the NAc.[4] This hypothesis is illustrated in FIGURE 2.

In the VTA, higher levels of tyrosine hydroxylase (which is the rate-limiting enzyme in the biosynthesis of dopamine) in the drug-addicted/drug-preferring state would be expected to be associated with altered dopaminergic transmission in the mesolimbic dopamine system. Given the evidence for an important role of dopamine

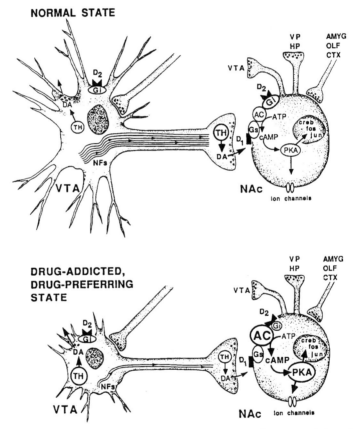

FIGURE 2. *Schematic summary of similar biochemical manifestations of the "drug-addicted" and "genetically drug-preferring" state.* The *top panel* depicts a normal VTA neuron projecting to an NAc neuron. Shown in the VTA neuron are tyrosine hydroxylase (TH), dopamine (DA), presynaptic dopamine receptors (D2) coupled to G-proteins (Gi), and neurofilaments (NFs). Shown in the NAc neuron are dopamine receptors (D1 and D2), G-proteins (Gi and Gs), components of the intracellular cyclic AMP system (AC, adenylate cyclase; PKA, cAMP-dependent protein kinase; and possible substrates for the kinase-ion channels and the nuclear transcription factors, CREB, fos, and jun) as well as major inputs and

function in drug reward, altered levels of tyrosine hydroxylase could be one molecular mechanism by which drugs of abuse influence reward processes and by which individuals have different inherent responses to these drugs.

The physiological consequences of lower levels of neurofilament proteins in the VTA in the drug-addicted/drug-preferring state are less certain, because the precise functional role of neurofilament proteins remains unknown. Neurofilament proteins are important determinants of neuronal morphology and have been associated with axonal caliber and transport and with dendritic sprouting.[28–31] Different levels of neurofilament proteins in the VTA might be associated with any of these parameters. Slow axonal transport, in particular, would be expected to be decreased in the mesolimbic dopamine system of drug-addicted or drug-preferring animals. A variety of *in vivo* manipulations, including axotomy,[32] aluminum intoxication,[33] and chronic β,β′-iminodiproprionitrile intoxication,[34] are known to lead to both decreased neurofilament protein levels and decreased slow axonal transport. More recently, we found that chronic morphine treatment impairs axonal transport in the VTA-NAc pathway, but not in other neural pathways studied.[35] Clearly, it would be interesting in future studies to compare axonal transport in the mesolimbic dopamine system of cocaine- and alcohol-treated animals and in Lewis *versus* Fischer rats. It would also be important to explore other possible mechanisms responsible for differences in tyrosine hydroxylase and neurofilament protein levels, such as a difference in the number or size of cell bodies, dendrites, or axons in this brain region. In any event, drug-induced or inherent differences in tyrosine hydroxylase and neurofilament protein levels suggest that prominent structural and functional differences exist in the VTA in the drug-addicted/drug-preferring state.

In the NAc, different levels of G-proteins and the cyclic AMP system would be expected to have important functional consequences on NAc neurons. Altered responsiveness of the NAc could then exert a powerful influence on cortical to subcortical neural pathways, critical to reward mechanisms, that course through this brain region. However, it is more difficult to study directly the physiological role of these intracellular adaptations in the NAc, compared to the LC for example, because

---

outputs of this region (VP, ventral pallidum; HP, hippocampus; AMYG, amygdala; OLF, olfactory cortex; CTX, other cortical regions). The *bottom panel* depicts a VTA neuron projecting to the NAc after chronic administration of morphine or cocaine or in an untreated Lewis (genetically drug-preferring) rat as compared to a relatively nonpreferring Fischer rat. In the drug-addicted or drug-preferring animal, TH levels are increased in the VTA and decreased in the NAc (due to either decreased phosphorylation as for morphine and cocaine or decreased enzyme levels as in Lewis *vs* Fischer rats). In addition, NF levels are decreased in the VTA in the drug-addicted and drug-preferring animal. This decrease in NFs may be associated with alterations in neuronal structure, decreases in axonal caliber, and/or decreases in axonal transport rate in these cells. This hypothetical decrease in axonal transport may account for the lack of correspondingly increased levels of TH in dopaminergic terminals in the NAc. Decreased TH levels imply reduced synthesis and may result in reduced dopaminergic transmission to the NAc. In the NAc of the drug-addicted or drug-preferring animal, Gi is decreased, and adenylate cyclase and cAMP-dependent protein kinase activities are increased, changes that could account for D_1 receptor supersensitivity observed electrophysiologically. It should be noted that alterations in dopaminergic transmission probably influence many cell types within the NAc as well as other nerve terminals in the NAc. Similarly, altered local dopaminergic transmission in the VTA would influence other VTA neurons as well as nerve terminals that innervate this brain region. Thus, biochemical alterations in the mesolimbic dopamine system could potentially lead to altered neuronal function in many other brain regions as well. Modified from ref. 42.

the NAc is a much more complex brain region in that it contains multiple types of neurons that show different responses to various drugs of abuse. There have also been no reports of the *chronic* effects of opiates or alcohol on the electrophysiological properties of NAc neurons. However, the chronic actions of cocaine on the NAc have been addressed in a recent study; chronic cocaine administration was shown to produce supersensitivity of NAc neurons to D_1-dopamine receptor activation.[36] As this functional supersensitivity occurs in the absence of detectable changes in D_1 receptor number or affinity,[37] and as D_1 receptors produce their physiological effects through activation of the cyclic AMP pathway (see ref. 1), it is likely that post-receptor intracellular mechanisms are involved. Indeed, chronic cocaine-induced decreases in levels of $Gi\alpha$ (which acts to inhibit adenylate cyclase) and increases in levels of adenylate cyclase and cyclic AMP-dependent protein kinase could account for the D_1 receptor supersensitivity observed electrophysiologically. It will be important in future studies to investigate the chronic effects of morphine and alcohol on the physiological state of NAc neurons.

FURTHER EVIDENCE FOR A ROLE OF INTRACELLULAR MESSENGER PROTEINS IN THE DRUG-PREFERRING STATE

On the basis of recent reports that individual differences in drug preference among outbred Sprague-Dawley rats are highly correlated with the animals' locomotor response to the mild stress associated with a novel environment,[38] we determined whether such within strain differences in locomotor behavior are also associated with differences in specific intracellular messenger proteins in the mesolimbic dopamine system. It was found that rats with the lowest locomotor responses (designated L rats) display higher levels of tyrosine hydroxylase and lower levels of neurofilament proteins specifically in the VTA than do rats with the highest locomotor responses (designated H rats).[39] In more preliminary experiments, L rats also displayed higher levels of cyclic AMP-dependent protein kinase specifically in the NAc compared to H rats.[40] Recent data have confirmed that L and H rats show prominent differences in several cocaine-related behaviors, including cocaine preference. It will now be important to more completely study the behavioral responses of these animals to opiates and alcohol as well as cocaine.

We also recently compared levels of intracellular messenger proteins in the mesolimbic dopamine system of two rat lines that were selectively bred for a difference in alcohol preference, the alcohol-preferring (P) and nonpreferring (NP) rats.[41] It was found that the VTA of P rats contain lower levels of neurofilament proteins than does the VTA of NP rats (FIG. 3), similar to earlier findings in Lewis *versus* Fischer rats. This observation provides further support for an association between lower levels of neurofilament proteins in the VTA and an inherent alcohol-preferring state.

However, no P-NP differences were found in levels of tyrosine hydroxylase in the VTA in this study.[41] This observation means that altered levels of neurofilament proteins and tyrosine hydroxylase may not be necessary concomitants and that altered levels of each protein might occur in isolation of the other. Of course, measures of tyrosine hydroxylase immunoreactivity by immunoblotting procedures (the sole method used to date) offer a rather crude evaluation of the protein. Thus, it is conceivable that levels of tyrosine hydroxylase might well be different in the VTA of P and NP rats, but perhaps in more localized subsets of dopaminergic neurons compared to the Lewis-Fischer (or drug-treated) situations. It is also con-

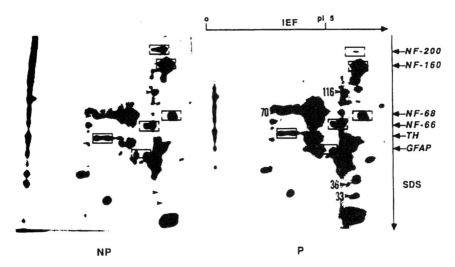

FIGURE 3. *Autoradiograms of two-dimensional gels showing differences in protein phosphorylation in the VTA between P and NP rats.* Extracts of VTA obtained from drug-naive P and NP rats were subjected to back phosphorylation with purified cyclic AMP-dependent protein kinase and [γ-³²P]ATP and to two-dimensional gel electrophoresis. Autoradiograms of representative gels are shown in their entirety. The figure shows that several neurofilament proteins (NF-200, NF-160, NF-68, and NF-66), indicated by *brackets*, are present in the VTA at different levels between P and NP rats. In contrast, no P-NP differences were observed in tyrosine hydroxylase (TH) or glial fibrillary acidic protein (GFAP), also identified by *brackets*. *Arrowheads* indicate the positions of four proteins of unknown identity, with M_r of 116, 70, 36, and 33 kD, which show consistent differences between P and NP rats. From ref. 41.

ceivable that certain functional properties of VTA dopaminergic neurons are similarly different between P *versus* NP rats and between Lewis *versus* Fischer (and drug-treated *versus* control) rats; however, the biochemical basis of the differences resides in a different protein(s) in these two situations.

COMMON MECHANISMS OF DRUG ABUSE AND ADDICTION

One corollary of our biochemical model of the drug-addicted/drug-preferring state (FIG. 2) is that the mechanisms underlying the actions of alcohol and inherent vulnerability for alcohol addiction are shared at least in part by other drugs of abuse and states of drug preference. This may seem surprising on the surface, inasmuch as alcohol, opiates, and cocaine initially influence the nervous system through disparate mechanisms of action. Alcohol seems to alter membrane fluidity and thereby disrupt the function of proteins particularly sensitive to their lipid milieu, for example, $GABA_A$ and NMDA-glutamate receptors. By contrast, opiates affect the nervous system acutely through activation of specific opioid receptors, whereas cocaine affects the nervous system acutely by inhibiting specific dopamine (as well as serotonin and norepinephrine) transporters and thereby reducing the reuptake of these neurotransmitters into their nerve terminals. Indeed, alcohol, opiates, and cocaine are known to produce many different effects on the nervous system both acutely and chronically. These differences are probably due to the different localizations of their initial protein targets (e.g., $GABA_A$ receptors, opioid receptors, and dopamine transporters) in the nervous system and therefore to different collections of neurons influenced by the drugs throughout the central and peripheral nervous systems. Several of these disparate actions could influence the rewarding properties of these drugs as well as inherent states of drug preference in some individuals.

Nevertheless, despite the different initial protein targets and disparate pharmacological actions, alcohol, opiates, and cocaine do produce some similar effects on the brain. Such common effects may be due to the colocalization of their initial protein targets in specific brain areas and the same downstream consequences (i.e., regulation of postreceptor mechanisms) initiated by drug regulation of the target proteins in these brain areas. A prominent example of a common acute action is increased levels of extracellular dopamine in the NAc. Moreover, because alcohol, opiates, and cocaine are psychologically addicting and the mesolimbic dopamine system plays an important role in psychological addiction, it makes sense that these drugs of abuse would induce *some* common functional changes in this neural pathway after chronic administration. The common chronic actions of alcohol, opiates, and cocaine described here could then be part of the biochemical basis of long-term functional changes in the VTA-NAc pathway that modify drug reward mechanisms. In a similar way, the observation that Lewis and Fischer rats exhibit prominent differences in preference for several drugs of abuse, including alcohol, opiates, and cocaine, supports the view that specific biochemical parameters in the mesolimbic dopamine system can indeed influence the rewarding effects of many types of drugs of abuse.

The studies described here demonstrate that through the study of intracellular messenger pathways it will be possible to build a progressively more complete understanding of the biochemical adaptations that underlie aspects of drug reward mediated by the mesolimbic dopamine system and that may also contribute to individual genetic vulnerability to drug addiction, including alcoholism.

REFERENCES

1. HYMAN, S. E. & E. J. NESTLER. 1993. The Molecular Foundations of Psychiatry. American Psychiatric Press. Washington, DC.
2. NESTLER, E. J. & P. GREENGARD. 1993. Protein phosphorylation and the regulation of neuronal function. *In* Basic Neurochemistry: Molecular, Cellular, and Medical Aspects, 5th ed. G. Siegel, B. Agranoff, R. W. Albers & P. Molinoff, eds.: 449–474. New York. Raven Press.
3. NESTLER, E. J. & R. DUMAN. 1993. G-proteins and cyclic nucleotides in the nervous system. *In* Basic Neurochemistry: Molecular, Cellular, and Medical Aspects, 5th Ed. G. Siegel, B. Agranoff, R. W. Albers & P. Molinoff, eds.: 429–448. New York. Raven Press.
4. NESTLER, E. J. 1992. Molecular mechanisms of drug addiction. *J. Neurosci.* **12**: 2439–2450.
5. KOOB, G. F., R. MALDONADO & L. STIMUS. 1992. Neural substrates of opiate withdrawal. Trends Neurosci. **15**: 186–191.
6. DUMAN, R. S., J. F. TALLMAN & E. J. NESTLER. 1988. Acute and chronic opiate-regulation of adenylate cyclase in brain: Specific effects in locus coeruleus. J. Pharmacol. Exp. Ther. **246**: 1033–1039.
7. NESTLER, E. J. & J. F. TALLMAN. 1988. Chronic morphine treatment increases cyclic AMP-dependent protein kinase activity in the rat locus coeruleus. Mol. Pharmacol. **33**: 127–132.

8. NESTLER, E. J., J. J. ERDOS, R. TERWILLIGER, R. S. DUMAN & J. F. TALLMAN. 1989. Regulation of G-proteins by chronic morphine in the rat locus coeruleus. Brain Res. **476**: 230–239.

9. GUITART, X. & E. J. NESTLER. 1989. Identification of morphine and cyclic AMP-regulated phosphoproteins (MARPPs) in the locus coeruleus and other regions of rat brain. Regulation by acute and chronic morphine. J. Neurosci. **9**: 4371–4387.

10. GUITART, X., M. D. HAYWARD, L. K. NISENBAUM, D. S. BEITNER-JOHNSON, J. W. HAYCOCK & E. J. NESTLER. 1990. Identification of MARPP-58, a morphine- and cyclic AMP-regulated phosphoprotein of 58 kD, as tyrosine hydroxylase: Evidence for regulation of its expression by chronic morphine in the rat locus coeruleus. J. Neurosci. **10**: 2635–2645.

11. KOGAN, J. H., E. J. NESTLER & G. K. AGHAJANIAN. 1992. Elevated basal firing rates and enhanced responses to 8-Br-cAMP in locus coeruleus neurons in brain slices from opiate-dependent rats. Eur. J. Pharmacol. **211**: 47–53.

12. HAYWARD, M. D., R. S. DUMAN & E. J. NESTLER. 1990. Induction of the c-fos proto-oncogene during opiate withdrawal in the locus coeruleus and other regions of rat brain. Brain Res. **525**: 256–266.

13. GUITART, X., M. A. THOMPSON, C. K. MIRANTE, M. E. GREENBERG & E. J. NESTLER. 1992. Regulation of CREB phosphorylation by acute and chronic morphine in the rat locus coeruleus. J. Neurochem. **58**: 1168–1171.

14. WISE, R. A. 1990. The role of reward pathways in the development of drug dependence. *In* Psychotropic Drugs of Abuse. D. J. K. Balfour, ed. 23–57. Pergamon Press. Oxford.

15. KUHAR, M. J., M. C. RITZ & J. W. BOJA. 1991. The dopamine hypothesis of the reinforcing properties of cocaine. *Trends Neurosci.* **14**: 299–302.

16. KOOB, G. F. 1992. Drugs of abuse: Anatomy, pharmacology and function of reward pathways. *Trends Pharmacol.* **13**: 177–184.

17. BEITNER-JOHNSON, D. & E. J. NESTLER. 1991. Morphine and cocaine exert common chronic actions on tyrosine hydroxylase in dopaminergic brain reward regions. J. Neurochem. **57**: 344–347.

18. BEITNER-JOHNSON, D., X. GUITART & E. J. NESTLER. 1992. Neurofilament proteins and mesolimbic dopamine system: Common regulation by chronic morphine and chronic cocaine in the rat ventral tegmental area. J. Neurosci. **12**: 2165–2176.

19. NESTLER, E. J., R. Z. TERWILLIGER, J. R. WALKER, K. A. SEVARINO & R. S. DUMAN. 1990. Chronic cocaine treatment decreases levels of the G-protein subunits Giα and Goα in discrete regions of rat brain. J. Neurochem. **55**: 1079–1082.

20. TERWILLIGER, R. Z., D. BEITNER-JOHNSON, K. A. SEVARINO, S. M. CRAIN & E. J. NESTLER. 1991. A general role for adaptations in G-proteins and the cyclic AMP system in mediating the chronic actions of morphine and cocaine on neuronal function. Brain Res. **548**: 100–110.

21. TREVISAN, L., J. ORTIZ, W. SHOEMAKER & E. J. NESTLER. 1993. Regulation of intracellular messenger proteins by chronic alcohol self-administration in specific regions of rat brain. Alcoholism: Clin. Res. **17**: 500.

22. GEORGE, F. R. & S. R. GOLDBERG. 1988. Genetic approaches to the analysis of addiction processes. Trends Pharmacol. Sci. **10**: 78–83.

23. SUZUKI, T., F. R. GEORGE & R. A. MEISCH. 1989. Differential establishment and maintenance of oral ethanol reinforced behavior in Lewis and Fischer 344 inbred rat strains. J. Pharmacol. Exp. Ther. **245**: 164–170.

24. GUITART, X., D. BEITNER-JOHNSON & E. J. NESTLER. 1992. Fischer and Lewis rat strains differ in basal levels of neurofilament proteins and in their regulation by chronic morphine. Synapse **12**: 242–253.

25. GARDNER, E. L. & J. H. LOWINSON. 1991. Marijuana's interaction with brain reward systems: Update 1991. Pharmacol. Biochem. Behav. **40**: 571–580.

26. BEITNER-JOHNSON, D., X. GUITART & E. J. NESTLER. 1991. Dopaminergic brain reward regions of Lewis and Fischer rats display different levels of tyrosine hydroxylase and other morphine- and cocaine-regulated phosphoproteins. Brain Res. **561**: 146–149.

27. GUITART, X., J. H. KOGAN, M. BERHOW, R. Z. TERWILLIGER, G. K. AGHAJANIAN & E. J. NESTLER. 1992. Lewis and Fischer rat strains display differences in biochemical, elec-

trophysiological, and behavioral parameters: Studies in the nucleus accumbens and locus coeruleus of drug naive and morphine-treated animals. Brain Res. **611**: 7–17.

28. HALL, G. F., V. M.-Y. LEE & K. S. KOSIK. 1991. Microtubule destabilization and neurofilament phosphorylation precede dendritic sprouting after close axotomy of lamprey central neurons. Proc. Natl. Acad. Sci. USA **88**: 5016–5020.

29. HOFFMAN, P. N. & R. J. LASEK. 1975. Identification of major structural polypeptides of the axon and their generality among mammalian neurons. J. Cell Biol. **66**: 351–366.

30. HOFFMAN, P. N., J. W. GRIFFIN & D. L. PRICE. 1984. Control of axonal caliber by neurofilament transport. J. Cell Biol. **99**: 705–714.

31. TYTELL, M., M. M. BLACK, J. A. GARNER & R. J. LASEK. 1981. Axonal transport: Each major component reflects the movement of distinct macromolecular complexes. Science **214**: 179–181.

32. GOLDSTEIN, M. E., H. S. COOPER, J. BRUCE, M. J. CARDEN, V.M.-Y. LEE & W. W. SCHLAEPFER. 1987. Phosphorylation of neurofilament proteins and chromatolysis following transection of rat sciatic nerve. J. Neurosci. **7**: 1586–1594.

33. TRONOSCO, J. C., P. N. HOFFMAN, J. W. GRIFFIN, K. M. HESS-KOZLOW & D. L. PRICE. 1985. Aluminum intoxication: A disorder of neurofilament transport in motor neurons. Brain Res. **342**: 172–175.

34. WATSON, D. F., J. F. GRIFFIN, K. P. FITTRO & P. N. HOFFMAN. 1989. Phosphorylation-dependent immunoreactivity of neurofilaments increases during axonal maturation and β,β'-iminodipropionitrile intoxication. J. Neurochem. **53**: 1818–1829.

35. BEITNER-JOHNSON, D. & E. J. NESTLER. 1992. Chronic morphine impairs axoplasmic transport in the rat mesolimbic dopamine system. Neuro Report, in press.

36. HENRY, D. J. & F. J. WHITE. 1991. Repeated cocaine administration causes persistent enhancement of D1 dopamine receptor sensitivity within the rat nucleus accumbens. J. Pharmacol. Exp. Ther. **258**: 882–890.

37. PERIS, J., S. J. BOYSON, W. A. CASS, P. CURELLA, L. P. DWOSKIN, G. LARSON, L.-H. LIN, R. P. YASUDA & N. R. ZAHNISER. 1990. Persistence of neurochemical changes in dopamine systems after repeated cocaine administration. J. Pharmacol. Exp. Ther. **253**: 38–44.

38. PIAZZA, P. V., J.-M. DEMINIERE, M. LE MOAL & H. SIMON. 1989. Factors that predict individual vulnerability to amphetamine self-administration. Science **1245**: 1511–1513.

39. MISERENDINO, M. J. D., X. GUITART, R. TERWILLIGER, S. CHI, T. A. KOSTEN & E. J. NESTLER. 1993. Individual differences in locomotor behavior are associated with levels of tyrosine hydroxylase and neurofilament proteins in the ventral tegmental area of Sprague-Dawley rats. Mol. Cell. Neurosci. **4**: 440–448.

40. MISERENDINO, M. J. D., T. A. KOSTEN, X. GUITART & E. J. NESTLER. 1992. Individual differences in vulnerability to drug addiction: Behavioral and biochemical correlates. Soc. Neurosci. Abstr. **18**: 1078.

41. GUITART, X., L. LUMENG, T.-K. LI & E. J. NESTLER. 1993. Alcohol preferring (P) and non-preferring (NP) rats display different levels of neurofilament proteins in the ventral tegmental area. Alcoholism: Clin. Exp. Res. **17**: 580–585.

42. BEITNER-JOHNSON, D., X. GUITART & E. J. NESTLER. 1992. Common intracellular actions of chronic morphine and cocaine in dopaminergic brain reward regions. Ann. N.Y. Acad. Sci. **684**: 70–87.

Platelet MAO Activity in Type I and Type II Alcoholism[a]

ERIC J. DEVOR,[b] CREED W. ABELL,[c]
PAULA L. HOFFMAN,[d] BORIS TABAKOFF,[d]
AND C. ROBERT CLONINGER[e]

[b]Department of Psychiatry
University of Iowa College of Medicine
Iowa City, Iowa 52242

[c]Division of Medicinal Chemistry
College of Pharmacy and Institute for Neuroscience
University of Texas
Austin, Texas 78712

[d]Department of Pharmacology
University of Colorado School of Medicine
Denver, Colorado 80262

[e]Department of Psychiatry
Washington University School of Medicine
St. Louis, Missouri 63110

Beginning with the familial observations early in the twentieth century and continuing through the twin and adoption studies of the last two decades,[1-6] an impressive array of evidence has been accumulated to support the view that alcoholism has a significant genetic component. As this evidence has been gathered through ever more sophisticated and careful study designs, it also became apparent that alcoholism is not a unitary disease. Rather, in the current view, alcoholism is a complex, heterogeneous disorder in which the role of genetic and nongenetic factors varies in both proportion and intensity.[7] Explicit recognition of this heterogeneity in alcoholism has taken the form of typologies in which alcoholics are grouped according to sets of clinical and social features.[8-10] One broadly recognized and well-studied classification scheme is the Type I/Type II dichotomy proposed by Cloninger[11,12] and based on the extensive Stockholm adoption studies of alcoholism. In this scheme a form of alcoholism exists, called Type I, in which males and females share relatively equal risk, the disease is less severe with a later age of onset and a better prognosis, fewer complications by other psychiatric illnesses such as personality disorders occur, and hereditary factors are less pronounced. Contrasted to this is Type II alcoholism in which males have a substantially higher risk than do females, the disease is more aggressive, has its onset at a much earlier age, is more refractory to treatment, and is associated with phenomena such as antisocial personality disorder, and in which genetic factors appear to play a major role.

Acceptance of a hereditary predisposition in alcoholism leads to a search for biological/genetic markers of risk that not only would mark the disease in general but also would correlate in a systematic way with the subtypes of the disease. In recent years a number of potential biological/genetic markers of alcoholism have

[a]This work was supported by grants MH-14109, MH-31302, AA-03539, and NS-24932 from the National Institutes of Health and by the Banbury Foundation.

119

been proposed.[13] Among these are enzymes, blood-group proteins, and physiologic phenomena that were at either increased or decreased frequency or activity among alcoholics compared with nonalcoholic controls. Throughout this search for biological/genetic markers, however, no potential candidate was more extensively studied and more consistently replicated than was the *in vitro* activity of the enzyme monoamine oxidase (MAO). Wiberg *et al.*[14] first described the phenomenon of lowered platelet MAO activity among alcoholics, and it has since been replicated in numerous independent studies.[15-20] More importantly, given the concept of multiple alcoholisms, several studies now indicate that MAO activity is correlated with alcoholic subtype as well.[21,22] This paper focuses on our studies of *in vitro* platelet MAO activity in families ascertained through alcoholism.[23,24] In particular, we discuss the issue of MAO activity as an indicator of genetic heterogeneity of risk in the alcoholisms.

SUBJECTS AND METHODS

The families used in our studies of MAO activity and the alcoholisms were ascertained through alcoholic probands as part of the St. Louis Family Interview Study of Alcoholism (directed by C.R.C.). A total of 21 families with 189 members made up the samples of two separate studies, one of eight families with 62 members and the other of 13 families with 127 members. All probands were Caucasian and were born within a 100-mile radius of the St. Louis Metropolitan area. Probands and other family members agreed under informed consent to undergo an extensive clinical and medical evaluation using several well-described psychiatric and psychometric instruments[25] and to provide a venous blood sample from which platelets were prepared. Determinations of affection status for alcoholism as well as for other psychiatric illnesses were made by blind ratings of the interview data and were based on the Feighner criteria.[26] The assignment of a diagnosis of definite alcoholism required the presentation of at least three symptoms in each of three or more symptom classes. These classes included medical problems such as organ damage, withdrawal symptoms, or blackouts; control problems such as morning drinking or use of nonbeverage alcohol; social problems including arrests resulting from drinking, job loss, or loss of friends; and prior identification including self-admission and guilt as well as identification of problem drinking by family members and friends. For the purpose of these studies, only a diagnosis of definite alcoholism was considered as affected. Any lesser diagnosis, such as probable or questionable alcoholism, was regarded as unaffected. Similarly, only definite diagnoses of other psychiatric illnesses were regarded as affected.

Platelet Preparation

Platelets were isolated from 20-ml samples of whole blood obtained via venipuncture in K_2-EDTA anticoagulant tubes. The isolation procedure followed the method of Corash[27] with minor modifications described elsewhere.[28] Blood samples were transferred in equal aliquots to 15-ml polypropylene centrifuge tubes to which 5 ml of "washout" buffer (0.15 M NaCl, 0.01 M Na_2HPO_4, 0.001 M NaH_2PO_4, and 0.01 M glucose) was added at RT. Tubes were then mixed by inversion and centrifuged at 600 × G for 3 minutes at RT without brake. The platelet-rich plasmas (PRPs) in washout buffer were transferred to a 50-ml Oak Ridge centrifuge tube in ice. This procedure was repeated several times to maximize platelet recovery. The ice cold PRPs were centrifuged at 2,500 × G for 20 minutes at 4°C. The supernatant was removed, and the platelet pellets were washed twice in 5 mM EDTA (pH 8.0), 0.15 M NaCl with spins of 2,500 × G for 20 minutes at 4°C. Finally, the washed pellets were resuspended in 0.5 ml of 0.05 M K_2HPO_4 and transferred to a sterile 1.5 ml microcentrifuge tube for storage at −70°C until needed.

MAO Assays

MAO activities in the platelets prepared on the smaller family sample ($n = 62$) were determined by a different method than were the activities in the platelets prepared on the larger family sample ($n = 108$). In the smaller sample the MAOB-specific substrate ^{14}C-benzylamine was used, and enzyme activity was measured by the method of Wurtman and Axelrod,[29] whereas in the larger sample the substrate ^{14}C-phenylethylamine (PEA) was used and enzyme activity was assayed according to the method described in Tabakoff *et al.*[30] Other details of the assay procedures are provided in the reports of the two studies.[23,24]

Inheritance of MAO Activity

Reich[31] suggested that a marker of genetic susceptibility to a disease must meet three specific criteria: (1) the marker itself must be highly heritable, (2) the marker must be present in as yet unaffected offspring of affecteds, and (3) there must be a demonstrable increase in risk among relatives of affecteds if they also carry the marker *versus* the relatives of affecteds who do not carry the marker.

Genetic studies of platelet MAO activity have consistently presented evidence of vertical transmission which could best be described by a single major gene locus model.[32–35] Complex segregation analysis of familial MAO activity data reported by Rice *et al.*[33] favored a transmission model in which the architecture of the single major locus was that of a partial recessive (D = 0.09) with a modest multifactorial background effect (H = 0.20).[36]

We carried out complex segregation analyses of both of our familial MAO data sets using the same POINTER program used in the Rice *et al.* study. The single major locus models recovered by us as the optimal solutions in both sets of data are in remarkable agreement with the earlier model (TABLE 1). The single major locus

TABLE 1. Genetic Architecture of Platelet MAO Activity from Segregation Analysis of Family Data

Study	Families	Parameter Estimates[a]			
		D	T	Q	H
Rice *et al.*[33]	154	0.09	2.44	0.22	0.20
Devor *et al.*[23]	8	0.17	2.58	0.40	0.00[b]
Devor *et al.*[24]	13	0.27	7.22	0.07	0.09

[a]Parameters of the mixed model of genetic transmission in POINTER are: D, the degree of dominance at a major gene locus; T, the deviation, in standard deviation units, between the homozygotes at the major locus; Q, frequency of the allele at the major locus leading to a high value in the trait; H, the relative strength of the multifactorial background effect (see ref. 36).
[b]This parameter reached the lower boundary value during the estimation procedure.

transmission models that best fit our own familial data are also partial recessives with modest multifactorial background effects. Given that the sample sizes in these studies were different and that three different assay systems with three different substrates were used, the three independent best-fit segregation analysis solutions are very consistent. Clearly, the generally held belief that platelet MAO activity is governed by a single major effect that is itself highly heritable is warranted.

MAO Activity and Alcoholic Subtype

In both of our familial studies of platelet MAO activity and alcoholism we obtained results that are in clear agreement with most extant studies indicating a relationship (association) between decreased enzyme activity and disease. In the earlier, smaller family study of 8 families ($n = 62$) and 20 unrelated, nonalcoholic controls,[23] we carried out a series of MAOB-specific assays that had recently been developed.[28,37,38] Although the sample was small, the comparisons between the 24 alcoholics in the families and the 20 controls did indicate significantly decreased enzyme activities regardless of how activity was assessed (TABLE 2). In contrast, our measures of MAOB-specific *concentration* did not display any significant differences (TABLE 2). We concluded from our results[23] that the decreased MAO activities observed among alcoholics compared with nonalcoholic controls were due to diminished catalytic activity of the individual MAOB molecules rather than to a decline in the number of molecules in the tissues.

Apart from the observations regarding overall differences between alcoholics and nonalcoholics, the initial familial study was too small to analyze in more detail. In the second, larger family study[24] we employed a point-biserial correlation[39] of the MAOB activity data to assess *co-segregation* of enzyme activity and alcoholism. As shown in TABLE 3, the point-biserial correlations between alcoholism and MAOB activities of $r = -0.34$ at K_m PEA substrate concentration (1.2 μM) and of $r = -0.29$

TABLE 2. Measures of MAOB Activities and Concentration in Eight Families and a Sample of Correlated Controls

Sample	n	MAOB Activities and Concentration Measures[a]			
		Whole Platelet Activity	Extracted Activity	Molecular Activity	Specific Concentration
All family members	62	66.6 ± 31.7	339.8 ± 125.6	180.0 ± 56.3	3.8 ± 1.4
All affected	24	66.0 ± 29.6	336.9 ± 122.2	180.8 ± 50.5	3.8 ± 1.3
Controls	20	86.1 ± 11.3[b]	420.5 ± 85.7[c]	213.0 ± 49.1[c]	3.9 ± 0.9[d]

[a]Whole platelet activity is expressed as nmol product/mg total protein/hour, extracted activity is expressed as nmol product/mg solubilized protein/hour, molecular activity is expressed as molecules of product/molecule of MAOB/minute, and specific concentration is expressed as μg MAOB/mg solubilized protein.

[b]Comparisons of alcoholic *versus* nonalcoholic mean are significant (t = 2.82, df = 42, p <0.01).

[c]Comparisons of alcoholic *versus* nonalcoholic mean are significant (t = 2.58, df = 42, p <0.05 and t = 2.11, df = 42, p <0.05, respectively).

[d]Comparisons of alcoholic *versus* nonalcoholic mean are not significant (t = 0.32, df = 42, ns).

TABLE 3. Mean and Standard Deviations of MAO Activities among Alcoholic Families by Affection Status and Alcoholic Subtype

Sample	n	PEA Concentration	
		1.2 µM	12.0 µM
Total	108	509.8 ± 182.7	1,172.0 ± 663.2
Unaffected	59	568.2 ± 209.4 ⎤ $r = -0.34**$	1,349.7 ± 749.2 ⎤ $r = -0.29**$
Alcoholic	49	443.0 ± 115.7 ⎦	958.0 ± 465.6 ⎦
Type I	33	465.2 ± 109.1 ⎤ $r = -0.28*$	1,006.1 ± 503.8 ⎤ $r = -0.16$
Type II	16	397.1 ± 119.0 ⎦	858.9 ± 369.9 ⎦

$*p < 0.05$; $**p < 0.01$.

at saturating PEA substrate concentration (12.0 µM) are both significant (t = 3.78, df = 106, p 0.01 and t = 3.32, df = 106, p <0.01, respectively). These results suggest significant co-segregation of low MAOB activities and illness in these families, which strengthens the extant evidence that MAOB activity is a genetic marker of risk to alcoholism.[23] Also shown in TABLE 3 is evidence that MAOB activity is related to clinical subtype within the alcoholics themselves. The point-biserial correlation of $r = -0.28$ between enzyme activity and alcoholic subtype at K_m PEA concentration is significant (t = 2.03, df = 47, p <0.05). Conversely, the correlation of $r = -0.16$ at the saturating PEA concentration is not significant (t = 1.06, df = 47, ns). Thus, at least at the K_m substrate concentration, MAOB activity co-segregates with respect to both disease and disease subtype. These results add more credence to the contention that the Type I/Type II alcoholism dichotomy is a valid clinical/biological concept.[22]

MAO Activity and Genetic Vulnerability

One aspect of the Type I/Type II alcoholism dichotomy that follows from the demonstrated relationship with MAOB activity is that if low MAOB activity is an indicator of genetic vulnerability to alcoholism in general, increased genetic vulnerability may exist in the families of Type II *versus* Type I alcoholics. As noted, in addition to diagnoses of alcoholism in these families, the structured psychiatric interviews permitted assessment of mental health status in general. Among the 13 families in this sample, 6 probands are Type I alcoholics and 7 are Type II alcoholics. We examined the mental health status of the relatives of the two types of probands for alcoholism, other psychiatric illnesses, and the two categories combined. The results of this assessment are presented in TABLE 4. A significant excess of alcoholism is present in the relatives of Type II probands compared with those of Type I probands even though a substantial number of relatives of Type I probands are affected. In addition to alcoholism, there are clear trends in these data toward higher levels of other psychiatric illness in the relatives of Type II alcoholics. Again, substantial levels of other psychiatric illness are present in the relatives of Type I alcoholics, but when the proband is classified as a Type II alcoholic, the families tend to have by comparison even higher levels of illness.

Perhaps even more important than simply the number of relatives of Type II probands compared with that of Type I probands diagnosed for a psychiatric illness

TABLE 4. Distribution of Alcoholism and Other Psychiatric Illness among the Relatives of Type I ($n = 6$) and Type II ($n = 7$) Probands

	Alcoholism Only		Other Psychiatric Illness[a]		Combined Illness[a]	
	Type I PB	Type II PB	Type I PB	Type II PB	Type I PB	Type II PB
Affected	12 (19.7%)	20 (37.7%)	18 (29.5%)	24 (45.3%)	26 (42.6%)	32 (60.4%)
Unaffected	49 (80.3%)	33 (62.3%)	43 (70.5%)	29 (54.7%)	35 (57.4%)	21 (39.6%)
Total	61	53	61	53	61	53
Significance (chi-square)	4.68[b]		3.04		3.50	

Note : PB = proband. Nine of the 12 alcoholic relatives of Type I probands are themselves Type I. Eleven of the 20 alcoholic relatives of Type II probands are Type II.
[a]These classes include individuals with multiple diagnoses. However, all affecteds are counted only once regardless of how many entities receive diagnoses.
[b]$p < 0.05$.

is the diversity of those illnesses. von Knorring *et al.*[22] reported a significant excess of both alcoholism and affective disorder in the parents of Type II alcoholics. In our families we find that this phenomenon is replicated in that 2 of 12 parents of the Type I probands received a diagnosis of either alcoholism (one father) or affective disorder (one mother), whereas 7 of 14 parents of Type II alcoholics had a diagnosis of either alcoholism (3 fathers, 1 mother) or affective disorder (3 mothers). Overall, however, affective disorder is diagnosed more among all relatives of Type I probands (16 of the 61 interviewed relatives, 26.2%) than among all relatives of Type II probands (11 of the 53 interviewed relatives, 20.8%). Conversely, more cases of phobia are noted among the relatives of Type II probands (8 cases) than among the relatives of Type I probands (4 cases), and all five cases of anxiety neurosis occur in relatives of Type II probands. Naturally, given the criteria applied to categorizing an alcoholic as Type I or Type II, it is not surprising that 9 of 13 cases of antisocial personality disorder occur in the relatives of Type II probands. Finally, the proportion of multiplex cases, that is, individuals receiving more than one diagnosis, is higher in the relatives of Type II probands. Among 32 relatives of Type II alcoholic probands who received any psychiatric diagnosis, 23 (71.9%) received two or more diagnoses, whereas among the 26 relatives of Type I probands, 15 (57.7%) received two or more diagnoses. These data indicate that in a carefully ascertained and evaluated sample, the families of alcoholics display a great deal of psychiatric illness. However, a clear trend suggests a higher loading of psychiatric illness in the families of Type II probands than of Type I probands.

As noted by Reich[31] and mentioned previously, one of the three crucial criteria that must be satisfied for a trait to be accepted as a biological/genetic marker of increased risk of illness is a demonstration that the relatives of affecteds (probands) who themselves carry the putative genetic marker display an excess of the disorder in question compared with the relatives who do not carry the putative genetic marker. In other words, the potential genetic marker must be shown to actually be associated with increased risk! Given the correlation of low MAOB activities with both alcoholism and alcoholic subtype in our families and evidence that there are increased levels of psychiatric illness both overall and in the relatives of Type II probands in particular, is there evidence in these families that low MAOB activities are associated with these increased levels of illness? To assess this question, we dichotomized MAOB assay data values obtained at the K_m PEA substrate concentration. As a reasonable threshold, we chose the mean proband MAOB activity plus 1 standard deviation, or 381.2 + 78.7 pmol product/min/mg protein. Family members having MAOB activity of less than 459.9 pmol product/min/mg protein were thus designated as having "low" MAOB activity, whereas family members having MAOB activity greater than 459.9 pmol product/min/mg protein were designated as having "high" MAOB activity.

Using this "high" *versus* "low" activity dichotomy, we again examined the mental health status of family members for alcoholism, other psychiatric illness, and both. The results, presented in TABLE 5, indicate a significant excess of alcoholism, as expected, in the relatives of alcoholics who themselves have lower MAOB activities compared with the relatives who do not have lower MAOB activities. In addition, evidence indicates that other psychiatric illnesses are also more frequent among relatives who have lower MAOB activities. These results are consistent with other reports that relate lower enzyme activities to increased risk of psychiatric illness in general.[40–43] If this is the case, then the statistical relationship between lower MAOB activity and alcoholism reflects but one aspect of a more basic relationship between lower MAOB activity and increased genetic vulnerability to psychiatric illness.

One interpretation of the more detailed analyses of the association of both lower MAOB activity and increased levels of psychiatric illness with Type II alcoholism is that families of Type II alcoholics have higher genetic risk loading than do families of Type I alcoholics. Mean MAOB activity at 1.2 μM PEA substrate concentration among relatives of Type I probands is 547.5 ± 202.4 pmol product/min/mg protein, whereas mean MAOB activity among relatives of Type II probands at the same substrate concentration is 496.2 ± 160.3 pmol product/min/mg protein. This difference is significant (t = 2.53, df = 96, p <0.05) and is consistent with other patterns observed in these familial data. Closer examination of these MAOB activity data shows that mean MAOB activities among the *affected* relatives are not different

TABLE 5. Distribution of Alcoholism and Other Psychiatric Illness among the Relatives of Alcoholic Probands by MAOB Activity Level[a]

	Alcoholism Only				Other Psychiatric Illness				Combined Illness			
	High MAO		Low MAO		High MAO		Low MAO		High MAO		Low MAO	
	n	%	n	%	n	%	n	%	n	%	n	%
Affected	21	(34.3)	19	(54.3)	24	(38.1)	20	(57.1)	27	(42.9)	24	(68.6)
Unaffected	42	(66.7)	16	(45.7)	39	(61.9)	15	(42.9)	36	(57.1)	11	(31.4)
Total	63		35		63		35		63		35	
Significance (chi-square)	4.19[b]				3.33				5.71[c]			

[a]High MAOB activity includes all individuals with a platelet MAOB activity above 459.9 pmol product/min/mg protein. Low MAOB activity includes all individuals with a platelet MAOB activity below 459.9 pmol product/min/mg protein.

[b]p <0.05.

[c]p <0.02.

TABLE 6. Mean Platelet MAOB Activities at 1.2 μM PEA Concentration among Relatives of Type I and Type II Alcoholic Probands by Affection Status for All Psychiatric Illness (Values Expressed as pmol Product/Min/Mg Protein)

	Type I Proband	Type II Proband
Affected	488.1 ± 146.3 ($n = 24$)	478.5 ± 157.5 ($n = 29$)
Unaffected	600.2 ± 228.9 ($n = 27$)	524.4 ± 181.3 ($n = 17$)

regardless of the proband alcoholism type, whereas mean MAOB activities among the *unaffected* relatives are different (TABLE 6). This difference, although not statistically significant (t = 1.05, df = 42, ns), would be expected if MAOB activity is in fact a genetic marker of overall risk of psychiatric illness and if the families of Type II alcoholics have a higher genetic risk loading than do the families of Type I alcoholics.

SUMMARY

Lowered activity of the enzyme MAOB in the platelets and other tissues of alcoholics than of nonalcoholics is the most replicated biological finding in genetic research in alcoholism. Data presented here and elsewhere also indicate that the relationship between MAOB activity and alcoholism extends to the clinical subtypes referred to as Type I and Type II alcoholism. A detailed examination of the relationship between *in vitro* platelet MAOB activity levels, alcoholic subtype, and general mental health status among the relatives of the probands suggests that low MAOB activity is a marker of increased risk overall and that the families of Type II alcoholics have a higher genetic risk loading than do the families of Type I alcoholics. This increased genetic loading is probably due to the classification of Type II alcoholics on the basis of features related to severity of illness and additional psychiatric features such as personality disorders. Although the families of alcoholics tend to have higher levels of psychiatric illness compared to the general population, the overall risk is compounded in the families of Type II alcoholics, and these differences in underlying risk are reflected in the observed differences in MAOB activities. Thus, MAOB is not a biological/genetic marker of alcoholism *sensu stricto* but is rather a biological/genetic marker of an underlying pathophysiologic process leading to alcoholism and other psychiatric illness. The task now before us is to understand this process and how the activity of MAOB is involved.

REFERENCES

1. GOODWIN, D. W., F. SHULSINGER, L. HERMANSEN, S. B. GUZE & G. WINOKUR. 1973. Alcohol problems in adoptees raised apart from alcoholic biological parents. Arch. Gen. Psychiatry **28**: 238–243.
2. GOODWIN, D. W., F. SHULSINGER, I. KNOP, S. A. MEDNICK & S. B. GUZE. 1977. Psychopathology in adopted and non-adopted daughters of alcoholics. Arch. Gen. Psychiatry **34**: 1005–1009.
3. GOODWIN, D. W., F. SHULSINGER, I. KNOP, S. A. MEDNICK & S. B. GUZE. 1977. Alcohol-

ism and depression in adopted-out daughters of alcoholics. Arch. Gen. Psychiatry **34**: 751–775.

4. CADORET, R. J. & A. GATH. 1978. Inheritance of alcoholism in adoptees. Br. J. Psychiatry **132**: 252–258.
5. CLONINGER, C. R., M. BOHMAN & S. SIGUARDSSON. 1981. Inheritance of alcohol abuse. Cross-fostering analysis of adopted men. Arch. Gen. Psychiatry **38**: 861–868.
6. CLONINGER, C. R., M. BOHMAN, S. SIGUARDSSON & A.-L. VON KNORRING. 1985. Psychopathology in adopted-out children of alcoholics. The Stockholm Adoption Study. *In* Recent Developments in Alcoholism. Galanter, M. ed.: 37–51. Plenum. New York, NY.
7. DEVOR, E. J. 1992. Why there is no gene for alcoholism. Behav. Genet., in press.
8. JELLINEK, E. M. 1960. The Disease Concept of Alcoholism. Hillhouse Press. New Haven, CT.
9. COID, J. 1982. Alcoholism and violence. Drug Alcohol Dependence **9**: 1–13.
10. STABENAV, J. R. 1984. Implications of family history of alcoholism, antisocial personality and sex differences in alcohol dependence. Am. J. Psychiatry **141**: 1178–1182.
11. CLONINGER, C. R. 1987. Neurogenetic adaptive mechanisms in alcoholism. Science **236**: 410–416.
12. CLONINGER, C. R. 1989. Clinical heterogeneity in families of alcoholics. *In* Genetic Aspects of Alcoholism. K. Kiianmaa, B. Tabakoff & T. Saito eds. Finnish Found. Alc. Studies. Helsinki, Finland.
13. DEVOR, E. J. & C. R. CLONINGER. 1989. Genetics of alcoholism. Ann. Rev. Genet. **23**: 19–36.
14. WIBERG, A., C.-G. GOTTFRIES & L. ORELAND. 1977. Low platelet monoamine oxidase activity in human alcoholics. Med. Biol. **55**: 181–186.
15. ALEXOPOULOS, G. S., K. W. LIEBERMAN, R. FRANCES & P. E. STOKES. 1981. Platelet MAO during the alcohol withdrawal syndrome. Am. J. Psychiatry **138**: 1254–1255.
16. ORELAND, L., C.-G. GOTTFRIES, K. KIIANMAA, A. WIBERG & B. WINBLAD. 1983. The activity of monoamine oxidase A- and B in brains from chronic alcoholics. J. Neurol Trans. **56**: 73–83.
17. MAJOR, L. F., P. F. GOYER & D. L. MURPHY. 1981. Changes in platelet monoamine oxidase activity during abstinence. J. Stud. Alcohol **42**: 1052–1057.
18. SCHUCKIT, M. A., E. SHASKAN & J. DUBY. 1982. Platelet MAO activities in the relatives of alcoholics and controls. Arch. Gen. Psychiatry **39**: 137–140.
19. FARAJ, B. A., J. D. LENTON, M. KUTNER & V. M. CAMP. 1987. Prevalence of low maintenance oxidase function in alcoholism. Alcohol Clin. Exp. Res. **11**: 464–467.
20. PANDEY, G. N., J. FAWCETT, R. GIBBONS, D. C. CLARK & J. M. DAVIS. 1988. Platelet monoamine oxidase in alcoholism. Biol. Psychiatry **24**: 15–24.
21. VON KNORRING, A.-L., M. BOHMAN, L. VON KNORRING & L. ORELAND. 1985. Platelet MAO activity as a biological marker in subgroups of alcoholism. Acta Psychiat. Scand. **72**: 51–58.
22. VON KNORRING, A.-L., J. HALLMAN, L. VON KNORRING & L. ORELAND. 1991. Platelet monoamine oxidase activity in Type 1 and Type 2 alcoholism. Alcohol Alcohol. **26**: 409–416.
23. DEVOR, E. J., C. R. CLONINGER, S.-W. KWAN & C. W. ABELL. 1992. A genetic/familial study of monoamine oxidase B activity and concentration in alcoholics. Alcohol Clin. Exp. Res. **17**: 263–267.
24. DEVOR, E. J., C. R. CLONINGER, P. L. HOFFMAN & B. TABAKOFF. 1992. Association of monoamine oxidase (MAO) activity with alcoholism and alcoholic subtypes. Neuropsychiat. Genet., in press.
25. GILLIGAN, S. B., T. REICH & C. R. CLONINGER. 1987. Etiologic heterogeneity in alcoholism. Genet. Epidemiol. **4**: 395–414.
26. FEIGHNER, J. P., E. ROBINS, S. B. GUZE, A. R. WOODRUFF, G. WINOKUR & R. MUNOZ. 1972. Diagnostic criteria for use in psychiatric research. Arch. Gen. Psychiatry **26**: 57–63.
27. CORASH, L. 1980. Platelet heterogeneity: Relevance to the use of platelets to study psychiatric disorders. Schizophrenia Bull. **6**: 254–258.
28. FRITZ, R. R., C. W. ABELL, R. M. DENNEY, C. B. DENNEY, J. D. BESSMAN, J. A.

BOERINGER, S. CASTELLANI, D. A. LANKFORD, P. MALEK-AHMADI & R. M. ROSE. 1986. Platelet MAO concentration and molecular activity. I. New methods using an MAO B specific monoclonal antibody in a radioimmunoassay. Psychiatry Res. **17**: 129–140.

29. WURTMAN, R. J. & J. AXELROD. 1963. A sensitive and specific assay for the estimation of monoamine oxidase. Biochem. Pharmacol. **12**: 1439–1441.

30. TABAKOFF, B., J. M. LEE, F. DELEON-JONES & P. L. HOFFMAN. 1985. Ethanol inhibits the activity of the B form of monoamine oxidase in human platelet and brain tissue. Psychopharmacology **87**: 152–156.

31. REICH, T. 1988. Biologic marker studies in alcoholism. New Engl. J. Med. **318**: 180–182.

32. RICE, J. P., P. MCGUFFIN & E. SHASKAN. 1982. A commingling analysis of platelet monoamine oxidase activity. Psychiat. Res. **7**: 325–335.

33. RICE, J., P. MCGUFFIN, L. R. GOLDIN, E. G. SHASKAN & E. S. GERSHON. 1984. Platelet monoamine oxidase (MAO) activity: Evidence for a single major locus. Am. J. Hum. Genet. **36**: 36–43.

34. GOLDIN, L. R., E. S. GERSHON, C. R. LAKE, D. L. MURPHY, M. MCGINNISS & R. S. SPARKES. 1982. Segregation and linkage studies of plasma dopamine-beta hydroxylase (DBH), erythrocyte catecholamethyltransferase (COMT), and platelet monoamine oxidase (MAO): Possible linkage between the ABO locus and a gene controlling DBH activity. Am. J. Hum. Genet. **34**: 250–262.

35. CLONINGER, C. R., L. VON KNORRING & L. ORELAND. 1985. Pentameric distribution of platelet monoamine oxidase activity. Psychiat. Res. **15**: 133–143.

36. LALOUEL, J. M., D. C. RAO, N. E. MORTON & R. C. ELSTON. 1983. A unified model for complex segregation analysis. Am. J. Hum. Genet. **35**: 816–826.

37. DENNEY, R. M., N. T. PATEL, R. R. FRITZ & C. W. ABELL. 1982. A monoclonal antibody elicited to human platelet monoamine oxidase B but not A. Mol. Pharmacol. **22**: 500–508.

38. ROSE, R. M., S. CASTELLANI, J. A. BOERINGER, P. MALEK-AHMADI, D. A. LANKFORD, J. D. BESSMAN, R. R. FRITZ, C. B. DENNEY, R. M. DENNEY & C. W. ABELL. 1986. Platelet MAO concentration and molecular activity. II. Comparison of normal and schizophrenic populations. Psychiat. Res. **17**: 141–151.

39. GIBBONS, J. D. 1976. Nonparametric Methods for Quantitative Analysis. Holt, Rinehart and Winston. New York, NY.

40. WYATT, R. J., D. L. MURPHY, R. BELMAKER, S. COHEN, C. H. DONNELLY & W. POLLIN. 1973. Reduced monoamine oxidase activity in platelets: A possible genetic marker for vulnerability to schizophrenia. Science **179**: 916–918.

41. SULLIVAN, J. L., C. S. STANFIELD & C. DACKIS. 1977. Platelet monoamine oxidase in schizophrenia and other psychiatric illnesses. Am. J. Psychiat. **134**: 1098–1103.

42. VON KNORRING, L., L. ORELAND & B. WINBLAD. 1984. Personality traits related to monoamine oxidase activity in platelets. Psychiat. Res. **12**: 11–26.

43. ORELAND, L. & J. HALLUM. 1988. Monoamine oxidase activity in relation to psychiatric disorders: The state of the art. Nor. Psykiatr. Tidsskr. **42**: 95–105.

Thyrotropin Response to Thyrotropin-Releasing Hormone in Young Men at High or Low Risk for Alcoholism

JAMES C. GARBUTT, LINDA P. MILLER,
JENNIFER S. KARNITSCHNIG, AND GEORGE A. MASON

Center for Alcohol Studies
Brain and Development Research Center and
Department of Psychiatry
University of North Carolina at Chapel Hill and
Clinical Research Unit
Dorothea Dix Hospital
Raleigh, North Carolina 27603

The identification of biological markers of vulnerability for alcoholism is important for clinical and theoretical reasons. One strategy to identify such markers is to search for biological differences between abstinent alcoholics and control populations and then to determine if these differences precede overt alcoholism. On the basis of this strategy, we investigated a reduced thyrotropin (TSH) response to thyrotropin-releasing hormone (TRH) as a potential biological marker of vulnerability for alcoholism.

A reduced TSH response to TRH has been identified in abstinent alcoholics.[1] However, results of preliminary studies on the prevalence of this marker in non-alcoholic subjects at risk for alcoholism have been contradictory. Radouco-Thomas et al.[2] found that 3 of 10 family history positive (FHP) subjects aged 20–25 years exhibited a blunted TSH response compared to 1 of 26 family history negative (FHN) subjects. Moss et al.[3] reported that FHP boys aged 8–17 years exhibited a higher basal TSH and a higher peak TSH response after TRH administration than did FHN boys. Girls, when compared by family history, did not differ in basal or TRH-stimulated TSH levels. Finally, Monteiro et al.,[4] in a comparison of 10 FHP and 10 FHN men aged 18–25 years, found no differences in basal TSH or TRH-induced TSH responses.

The TRH test findings in high-risk subjects are provocative but are also incomplete and require further study. To further test the possibility that a reduced TSH response to TRH is a marker of vulnerability for alcoholism we tested 24 FHP and 26 FHN subjects matched for alcohol consumption. Our results indicate that a reduced TSH response to TRH may be a marker of vulnerability for alcoholism.

METHODS

Subjects

Male subjects were recruited from a local university and evaluated for a familial and personal history of alcoholism using Research Diagnostic Criteria (RDC) and the Family History RDC. For the purposes of the study, subjects were considered FHP if their father was alcoholic. Subjects with an alcoholic mother were excluded. Subjects were identified as FHN if they had no first or second degree relatives with alcoholism. Alcohol and drug use was initially determined by a questionnaire and later verified during a clinical interview. FHP and FHN subjects were matched on age, weight, and alcohol and drug use. All subjects were less than 31 years of age, Caucasian, and in good medical and psychiatric health. TABLE 1 shows the demographic features of FHP and FHN subjects. There were no significant differences in age, alcohol intake, or drug intake between FHP and FHN subjects.

Test Procedure

A standard TRH test using a dose of 500 μg of TRH was performed in all subjects. Following an overnight fast, subjects reported to our research unit where an intravenous line was started at 8:30 AM. At 8:45 and 9:00 AM blood samples were withdrawn. At 9:00 AM TRH was injected over 1 minute. At 9:15, 9:30, 9:45, and 10:00 AM blood samples were withdrawn. Serum was separated and stored at −70°C until assayed. Samples were assayed in duplicate and without knowledge of the FHP or FHN status of the subject. Baseline samples were drawn for TSH, prolactin (PRL), thyroxine (T_4), triiodothyronine (T_3), and cortisol. Postinfusion samples were measured for TSH and PRL. Assays were completed using commercial immunoradiometric kits. Δ max TSH and Δ max PRL represent the maximum hormone level after TRH injection minus the average of the two baseline values.

RESULTS

Baseline hormone and Δ max TSH and Δ max PRL values did not differ between FHP and FHN subjects, as shown in TABLE 2. The proportion of FHP and FHN subjects with specific Δ max TSH and Δ max PRL values is shown in FIGURES 1 to 4. Statistical analysis of Δ max TSH by FHP or FHN status was performed using the Wilcoxon 2-sample test with $p = 0.056$. Using a Δ max TSH of <6 mU/L to define a low response, 9 of 24 (38%) FHP subjects had low Δ max TSH values compared to 2 of 26 (8%) FHN subjects (chi-square = 6.4, $p = 0.011$). Examination within the FHP group for a relation between the number of first and second degree relatives with alcoholism and Δ max TSH revealed no significant association (Spearman $r = 0.02$).

Examination of Δ max PRL by FHP or FHN status using the Wilcoxon 2-sample test yielded $p = 0.788$. Using a Δ max PRL of <14 g/L to define a low response, 3

TABLE 1. Demographic Characteristics of FHP and FHN Subjects

Characteristics	FHN ($n = 26$)	FHP ($n = 24$)
Age (yr.; mean ± SD)	21 ± 2	23 ± 3
Weight (lb)	160 ± 24	172 ± 32
Number of drinks/mo.	29 ± 23	33 ± 24
Marijuana, (no. of times/last 6 mo.)	2 ± 8	7 ± 21
Cocaine (no. of times/last 6 mo.)	0.2 ± 0.6	0.5 ± 2.0

TABLE 2. Hormone Values of FHP and FHN Subjects

Hormones	FHN (n = 26)	FHP (n = 24)
Thyrotropin (mU/L)	2.1 ± 1.1	1.8 ± 1.0
Prolactin (g/L)	5.4 ± 2.8	4.3 ± 2.8
T_4 (µg/dl)	6.9 ± 1.5	6.7 ± 1.5
T_3 (ng/dl)	102 ± 27	90 ± 23
Cortisol (µg/dl)	13 ± 4	12 ± 5
Glucose (mg/dl)	79 ± 16	85 ± 20
Δ max TSH (mU/L)	11.1 ± 4.1	9.8 ± 6.0
Δ max PRL (g/L)	19.5 ± 9.3	17.7 ± 5.2

of 24 (13%) FHP subjects had low Δ max PRL values compared to 6 of 26 (23%) FHN subjects (chi-square = 0.9, p = 0.331).

To test whether alcohol or drug use was related to hormonal responses, Spearman correlations were completed. No relation was found between ethanol intake or marijuana or cocaine use and Δ max TSH or Δ max PRL in the entire sample when testing was completed or when FHP and FHN subjects were evaluated separately. When studied, season, a potential confounder for TSH values, was not different across FHP and FHN groups and was not related to Δ max TSH.

COMMENTS

Results of the present study suggest that young, nonalcoholic men with an alcoholic biological father are more likely to have a low TSH response to TRH than are young FHN men. Alcohol use, drug use, age, or weight did not explain the difference in TSH response between groups. In fact, no effect of alcohol or drug use on TSH response was detected. No effect of thyroid hormones, cortisol, or glucose levels on TSH response was evident either. Therefore, on the basis of the present

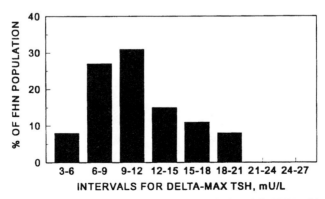

FIGURE 1. Distribution of delta-max thyrotropin (TSH) in FHN subjects.

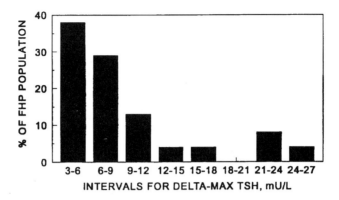

FIGURE 2. Distribution of delta-max thyrotropin (TSH) in FHP subjects.

data, a positive familial history of alcoholism may be associated with an increased prevalence of low TSH responses to TRH in young men. These findings are consistent with the report of Radouco-Thomas *et al.*[2] Because the present results suggest an abnormality in the hypothalamic-pituitary-thyroid axis, they are potentially compatible with the data from Moss *et al.*[3] who found exaggerated TSH responses in young FHP boys.

Interestingly, although a low TSH response to TRH was evident in our FHP subjects, PRL responses did not differ between FHP and FHN groups. This observation is in keeping with what we[5] and others[1] found in abstinent alcoholics, that is, a reduced TSH response in the presence of a normal PRL response. At this point in time the underlying mechanism for a reduced TSH response is not known. Possible explanations include compensatory down-regulation of thyrotroph TRH receptors secondary to TRH release from the hypothalamus or disturbances at some other site in the thyrotroph cell. Pharmacokinetic explanations are not likely, given the normal PRL response.

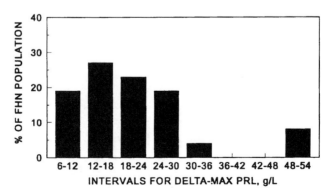

FIGURE 3. Distribution of delta-max prolactin (PRL) in FHN subjects.

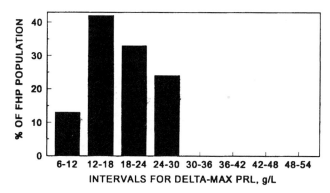

FIGURE 4. Distribution of delta-max prolactin (PRL) in FHP subjects.

In summary, a reduced TSH response to TRH may be a biological marker of the risk for alcoholism. Further studies in abstinent alcoholics and their at-risk offspring combined with longitudinal follow-up of at-risk subjects with or without the marker will permit this hypothesis to be tested more fully.

REFERENCES

1. LOOSEN, P. T., I. C. WILSON, B. W. DEW, & A. TIPERMAS. 1983. Thyrotropin-releasing hormone (TRH) in abstinent alcoholic men. Am. J. Psychiatry **140:** 1145–1149.
2. RADOUCO-THOMAS, S., F. GARCIN, M. R. V. MURTHY, N. FAURE, A. LEMAY, J. C. FOREST & C. RADOUIO-THOMAS. 1984. Biological markers in major psychoses and alcoholism: Phenotypic and genotypic markers. J. Psychiatric Res. **18:** 513–539.
3. MOSS, H. B., S. GUTHRIE & M. LINNOILA. 1986. Enhanced thyrotropin response to thyrotropin-releasing hormone in boys at risk for alcoholism. Arch. Gen. Psychiatry **43:** 1137–1142.
4. MONTEIRO, M. G., M. IRWIN, R. L. HAUGER & M. A. SCHUCKIT. 1990. TSH response to TRH and family history of alcoholism. Biol. Psychiatry **27:** 905–910.
5. GARBUTT, J. C., J. P. MAYO, G. M. GILLETTE, K. Y. LITTLE, R. E. HICKS, G. A. MASON & A. J. PRANGE, JR. 1991. Dose-response studies with thyrotropin-releasing hormone (TRH) in abstinent male alcoholics: Evidence for selective thyrotroph dysfunction? J. Stud. Alcohol **52:** 275–280.

Developmental Evidence for at Least Two Alcoholisms

I. Biopsychosocial Variation among Pathways into Symptomatic Difficulty[a]

ROBERT A. ZUCKER, DEBORAH A. ELLIS,
AND HIRAM E. FITZGERALD

Department of Psychology
Michigan State University
Psychology Research Building
East Lansing, Michigan 48824–1117

The phenomena variously described as alcoholism or alcohol dependence do not emerge abruptly in adulthood; like most other chronic behavior disorders, they show themselves gradually, in bits and pieces over time. To put this matter another way, developmental variation is an intrinsic part of the etiologic process. In fact, the very concept of etiology implies a process of moving from a state involving the presence of risk to a state involving the emergence of the full-blown disease entity. The prospective study upon which we report here has as its focus the charting of this process from early childhood onward, with the expectation that tracking developmental variation within the context of a broadly drawn, multidomain database will provide clues to etiologic mechanisms for alcoholism that would not be likely to emerge without such breadth and without close monitoring of time-based variations in causal patterning over significant portions of the life course. The present paper provides an overview of one important strand of this research, which pursues the issue of multiple pathways into alcoholic risk in a variety of ways. We review the issues that led to our interest in this problem,[1-3] present evidence that strongly encouraged continuation of this line of inquiry, and last, report some preliminary findings among both alcoholic adults as well as their offspring which indicate that the conception of a single alcoholism is no longer a tenable etiologic hypothesis.

DEVELOPMENTAL THEORY AND ETIOLOGIC PROCESS — MATURATION + CONTEXT

The application of a developmental lens to the course of psychiatric disorder can inform us in a variety of ways that have until recent times been overlooked in traditional etiologic research. The developmental method is concerned with mapping trajectories over long spans of the life cycle and examining the growth, maturation, and decline of symptoms all the way from conception (wherein genetic variation is transmitted), through the period of intrauterine existence, through all stages of the life cycle. Whether used in biology[4] or in the behavioral sciences,[5,6] it is concerned with tracing out multiple interactive processes, documenting the relationship of maturation to environmental trigger events, establishing the extent to which the basic

[a]This work was supported in part by grant RO1 AA07065 to R. A. Zucker and H. E. Fitzgerald from the National Institute on Alcohol Abuse and Alcoholism.

134

structure of such processes has the plasticity to be modified by environmental experiences, and describing the organism-environment interchanges that are necessary to either bring about such change or resist it.

Patterns of change are considered to be as important as patterns of stability, but the operation of either trajectory, within this framework, needs to be viewed as a dynamic system, operating in multiple contexts through time.[7] Given the inconsistencies in the environment and the large variety of social and biological events to which we are exposed, the attainment of stability can be taken as evidence of the operation of internal mechanisms that regulate and insure the constancy. Nonetheless, contextual embeddedness is also considered critical to what takes place. Thus, the importance of individual factors, whether they pertain to variations in genetic vulnerability, to differences in socialization practices by parents, or to differences in exposure to peer pressure for heavy or problematic drinking are likely to be misinterpreted if a system analysis that examines the interplay of these factors over maturational time is not also examined.[8] What may be a distal and insignificant influence at birth (e.g., peer effects) may be a proximal one at a later time (e.g., adolescence). Finally, the notion that systems adjust, shift, and not only develop momentums of their own, but also enter into and perhaps even seek out contexts that drive and encourage constancy is very much a strategic part of this analytic strategy.

INTRODUCTION TO THE MICHIGAN STATE UNIVERSITY LONGITUDINAL STUDY

The Michigan State University Longitudinal Study has as its focus the charting of these processes from early childhood onward, with the expectation that this scrutiny will provide clues to etiologic mechanisms for alcoholism that would not be likely to emerge without such close monitoring over such a long span of time. Starting with a sampling plan that would allow the recruitment of a population sample of alcoholic men and a community contrast group also generated by a population net, the study's aim was to commence the charting of developmental progression among initially preschool age children and their parents and follow them at 3-year intervals. Some of these children would be at substantially heightened risk for alcohol abuse/dependence in adult life, and some would not. The extreme youth of the sample was chosen on the basis of pilot data that had already indicated evidence for putative risk differences, which could be detected even in the preschool years.[9,10]

Pilot work for this project began in 1982, and the beginnings of the study as it exists today were in 1985. TABLE 1 provides detail on the core study design which involved a high risk group of alcoholic families and a yoked group of community control families, all of whom have data collected upon fathers, mothers, and the male target child. When the parents separate, the study follows both parents as well as doing add-on assessments on stepparents.

The project begins with a largely preschool age cohort of high risk and contrast children; the broadness of the database as well as the children's very young age contributes to the study's potential power to detect emerging causal processes. The variable network involves the assessment of seven domains of psychosocial functioning collected during a 9–10 session protocol involving both in-home visits as well as laboratory interaction data collected at the university. The domains assessed include evaluations of parental psychopathology, child developmental adaptation, early symptomatic manifestations, child cognitive and control functions, child-rear-

TABLE 1. Study Design: Michigan State University Longitudinal Study

High Risk Families	Comparison (Control) Families
Fathers arrested for driving while impaired (DWI) with blood alcohol levels of 0.15% or higher when apprehended (0.12 if multiple arrest history)	Parents reside in same neighborhoods as high-risk families and are recruited by door-to-door canvass
Fathers meet Feighner diagnostic criteria at least for "probable alcoholism" (virtually all meet "definite alcoholism" criterion); mothers' substance abuse status not a basis for selection	Neither parent has history of either alcohol or other drug diagnosis
Families are intact and have a male target child between the ages of 3.0 and 6.0	Same criteria as high-risk families; in addition, comparison families are yoked to high-risk families for age of target child (± 6 months) and where possible have parallel sibling composition (number, sex, age)

NOTE: Initial data collection done on father, mother, and target child involves approximately 15 hours of family contact. During this contact time (partially at the university and partially at family's home) 20 hours of questionnaire, laboratory interaction data, developmental assessments, and extensive observer ratings are collected (in the blind) on all families. Follow-up with parallel data collection is at 3-year intervals.

ing practices, and an extensive array of measures of family history and of the broader social environment. Children and adults are assessed at 3-year intervals, and plans call for following both the child and parent cohorts at least until the children reach adulthood.

Although the ultimate scientific significance of such a project must await adulthood for these children, several features of the work allow it to be useful along the way. First, it needs to be emphasized that the project, as designed, has a commitment to follow the children and both their parents as development proceeds. Thus, it is also a longitudinal study of young alcoholic men (M = age 31 at Wave 1) and their spouses as well as of a set of contrast families in which risk is lower, but still not negligible, wherein different patterns of alcoholism might be anticipated to emerge. Second, given that some of the women in these families are also alcoholic (44%), it is also a study of the evolution of alcoholism among women, albeit of a special subset who are married to alcoholic men. Third, the focus on such an initially young child cohort requires that, for the study to be useful, it must establish a base of proxy measures pertaining to later alcohol use because the study starts too early to actually utilize alcohol involvement as a measure of risk status. This has encouraged the development of a theoretical base that has articulated the case for the interplay between what are termed *alcohol-specific* and *nonalcohol-specific* factors.[11,12] This theory in turn formed the basis for the hypothetico-deductive framework of the study pertaining to the interplay between genetic diathesis, aggressive behavior and negative mood, and the anticipated potentiating or moderating effect to be played by social relationships (initially by way of parent modeling and socialization effects and later through peer relations and other contextual influences [e.g., social status]. Analyses on the alcoholic men from Wave 1 of the study provide the basis for positing multiple alcoholisms among the adults. Other analyses on the child data, indicating multiple etiologic processes leading into risk, that in turn are tied to the adult subtypes, have provided additional validation for the typological differentiation. The present report provides a brief overview of that work.

EVIDENCE FOR MULTIPLE ALCOHOLISMS AMONG ADULTS: STUDY ONE[a]

Although subtyping of alcoholism has a long history,[13] the degree to which such subtypes would have different patterns of etiology has received considerably less attention. For these reasons, findings from the Swedish Adoption Study data set[14,15] and the collateral theoretical framework proposed to account for the observed differences[2] have received so much attention. From a heuristic standpoint, the Cloninger group's typology, involving the distinction between early onset, antisocial alcoholism (Type II or "male-limited alcoholism") and later onset, nonantisocial alcoholism (Type I or "milieu-limited alcoholism"), has the additional advantage that it specifically articulates differences in genetic and environmental sources of effect across the two types. Although this work has received significant criticism in recent years, it presented a potentially useful starting place in approaching the multiple pathways question. Thus, in our own work we set as a beginning point the question of whether we could replicate the core finding of the Cloninger group's work, namely, that biological risk might be differentially implicated in the symptomatic outcomes for Type Is and Type IIs. As a second question we had a special interest in exploring the degree to which other, not necessarily genetically mediated mechanisms might also play a role in the apparently nonenvironmentally mediated Type II outcomes. As noted in earlier reviews in this area,[16,17] there is strong reason to question the early conclusion of the lack of environmental contribution, given the exceptionally sparse and inadequate sampling of environmental variables already known to influence problematic alcohol involvement. The rich nature of the MSU study's variable network would allow a broader test of other potential paths into alcoholic outcome.

METHOD

Subjects

The subjects for the present study were 102 men participating in the MSU study.[10,18] Subjects are a community-based sample of alcoholic men who were identified from the population of all males convicted of drunk driving in a four-county mid-Michigan area. They were required to have had a blood alcohol concentration (BAC) of 0.15% (150 mg/100 ml) or higher when arrested or a BAC of 0.12% but also a history of multiple alcohol-related driving offenses. In addition, subjects met the following requisite demographic characteristics for the purpose of the larger study: they had a male offspring between 3 and 5 years of age, and both father and son resided with the child's mother at the time of first contact. At recruitment, participants were screened for a positive alcoholic diagnosis using the SMAST. The diagnosis was later verified with the Diagnostic Interview Schedule[19]; participants needed to meet Feighner criteria for either probable or definite alcoholism. Eighty-eight percent of subjects met a definite diagnosis. Thereafter, DSM-III-R diagnoses were also established. In fact, well over half (73%) met the criteria for a diagnosis of either moderate or severe dependence. The mean age of the sample was 31.2 years (SD = 4.9). Mean number of years of education was 12.2 (SD = 1.9) and the mean yearly family income was $26,270 (SD = $16,878).

MEASURES

Alcoholic Subtype

The sample was divided into Type Is and Type IIs using the criteria laid out by Cloninger's group.[2,20] Subjects who met a Feighner alcoholic diagnosis, whose drinking had not incurred social consequences (such as family arguments, violence while drinking, loss of job, and other legal difficulties) and who had suffered psychological distress for their drinking were coded as Type I. Subjects who met alcoholic criteria, whose drinking began before age 25, and whose drinking had incurred these social consequences were coded as Type II. Possible Type Is subjects who failed the positive social complications screen but who did not pass the psychological complications screen were coded as indeterminates; possible Type IIs who met the social complications but not the early onset criterion were also coded as indeterminate.

Family Expression of Alcoholism

Information on familial alcoholism was obtained via a family history interview in which subjects provided data on psychiatric and physical disorders in other family members. Respondents and their mates were first asked to produce a family tree extending back to the grandparental generation which included such second-degree relatives as aunts, uncles, and first cousins[b]; they then were given a list of various physical and psychological disorders, including alcoholism, and were asked to identify any relatives who were affected by any of the listed disorders, thus creating a genogram.

The family expression of alcoholism (FEA) score was computed by first allotting a weight equivalent to the coefficient of relationship value to each alcoholic family member (first-degree relatives = 0.5; second-degree relatives, 0.25; etc.). FEA scores were then calculated by (1) summing within each generation, the weightings for all alcoholic relatives; (2) multiplying this sum by the ratio of alcoholics to total number of family relatives in that generation; and (3) summing the subscores across generations. As FEA includes points assigned to alcoholic parents who both raised the index subject as well as contributed to his genetic makeup, the score cannot be considered a pure measure of genetic loading for alcoholism as separate from rearing effects. However, FEA reflects the density of alcoholism in the subject's extended family as well as the degree of relatedness of these family members to him. Inasmuch as many of the alcoholic relatives contributing to this score would not have contributed to day-to-day rearing, it is also appropriate to regard it as an index of biological risk.

Alcoholism Load and Alcohol Involvement Measures

To index the magnitude of alcoholic disorder over the life course, the Lifetime Alcohol Problems Score (LAPS)[21] was used as one of the core dependent variables. LAPS is a dimensional measure that incorporates information on the primacy (onset), variety, and degree of life invasiveness of drinking problems. Information from which LAPS was coded was provided by the DIS, Drinking and Drug History, and SMAST. LAPS consists of three component subscores: (1) the primacy component, or squared inverse of age of first drunkenness; (2) the variety component, or number of areas in which drinking problems are reported; and (3) the life percent component, or measure of the interval between most recent and earliest drinking problem corrected for current age. The measure effectively distinguishes between alcoholics and nonalcoholics, among levels of severity of DSM-III-R alcohol dependence, is unrelated to current alcohol consumption in problem-drinking samples, and is correlated with a wide range of external measures of alcohol-related difficulty such as BAL at arrest and treatment involvement.[21]

Current alcohol consumption was coded from a drinking and drug history questionnaire which assessed the amount of alcohol intake in the last 6 months using an extended version of the standardized survey questions developed by Cahalan, Cisin and Crossley[22] and also contained questions assessing the presence, onset, and duration of an extensive list of alcohol-related problems. Data were coded for quantity-variability and frequency of alcohol consumption as well as for a revised version of Cahalan, Cisin, and Crossley's Alcohol Consumption Index, the QFV-R[23] which provides a more extended range for the measure. Subjects who were no longer drinking at the time of data collection were not included in computations that involved current consumption measures.

Background Characteristics

Information from a demographic questionnaire was used to calculate childhood and adulthood socioeconomic status (SES) using the Duncan TSE12 Socioeconomic Index,[24] an occupationally based measure of social prestige. This instrument also had items pertaining to marital history (*number of divorces*) and *environmental exposure to alcoholism*. Items identified people involved in the primary caretaker role for respondents over the course of childhood, and the genogram provided a means of assessing drinking status of parents or other primary caretakers. Degree of exposure to an alcoholic environment was rated on a scale which assigned one point for each alcoholic caretaker involved in raising the respondent. Guided by the rationale that the pairing of alcoholic caretakers would have a potentiating socialization effect,[25] an additional point was added for any pairing of alcoholic caretakers (i.e., two alcoholic parents). Scores thus could vary between 0 and 3.

Life Difficulty Variables

Antisocial Behavior (Childhood and Adult). The DIS was used to assess child and adult antisociality. A count was taken of the number of DSM-III antisocial symptoms experienced by each respondent during childhood/adolescence and in adulthood (age 18 and after).

Depression. The short form of the Beck Depression Inventory (BDI)[26] was used to evaluate self-reported depression. The BDI focuses on various areas of functioning known to be affected by depression, such as mood, appetite, and sleep.

RESULTS[c]

A series of analyses, beginning with tests of difference on demographic and life difficulty variables between Type Is and IIs, followed by an examination of differences in the correlational pattern of the alcohol and other life difficulty variables, and concluding with a test of the appropriateness of using different path models to characterize the two types, were performed to explore the set of questions we had posed. A first pass of relationships among the entire male risk study sample showed that the index of biologic risk, based upon the family expression of alcoholism measure, was significantly related to earlier age of first drunkenness ($r = -0.28$; $p < 0.01$), to greater severity of alcohol problems over time ($r = .32$; $p < 0.001$), as well as to more frequent current alcohol consumption ($r = .26$, $p < 0.01$). When we coded for the Cloninger subtypes, of 102 alcoholic men in the sample, 25 were classifiable as Type Is, 60 as Type IIs, and 17 (17%) as indeterminate because the schema would not allow placement in either category. This indeterminate group was not included in further work.

A second set of analyses examined differences between the two types on demographic, alcohol-related, and nonalcohol-related life difficulty variables (TABLE 2). Type IIs were higher in extent of antisocial behavior in both childhood and adulthood, on the family alcoholism load measure, and in number of separations/divorces, and they were lower in achieved socioeconomic status. However, no differences were found on measures of current alcohol consumption or on the Beck Depression index.

Patterns of correlation between current and lifetime problem alcohol involvement and the family expression of risk measures were highly similar among Type Is and IIs *when the degree of exposure to an alcoholic parenting environment was not partialed out.* When rearing environment effects were controlled, a significant difference ($p < 0.05$) in relationships to LAPS was found; Type IIs showed a positive and significant relationship between biological risk and level of lifetime alcohol problems ($r = 0.28$; $p < 0.05$), and Type Is showed a negative relationship ($r = -0.21$, *ns*).

Last, a stacked LISREL analysis was performed to test the appropriateness of using different path models for the two groups. This analysis is first estimated with effect coefficients constrained to be equal between groups; the model is then reestimated with effect coefficients allowed to vary by group. The difference chi-square

TABLE 2. Differences between Type I ($n = 25$) and Type II ($n = 60$) Alcoholics on Alcohol Involvement, Demographic, and Life Difficulty Variables

	M (SD)		
	Type I	Type II	t
Family expression of alcoholism	0.137 (.19)	0.277 (0.36)	−2.32*
Lifetime alcohol problems score	9.8 (1.4)	11.2 (2.0)	−3.28**
QFV-R frequency classification	6.2 (2.6)	5.9 (2.8)	0.54
QFV-R quantity-variability classification	13.4 (6.7)	15.0 (6.9)	−0.84
Socioeconomic status (Duncan TSE12)	35.5 (16.4)	26.2 (12.0)	2.58*
Number of separations/divorces from partner(s)	0.9 (1.0)	2.2 (1.8)	−4.23***
Number of child antisocial behavior symptoms from DIS	1.7 (1.5)	4.4 (2.8)	−5.86***
Number of adult antisocial behavior symptoms from DIS	2.5 (1.0)	4.4 (1.7)	−6.43***
Beck Depression Inventory Score	2.6 (2.9)	3.8 (3.7)	−1.45

NOTE: Figures in parentheses are standard deviations.
*$p < 0.05$; **$p < 0.01$; ***$p < 0.001$.

of combined *versus* separate models then provides a test of the goodness of fit when models for the two groups are allowed to differ. These analyses showed that the path models predicting the severity of alcohol-related difficulty for the subtypes were best described by way of two developmentally distinct sets of processes. Out of four potential distal predictors (i.e., the index of family expression of alcoholism, a childhood history of antisocial behavior, family socioeconomic status during childhood, and extent of rearing in an alcoholic environment during childhood) and one proximal predictor of later alcoholic difficulty (extent of antisocial behavior during adult life), only exposure to an alcoholic rearing environment was predictive of later alcoholic difficulty for Type Is (Fig. 1). In contrast, family expression of alcoholism and childhood antisocial behavior were both distal precursors of alcoholic severity for Type IIs, and adult antisociality was also a proximal contributory factor (Fig. 1). In other words, for this subtype, delinquent or conduct-disordered behavior in childhood is predictive of antisociality in adulthood, which in turn drives adult alcohol-related difficulty. This is in addition to the separate, direct path to which childhood antisociality and family risk also contribute. The difference chi-square testing the appropriateness of a combined model *versus* two separate models for Type Is *vs* IIs was 10.17 ($p < 0.05$), indicating that the data fit significantly better when two models are allowed.

PROBLEMS WITH THE TYPE I/TYPE II CATEGORIZATION

Although the evidence from these analyses clearly indicates that the Type I/Type II differentiation provides a beginning base for postulating multiple etiologic pro-

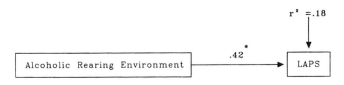

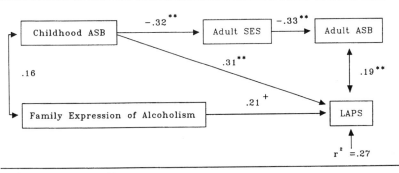

$^{+}$ P<.10 * p<.05 ** p<.01

FIGURE 1. Life course path models predicting magnitude of alcoholic load in adulthood for Type I (**top**) and Type II (**bottom**) alcoholics. $^{+}p < 0.10$; $^{*}p < 0.05$; $^{**}p < 0.01$.

cesses, a number of problems with it led us to move in another direction. For one, the schema provides a substantial amount of indeterminacy,[28,29] thereby limiting its utility as a clinical tool. In addition, evidence for an initially high promising heuristic model linking the typology to temperament and ultimately to neurochemical mechanisms[2] has failed to find support in recent studies.[29,30] For another, the extent to which early onset can unequivocally be tied to this classification has also been questioned.[31]

More importantly, the schema is open to misinterpretation and confusion on several different grounds, involving the inappropriate equation of antisocial comorbidity with the Type II categorization and the utilization of severity of alcoholic course as one of the markers of the male limited type,[20,33] even when the original typological distinction[14] explicitly linked *moderate* severity of alcoholism with Type II (and mild *or* severe outcome with Type I, depending on the quality of environmental risk exposure).

A DEVELOPMENTAL APPROACH TO SUBTYPING: ESTABLISHING TYPOLOGICAL VARIATION IN ADULTHOOD BY A CLASSIFICATION SCHEME WHICH USES HISTORY AS WELL AS CURRENT FUNCTIONING TO CATEGORIZE

For all these reasons we decided to use another approach to classification. We started with an observation and a developmental hypothesis. The observation was that the comorbid occurrence of both antisociality and alcohol problem symptoms in adulthood was the most prominent symptomatic covariation known among alcoholics.[33,34] The developmental hypothesis was that, given the already established nature of these behaviors, they needed to be regarded as later stage phenomena, for which earlier stage covariation had to exist. On those grounds, an index that captured continuity of process for both earlier (childhood) as well as later (adolescent and adult) antisociality should simultaneously capture continuity of process (and likely greater severity in development) for the other symptom set. Given that the development of antisociality typically occurs prior to the development of problem drinking,[35] it made sense to begin looking for differences in phenotypic variation in alcoholism among individuals who vary in the extent to which they display antisocial symptoms both in childhood as well as in adulthood.

STUDY II

In pursuing this hypothesis, we used a dimensional approach that characterizes antisocial variation by level and severity of antisocial symptoms, even when symptoms are not sufficient to warrant an ASP diagnosis. The sample was divided into two alcoholic subtypes; antisocial alcoholics (AALs) and non-antisocial alcoholics (NAALs). To categorize individuals as an AAL or NAAL, subjects' scores on the Antisocial Behavior Inventory[36] were summed over both child and adult domains, and a median split on the summary score was used to divide the men. Those with scores of 24 or higher were classified as AALs; with this cut-off score, the instrument's sensitivity is 0.85 and its specificity is 0.83 for a diagnosis of antisocial personality disorder using DSM-III-R criteria.

By using a measure that assesses *both* childhood and adulthood antisocial involvement to determine alcoholic subtype, this classification scheme is able to chart the degree to which a developmental trajectory has been established that is continuous across childhood and adulthood. Higher scores thus indicate a trajectory that begins early in life with aggressive/antisocial behavior and that thereafter crystallizes in alcoholism and antisociality during adulthood, rather than providing a categorical classification based solely on adult antisocial status. Using this schema we then evaluated whether or not a sustained history of high levels of antisocial involvement concurrent with alcoholism would yield a different type of alcoholism, both symptomatically and etiologically from one that does not share such developmental covariation. The same alcoholic men used in Study I were involved in these analyses. However, the variable network was broadened to include not only the antisociality measure, but also a measure of depression that would evaluate lifetime experience in addition to the Beck index, which only assesses depressive experience in the immediately past year.

Space limitations do not allow a detailed presentation of these results; what follows is a brief summary.[d] Findings provide strong evidence for two subsets of alcoholic men who differ in age of onset of first alcoholic problems (AALs begin earlier; $p < 0.001$), in the way their alcoholic symptoms are expressed (the AALs have a heavier load of alcohol-related difficulty and have had a greater variety of alcohol-related symptoms than do the NAALs [$ps < 0.01$]); in measures of their ongoing life positional status as assessed via demographic variables (AALs are lower in achieved socioeconomic status [$p < 0.01$] although childhood level of SES does not differ between the two groups), in type and severity of psychopathology that co-occurs with the alcoholism (AALs have greater depression both currently and over the life course [$p < 0.01$]) and in importance of family history load of alcoholism (AALs have a greater load). Most importantly, these two groups differ in the manner in which these contributory sources of variation interrelate, as assessed by path models predictive of the magnitude of the alcoholic load. For AALs, the best fit model involves contributions from both child and adult antisociality as well as a separate path involving the measure of Family Expression of Alcoholism, which in a crude way indexes family pedigree load. In contrast, for NAALs, only lifetime variation in the level of depressive symptoms (based on Hamilton Rating Scale for Depression scoring of worst ever depressive episode) is predictive of alcoholic outcome.

EARLY EVIDENCE OF DIFFERENCES IN CAUSAL PROCESS FOR THE DEVELOPMENT OF ANTISOCIAL AND NONANTISOCIAL ALCOHOLISM IN CHILDHOOD — STUDY III

Last, also using Wave 1 data from the MSU study, other just completed analyses utilizing this schema[38] show that it is able to significantly differentiate the degree of risk load in early childhood for these different subtypes. Childhood risk was evaluated by way of the Achenbach and Edelbrock[40] Child Behavior Checklist (CBCL) measure of externalizing behavior, using a composite index based on both mother and father ratings. When we examined variations in child outcome among 3- to 5-year olds as related to patterns of relationship among predictor variables in the children of AALs, NAALs, and nonalcoholic controls, both tests of difference among outcome variables in the three groups as well as path analyses showed (1) differences in the level of functioning and in the amount of contextual trouble among the groups, and (2) varying patterns of relationship between child risk and child outcome variables, suggesting that different factors play a role in the emergence of child behavior problems among these groups. Whereas heritable factors appear to play a role in the development of externalizing behavior problems for children of AALs, this is not true for children of NAALs or nonalcoholic controls. Moreover, the impact of being raised by an alcoholic parent appears less germane in the emergence of psychopathology among children of NAALs, possibly because their parents are less troubled. There is some suggestion as well that processes such as marital conflict, which would be anticipated to lead to poorer parenting, and differences in child temperament as well as parent-child temperament mismatch may play a greater role in the appearance of child risk among both NAALs and Controls.

DISCUSSION

There appear to be several advantages of the current category system that contrasts it with earlier approaches to typological variation; although it retains the capacity to differentiate among subsets of alcoholics in relative salience of family history as a predictor of outcome, it is (1) more differentiating of alcoholic severity in adulthood, (2) more differentiating of comorbid psychopathology in adulthood; and (3) the pattern of suggested etiology, although currently based on retrospective data, is more clearly consistent with the existing child and adolescent longitudinal literature on the early origins of problem alcohol involvement than has been true of other models.[12,41] The present study differs from other recent typological work[42] insofar as it uses a population based, rather than a clinically assessed set of adult alcoholics; despite this significant sampling disparity, the degree of overlap of findings is very strong.

Similarly, the child findings are based on a population sample of children with very substantially elevated risk, who have not yet shown signs of the clinical disorder, as well as a sample of nonalcoholic families who also are drawn from the community. The models obtained are indicative of variations in etiologic process that are consistent with the models derived from the adult report data. They also run strongly in parallel with other currently ongoing studies of early etiology.[43] Despite these very substantial correspondences, the child findings still need to be regarded as suggestive. Until the longitudinal study proceeds far enough to be able to connect child observations to firm adolescent and early adult outcomes, the early linkages are best viewed as increasingly more plausible hypotheses rather than as definitive proof.

END NOTES

[a]The present discussion is focused exclusively on subtyping among men. Although our project is also examining alcoholic variability and etiology among women, this work is not yet far enough advanced to report.

[b]Cousins were excluded from later analyses using this measure because it became clear that subjects' familiarity with them, as more distant relatives, was insufficient to allow accurate labeling of illness data.

[c]These data are based on Deborah Wynblatt Ellis' M.A. research,[27] conducted under the direction of the other authors.

[d]A series of more lengthy papers[37-39] report these findings in detail.

REFERENCES

1. BABOR, T. F. & Z. S. DOLINSKY. 1988. Alcoholic typologies: Historical evolution and empirical evaluation of some common classification schemes. *In* Alcoholism: Origins and Outcome. R. M. Rose & J. Barrett, eds.: 245–266.

2. CLONINGER, R. 1987. Neurogenetic adaptive mechanisms in alcoholism. Science **236:** 410–416.
3. HESSELBROCK, M. N., V. M. HESSELBROCK, T. F. BABOR, J. R. STABENAU, R. E. MEYER & M. WEIDENMAN. 1984. Antisocial behavior, psychopathology, and problem drinking in the natural history of alcoholism. *In* Longitudinal Research in Alcoholism. D. W. Goodwin, K. T. VanDusen & S. A. Mednick, eds.: 197–214. Kluwer-Nijhoff. Boston.
4. WADDINGTON, C. H. 1968. The basic ideas of biology. *In* Towards a Theoretical Biology; Vol. 1. Prolgeomena. C. H. Waddington, ed. Adline. Boston.
5. GOTTLIEB, G. 1991. Experimental canalization of behavioral development. Theory & Dev. Psychol. **27:** 4–13.
6. LERNER, R. M. 1991. Changing organism-context relations as the basic process of development: A developmental contextual perspective. Dev. Psychol. **27:** 27–32.
7. FITZGERALD, H. E., W. H. DAVIES, R. A. ZUCKER & M. T. KLINGER. 1994. Developmental systems theory and substance abuse: A conceptual and methodological framework for analyzing patterns of variation in families. *In* Handbook of Developmental Family Psychology and Psychopathology. L. L'Abate, J. Alexander, A. Crouder & E. Menaghan, eds. Chapt. 17: 350–371. Wiley. New York.
8. TARTER, R. E., A. I. ALTERMAN & K. L. EDWARDS. 1985. Vulnerability to alcoholism in men: A behavior-genetic perspective. J. Stud. Alcohol **46:** 329–355.
9. ZUCKER, R. A., J. A. BAXTER, R. B. NOLL, D. P. THEADO & C. M. WEIL. 1982. An alcoholic risk study: Design and early health related findings. Presented at the American Psychological Association Meetings, Washington, D.C.
10. ZUCKER, R. A. 1987. The four alcoholisms: A developmental account of the etiologic process. 1986. *In* Alcohol and Addictive Behaviors: Nebraska Symposium on Motivation. P. C. Rivers, ed. Vol. **34:** 27–83. University of Nebraska Press. Lincoln, NE.
11. ZUCKER, R. A. 1979. Developmental aspects of drinking through the young adult years. *In* Youth, Alcohol and Social Policy. H. T. Blane & M. E. Chafetz, eds. Chapter 4: 91–146. Plenum. New York, NY.
12. ZUCKER, R. A. & H. E. FITZGERALD. 1991. Early developmental factors and risk for alcohol problems. Alcohol Hlth. & Res. World. **15:** 18–24.
13. BABOR, T. F & R. LAUERMAN. 1986. Classification and forms of inebriety: Historical antecedents of alcoholic typologies. *In* Recent Developments in Alcoholism. M. Galanter, ed. Chapt. **5:** 113–144. Plenum. New York, NY.
14. CLONINGER, C. R., M. BOHMAN & S. SIGVARDSSON. 1981. Inheritance of substance abuse: Cross-fostering analysis of adopted men. Arch. Gen. Psychiatry **38:** 861–867.
15. CLONINGER, C. R., M. BOHMAN, S. SIGVARDSSON & A. L. VON KNORRING. 1985. Psychopathology in adopted-out children of alcoholics: The Stockholm adoption study. *In* Recent Developments in Alcoholism. M. Galanter, ed. Vol. 3. Plenum. New York, NY.
16. SEARLES, J. S. 1988. The role of genetics in the pathogenesis of alcoholism. J. Abnormal Psychol. **97:** 153–167.
17. ZUCKER, R. A. 1989. Is risk for alcoholism predictable? A probabilistic approach to a developmental problem. Drugs & Soc. **4:** 69–93.
18. ZUCKER, R. A., R. B. NOLL & H. E. FITZGERALD. 1986. Risk & Coping in Children of Alcoholics. Grant application: National Institute on Alcohol Abuse & Alcoholism. RO1 AA 07065.
19. ROBINS, L. N., J. E. HELZER, J. CROUGHAN & K. S. RATCLIFFE. 1980. The NIMH diagnostic interview schedule: Its history, characteristics, and validity. Arch. Gen. Psychiat. **38:** 381–389.
20. VON KNORRING, L., V. PALM & H. ANDERSON. 1985. Relationship between treatment outcome and subtype of alcoholism in men. J. Stud. Alcohol **46:** 388–391.
21. ZUCKER, R. A. 1991. Scaling the developmental momentum of alcoholic process via the Lifetime Alcohol Problems Score. Alcohol Alcohol. (Suppl. 1): 505–510.
22. CAHALAN, D., I. CISIN & H. CROSSLEY. 1969. American drinking practices: A national study of drinking behavior and attitudes. Publications Division, Rutgers Center of Alcohol Studies. New Brunswick, NJ.
23. ZUCKER, R. A. & W. H. DAVIES. 1989. The revised quantity-frequency-variability index: Rationale and formulae. Unpublished paper. Department of Psychology, Michigan State University. East Lansing, MI.

24. STEVENS, G. & D. L. FEATHERMAN. 1981. A revised socioeconomic index of occupational status. Social Science Res. **10:** 364–395.
25. REIDER, E. E., R. A. ZUCKER, G. MAGUIN, R. B. NOLL & H. E. FITZGERALD. August, 1989. Alcohol involvement and violence toward children among high risk families. Paper presented at the American Psychological Association Meetings. New Orleans, LA.
26. BECK, A., W. RIAL & K. RICKELS. 1974. Short form of the Beck Depression Inventory: Cross validation. Psychol. Rep. **34:** 1184–1186.
27. WYNBLATT, D. A. 1990. Genetic loading for alcoholism: New evidence for subtypes. Unpublished Master's Thesis, Department of Psychology, Michigan State University. East Lansing, MI.
28. PENICK, E. C., B. J. POWELL, E. J. NICKEL, M. R. READ, W. F. GABRIELLI & B. I. LISKOW. 1990. Examination of Cloninger's type I and type II alcoholism with a sample of men alcoholics in treatment. Alcohol. Clin. Exp. Res. **14:** 623–629.
29. PICKENS, R. W., D. S. SVIKIS, M. McGUE, D. T. LYKKEN, L. L. HESTON & P. J. CLAYTON. 1991. Heterogeneity in the inheritance of alcoholism. Arch. Gen. Psychiat. **48:** 19–28.
30. NIXON, S. J. & O. A. PARSONS. 1990. Application of the tridimensional personality questionnaire to a population of alcoholics and other substance abusers. Alcohol. Clin. Exp. Res. **14:** 513–517.
31. IRWIN, M., M. SCHUCKIT & T. SMITH. 1990. Clinical importance of age at onset in type 1 and type 2 primary alcoholics. Arch. Gen. Psychiat. **47:** 320–324.
32. BUYDENS-BRANCHEY, L., M. H. BRANCHEY & D. NOUMAIR. 1989. Age of alcoholism onset, I. Relationship to psychopathology. Arch. Gen. Psychiatry **46:** 231–236.
33. HELZER, J. E., A. BURNAM & L. T. McEVOY. 1991. Alcohol abuse and dependence. *In* Psychiatric Disorders in America: The epidemiologic Catchment Area Study. L. N. Robins & D. A. Regier, eds.: 81–115. The Free Press. New York, NY.
34. ZUCKER, R. A. 1994. Pathways to alcohol problems and alcoholisms: A developmental account of the evidence for multiple alcoholisms and for contextual contributions to risk. *In* The Development of Alcohol Problems: Exploring the Biopsychosocial Matrix of Risk. R. A. Zucker, J. Howard & G. M. Boyd, eds. National Institute on Alcohol Abuse and Alcoholism. Rockville, MD. In press.
35. PATTERSON, G. R., B. D. DEBARYSHE & E. RAMSEY. 1989. A developmental perspective on antisocial behavior. Am. Psychol. **44:** 329–335.
36. ZUCKER, R. A. & R. B. NOLL. 1980. Assessment of antisocial behavior: Development of an instrument. Unpublished manuscript. Michigan State University. East Lansing, MI.
37. ZUCKER, R. A., D. A. ELLIS & H. E. FITZGERALD. 1993. Other evidence for at least two alcoholisms. II. The case for lifetime antisociality as a basis of differentiation. Unpublished manuscript. Michigan State University. East Lansing, MI.
38. ELLIS, D. A. 1993. Typological differences in patterns of risk among young children of alcoholics. Unpublished doctoral dissertation, Department of Psychology, Michigan State University.
39. ELLIS, D. A., R. A. ZUCKER & H. E. FITZGERALD. 1993. Other evidence for at least two alcoholisms. III. Typological differences in patterns of risk among young children. Michigan State University. In preparation.
40. ACHENBACH, T. M. & C. EDELBROCK. 1983. Manual for the child behavior checklist and revised child behavior profile. University of Vermont. Burlington, VT.
41. KANDEL, D. B. 1978. Convergences in prospective longitudinal surveys of drug use in normal populations. *In* Longitudinal Research on Drug Use. D. B. Kandel, ed. Hemisphere. Washington, DC.
42. BABOR, T. F., M. HOFMANN, F. K. DELBOCA, V. HESSELBROCK, R. E. MEYER, Z. S. DOLINSKY & B. ROUNSAVILLE. 1992. Types of alcoholics. I. Evidence for an empirically derived typology based on indicators of vulnerability and severity. Arch. Gen. Psychiatry **49:** 599–608.
43. TARTER, R. E. & M. M. VANYUKOV. 1993. Stepwise developmental model of alcoholism etiology. *In* The Development of Alcohol Problems: Exploring the Biopsychosocial Matrix of Risk. R. A. Zucker, J. Howard & G. M. Boyd, eds. National Institute on Alcohol Abuse and Alcoholism. Rockville, MD. In press.

Environmental Differences in Young Men with and without a Family History of Alcoholism

JOHN S. SEARLES[a] AND ARTHUR I. ALTERMAN[b]

[a]Vermont Alcohol Research Center
and
University of Vermont
2000 Mountain View Drive
Colchester, Vermont 05446

[b]Addiction Research Center
University of Pennsylvania
Veterans Affairs Medical Center
Philadelphia, Pennsylvania

In the United States, alcoholism remains a singularly devastating problem with immense societal, family, and personal consequences. Although much research has been focused on causal mechanisms, the etiology of alcoholism is still not yet well specified. Studies on the genetics of the disorder have contributed much to the current understanding of the transmission of some forms of alcoholism. However, two very important considerations have been overlooked in most of the research that investigates the genetics of alcoholism. First, most individuals with a biological family history of alcoholism do not themselves become alcoholic. In fact, a rather small proportion of these individuals go on to abuse alcohol. For example, in the Danish adoption study,[1] 82% of adopted-out sons of alcoholics did *not* develop alcoholism. In the Iowa Adoption Studies,[2] 86% of the adoptees with a definite or possible alcohol diagnosis in a first-degree biological relative did not develop alcoholism. Second, the majority of individuals from studies on the genetics of alcoholism who are diagnosed as alcoholic do not have a biological family history of alcoholism. For example, in the Stockholm adoption study,[3] 62% of the individuals diagnosed as alcoholic (mild, moderate, or severe) did *not* have a family history of alcoholism.

These considerations suggest an important contribution of unspecified environmental influences in the development of alcohol abuse *even within individuals having an elevated genetic risk for alcoholism.* That is, even if a genetic basis for alcoholism were established beyond doubt and an effective treatment formulated, the problem would persist in its more prevalent nongenetic form. They also suggest a moderating effect on risk status, possibly mediated by personality factors.

Nonetheless, clear empirical support exists for the notion that the principal risk factor for male alcoholism is a familial history of alcohol abuse in general and paternal alcohol abuse in particular. For example, in a community sample, alcoholics and problem drinkers who had a positive family history (FH+) of alcohol abuse/dependence were twice as prevalent as were those without a familiar alcohol problem (FH−) (22% FH+ *vs* 11% FH−).[4]

The methodologies of the three major adoption studies whose results converge on a genetic mechanism for the development of alcoholism have not been accepted uncritically. We briefly highlight some of the major criticisms below.

THE DANISH ADOPTION STUDY[1]

First, as also pointed out by Murray et al.,[5] the entire genetic effect depends on a rather artificial distinction between what Goodwin et al.[1] call problem drinking and alcoholism. The percentage of heavy and problem drinkers in the control group (those adoptees without alcoholism in a biological relative) is actually higher than in the adoptee group. When the problem drinkers and alcoholics are combined, there is no statistical difference in the rates of alcoholism between the probands and controls.

The second significant problem with the Danish study is one primarily of omission, that is, over 60% of both probands and controls were under 30 years of age at the time of the study, which may be too young to be considered through the lifetime risk for alcoholism. Thus, a reevaluation of these individuals seems warranted. To date, no such follow-up has been published.

STOCKHOLM ADOPTION STUDY[3,6]

The most problematic issue of this study involves how the adoptees were classified as mild, moderate, or severe alcohol abusers. Individuals were classified as mild abusers if they had a single registration with the Temperance Board but had never been treated for alcoholism. Moderate abusers had two or three registrations with the Temperance Board and had never been treated for alcoholism. Severe abusers had four or more Temperance Board registrations, compulsory treatment, or a psychiatric hospitalization with a diagnosis of alcoholism.

The Temperance Boards, one for each county, record instances of alcohol-related family violence and traffic offenses as well as behavior and social problems associated with alcohol abuse. It is not clear who may report abuses (law enforcement persons, health professionals, and family and friends) and what the criteria are for reporting or recording abuses. The Temperance Boards have no American counterpart, making direct diagnostic comparisons difficult. Of those classified as abusers, 7.4% were classified as mild abusers, 4.2% as moderate abusers, and 5.9% as severe abusers. It is not clear if any but those meeting the severe abuse criteria could be considered alcohol dependent by standard criteria in the United States (e.g., DSM-III, DSM-III-R, and DSM-IV). It is possible that most American college students could qualify as at least mild abusers by the Swedish criteria.

The second major problem is also related to and a consequence of the classification system. Specifically, the results of this study suggest that both mild *and* severe abusers cluster into the environmentally mediated Type 2 alcoholic, whereas the Type 2, highly heritable, male-limited subtype consists of the moderate abusers. Thus, the hypothesized genetic liability is greatest with the moderate abusers and less robust with *both* mild and severe abusers. This seems to be a rather implausible mechanism and will need further explanation by the Cloninger group.

A serious and ubiquitous misunderstanding has arisen regarding this typology. The Type 2 abuser is often described as having an extensive alcohol treatment history and a high level of criminality. Actually, these characteristics are associated with the *fathers* of the identified Type 2 abusers. The abusers themselves are classified as moderate (i.e., 2–3 registrations), *with no history of treatment or criminality.* This misconstrual of the Swedish data has obvious and dramatic implications for treatment and research.

It is particularly difficult to interpret the criteria used in the Swedish Adoption

Study as there is no American counterpart to the Temperance Boards. Because this study has played such a critical role in the development of theory in the etiology of alcoholism, it is appropriate to delve deeper into this criteria problem. Data are available from this study that may shed some light on this important issue. von Knorring et al.[7] provided data on an expanded sample of the Stockholm Adoption Study. The significance of the von Knorring et al.[7] study is that recognized and established criteria were employed in the diagnosis of alcohol abuse. The background of all 2,966 adoptees (males and females) in the project was examined for psychiatric illness that resulted in missing work for a minimum of 2 weeks (including homemakers and students) and with a psychiatric diagnosis according to the International Classification of Diseases, eighth edition (ICD-8). Both inpatient and outpatient cases were included in the selection. With these criteria, 218 adoptees qualified for a psychiatric diagnosis, 55 with a primary diagnosis of alcoholism and 2 with an alcoholism diagnosis secondary to drug abuse. Primary and secondary diagnoses were determined by age of onset of the disorder. Alcoholism was diagnosed in 35 males and 22 females.

Although not specified, it is likely that the sample was roughly the same gender composition of the earlier studies reported by Cloninger et al.[3] and Bohman et al.[6] which was 51.4% female and 48.6% male. Based on the same sample partition for gender, the rate of alcoholism diagnosed with standard criteria was 2.4% for male and 1.4% for female adoptees. The rate reported in the previous studies employing nonstandard criteria was 17.5% for male[3] and 3.4% for female[6] adoptees. Thus, when using standard psychiatric criteria implemented by von Knorring et al.,[7] the rate of alcoholism reported by Cloninger et al.[3] for males is over seven times greater and the rate for females reported by Bohman et al.[6] is over twice as great. Furthermore, the severe abuse category for males assessed by nonstandard criteria reported by Cloninger et al.[3] was more than twice that when standard criteria were applied (5.9% /2.4%).

Von Knorring et al.[7] reported no clear genetic link between substance abuse in the biological mothers or fathers and substance abuse in the adoptees. In fact, the largest effect reported in this study was a purely environmental one. Adoptees who had been treated by a psychiatrist were five times more likely to have an adoptive father who was treated by a psychiatrist compared to a control group of adoptees who had no diagnosed psychopathology. The size of this effect is even more impressive when we consider that prospective adoptive parents are intensively screened for physical and psychological health before being allowed to adopt a child.

Finally, von Knorring et al.[7] suggested that only milder forms of psychopathology, including alcoholism, may be genetically controlled[a]: "The implication of our current data is that researchers seeking to identify traits that are more fully determined by genetic factors will have to use measures of impairment besides occupational disability or psychiatric treatment. The criterion for disability we used here seems to depend more on nongenetic factors than genetic factors" (p. 948). Therefore, when objective, standard criteria were implemented for psychiatric diagnoses, a genetic effect was difficult to establish and environmental factors in the family predominated.

[a]It should be noted that in referring to the earlier Cloninger et al.[3] study of alcohol abuse in men, von Knorring et al.[7] misstate their own findings: "In our earlier study of alcohol abuse in men we found that susceptibility to mild or untreated alcohol abuse was stronger than the heritability of severe or treated alcohol abuse" (p. 948). What Cloninger et al.[3] actually reported was that mild *and* severe abuse was less heritable than was moderate abuse. See Searles[13] for a discussion of this issue.

IOWA ADOPTION STUDIES

Cadoret *et al.*[2,8] reported on two separate cohorts of adoptees in Des Moines, Iowa. These have been among the most influential American adoption studies of the genetics of alcoholism and antisocial personality reported to date.[b] There are two major problems with the method of these investigations. First, these studies suffer from a lack of objective information about the biological parents, particularly the biological father. Much of the critical information that was used to derive an alcoholism diagnosis in the biological father was not obtained directly and may even have been based on information provided by the biological mother. The veracity and reliability of such data must be considered suspect unless confirmed by an additional source.

The second issue in the Iowa studies concerns the diagnosis of alcoholism in the adoptees. Diagnoses were made by Cadoret himself, blind to the biologic background of the adoptees. The diagnosis was based primarily on interviews with the adoptee and the adoptee's adoptive parents. Additional data were garnered from available records of treatment for psychiatric or behavioral problems. From this information, Cadoret was able to diagnose alcoholism in accordance with DSM-III criteria, which is a real strength of these studies. However, data analyses were conducted not just on the individuals who met the criteria, but also on those who were "possible" alcohol abusers defined as those in whom "one criterion behavior was lacking to make the definite diagnosis." Given the rather restrictive nature of the DSM-III criteria, it is not clear how the "possible" diagnosis was made.

This problem is of some consequence because the 1985 study reported on 14 men who met definite criteria for alcohol abuse or dependence and an additional 23 males were classified as "possible" alcohol abusers. However, the data were analyzed and reported on the combined sample of 37 definite *or* possible alcohol abusers. The basic diagnostic data were not reported in the 1987 paper, but they presumably were similar to the 1985 description.

New data presented by Cadoret (this volume) on an expanded and better assessed sample of adoptees and their biological parents (especially the fathers; see above) indicate that the genetic effect is substantially weaker than originally thought. In fact, when the independent samples are combined, the genetic effect disappears entirely.

Despite theoretical and methodological criticism from a number of sources,[5,9–16] interest in an underlying conceptual notion of a direct genetic causal mechanism in the etiology of alcoholism has been remarkably robust.

It might be said that despite some minor methodological and statistical anomalies, these three adoption studies all converge on a similar result. Given the widely varying diagnostic criteria that the studies employed, this is a difficult argument to sustain. In the three studies, diagnoses of alcohol abuse that was genetically influenced ranged from 7.1 to 29.7%; diagnoses of environmentally induced alcoholism ranged from 4.1 to 32.4%; and gene X environment interaction diagnoses ranged from 0 to 26.7%. These findings suggest that the samples may be too diverse to draw any general conclusions or that qualitatively different phenomena are being studied. It is crucial to be able to compare these studies with a common gauge rather than with criteria of convenience that are study specific.

Although several studies are currently investigating the genetic contribution to

---

[b]Recent reports from the Minnesota sample of twins recruited from treatment facilities[20,21] are arguably more methodologically sound. On the other hand, we note that clinical samples may be qualitatively different from nonclinical samples.

alcohol abuse, the role of early and sustained negative environmental factors in the pathogenesis of alcoholism has been relatively neglected. Although offspring of alcoholic fathers are at an increased risk of developing alcohol problems themselves, it is not an inevitable or even a common outcome. In fact, West and Prinz[17] concluded that most children who are reared by one or more alcoholic parents do *not* develop psychopathologies and are as psychologically and emotionally adjusted as their peers who are reared by nonalcoholic parents. Windle and Searles[14] have also pointed out that most children of alcoholics function within a normal range in the cognitive, neuropsychological, and personality domains, even if they exhibit statistically significant differences from the non-children of alcoholics control groups.

Although genetic factors may (and probably do) play a role, as yet unspecified environmental factors are likely to be at least as important in the development of alcoholism.[13-15] Both direct and indirect influences may be implicated. Adverse life events or inadequate role models (e.g., growing up in an "alcoholic home") may have a direct causal impact. For example, Richardson et al.[18] showed that children who are left to take care of themselves more than 11 hours per week were at a significantly increased risk for substance abuse compared to children who were supervised after school, *and* this relationship obtained across all sociodemographic levels.

In another study of potentially significant environmental influences, Holmes and Robins[19] reported that retrospectively recalled differential parental disciplinary practice discriminated between siblings who did and those who did not become alcoholics. This effect is likely to be extremely complex and probably is more adequately described in terms of gene-environment correlation models rather than simple main effects models. Nonetheless, it demonstrates that early environmental influences can have at least an association with long-term adverse outcomes in the development of alcoholism.

Indirectly, certain biological vulnerabilities may require specific environmental releasers for alcoholism to develop. The more prevalent type of alcoholism identified by Cloninger et al.[3] presumably requires both a genetic predisposition and an environmental catalyst (but see Searles[13] for a critical review of the methodology and results of this and other genetic studies of alcoholism).

At the same time, because most individuals in epidemiological studies such as those of Russell et al.[4] do not exhibit alcoholic/problem-drinking symptoms regardless of family background, it would be important to identify those factors, whether constitutional/personological or environmental, that may differentiate between individuals at risk who do and those who do not eventually engage in alcoholic behavior.

Pickens et al.[20] recently reported that the genetic influence for alcoholism in male and female twins was modest ($h^2 = 0.35$ for males and 0.24 for females). In a companion study, McGue et al.[21] found substantial genetic effects only for symptoms of early onset alcoholism (prior to age 20) in males and a small genetic effect for symptoms of late onset alcoholism in males. No genetic influence was noted for either early or late onset of symptoms in females. McGue et al.[21] concluded that: "In any case, our findings suggest that in the headlong rush to identify molecular genetic processes, researchers may be ignoring the significant influences that the environment has in the origins of alcoholism" (p. 15).

The present study further elaborates the role of a subset of potentially important environmental factors that may influence the development of alcohol-related problems and other psychopathologies in young men.

METHOD

We screened introductory psychology, mathematics, and english classes at a public university in the Philadelphia area in an effort to recruit 18–20-year-old men who would self-report paternal alcohol abuse symptoms. In addition, we recruited similar age males from presumed low SES sources (e.g., trade schools, smaller junior colleges, residential facilities, as well as ads in newspapers targeted to specific areas of Philadelphia that would yield subjects who fit our inclusion criteria). The screening questionnaire was entitled "Correlates of Drinking Behavior Project Screening Questionnaire." Screening was conducted for the most part in small discussion sections of 10 to 30 university students; questionnaires were mailed along with an addressed, postage-paid return envelope to individuals who responded to the newspaper ads and on an *ad hoc* basis for other sources (e.g., in one residential facility, questionnaires were left in individual mail slots, and a specific date and time were established to meet with research assistants on site). The potential subjects were asked to complete the questionnaire which asked about drinking activities for the last 30 days and last year. The last question asked the students to report on alcoholic symptoms for themselves, first-degree relatives (mother, father, brothers, and sisters) and second-degree relatives (uncles, aunts, and grandparents). Males who identified one or more symptoms of alcohol abuse for their biological father were selected for the study. This group was designated as Family History Positive (FH+). A second set of individuals who reported no first- or second-degree biological relatives with any symptoms of alcohol abuse were selected as controls, were matched for age and recent drinking history with the index group, and were designated as Family History Negative (FH−). Nondrinkers were excluded from the samples. None of the individuals was aware of the criteria for selection. Individuals in both groups were contacted by telephone and administered the Family History section of the Renard Diagnostic Interview to confirm alcohol abuse symptoms in their fathers (for FH+) and the absence of a family history of alcoholism (for FH−). All but three individuals contacted were concordant on both instruments and were asked to participate in the full study.

Individuals were scheduled at their convenience to come to our on-campus office, located in the building housing the psychology and counseling departments, or to our research offices at the Addiction Research Center of the University of Pennsylvania. Subjects were administered a 4-hour assessment battery consisting of interviews and self-report questionnaires that ranged from standardized personality tests to a general background questionnaire. The first instrument was the MAST-10[22] in order to reconfirm paternal symptoms of alcoholism. Subjects were not required to complete the battery in one sitting; most spread it out over 2 or 3 days. Each individual was compensated $50 for his efforts.

This report presents preliminary data from the Life Events Record (LER) as well as some relevant personality data. The LER is a refinement of the Social Readjustment Rating Scale developed by Holmes and Rahe.[23] There are four separate rating periods: preschool (30 events), elementary school (35 events), junior high school (40 events), and senior high school (42 events). Each of the time periods has its own associated life events, but there is considerable overlap (e.g., "death of a parent"). We further revised the instrument to include ratings of the positive or negative impact of each event. For each event the subject is asked to report if the event occurred, how many times it occurred, and the overall impact on the subject's life at the time of the event on a scale of +100 being the best possible impact and −100 indicating the worst possible impact. The instructions encouraged the subjects to employ the entire range of the rating scale for their evaluations.

RESULTS

Data were gathered on 465 individuals, 235 FH+ and 230 FH–. Both FH+ and FH– individuals were approximately 19 years of age. FH+ subjects were reared in the same home as their biological father approximately 3 years less than were FH– subjects. In addition, FH+ subjects drank on more occasions, drank more per occasion, and had more severe hangovers than did FH– subjects. Finally, the FH+ individuals reported considerably more illicit drug involvement than did the FH– group. It should be noted, however, that both groups showed a relatively high illegal drug use.

Personality

The California Psychological Inventory (CPI) yields 20 personality scales as well as three structural scales that represent broader, underlying personological themes measured by the instrument (internality, norm-favoring, and self-realization). The FH+ group scored significantly lower than did the FH– group on the Responsibility, Socialization, and Self-Control scales as well as less internal and less norm-favoring on the structural scales.

The Tridimensional Personality Questionnaire (TPQ) taps three major dimensions of personality consonant with Cloninger's[24] neurobiological theory of personality (Novelty Seeking, Harm Avoidance, and Reward Dependence). Several subscales are derived from each of the three overarching constructs. Three of the four subscales of the Novelty Seeking dimension (impulsiveness, extravagance, and disorderliness) as well as the overall scale and one subscale of the Harm Avoidance dimension (fear of uncertainty) significantly discriminated the FH+ and FH– groups.

Life Events

Impact scores for the four time periods (preschool, elementary, junior high, and senior high school) were computed for each time period as the sum of the negatively rated events and the sum of the positively rated event. Negative impact for all four time periods was significantly greater for the FH+ group than for the FH– group. The total number of events reported, either positive or negative for the time periods separately and the combined total across all four periods did not differ between the groups. A subset of the fathers in the FH+ group were explicitly treated for alcoholism, and it was thought that this indicator of severity might have a separate effect with respect to life events. However, no such effect appeared; LER scores for both FH+ groups were the same and significantly greater than those of the FH– group. Again, no differences for positive impact scores emerged across all three groups.

Effect sizes were calculated for both the personality and the LER data. The mean effect size for the significant personality scales was 0.25, whereas the mean for the LER was 0.49.

DISCUSSION

This study demonstrates that individuals at a putatively higher risk of developing alcoholism retrospectively report experiencing a greater negative impact of life events as far back as their preschool years than did the control subjects. Thus, sons of alcoholics at least perceive their rearing environment in more negative terms than do their peers who were not exposed to paternal alcohol problems while growing up. Although these individuals are too young to associate negative life events with the development of alcoholism, these results show that the environments of sons of alcoholics and sons of nonalcoholics are perceived differently. Since we are tracking these young men over the next several years, we will be able to assess the relative etiological significance of the impact of life events on psychopathology in general and alcohol abuse/dependence specifically.

Personality differences between the FH+ and FH− groups were surprisingly small, and the effect sizes associated with those differences were about half as large as those associated with the environmental variables (as measured by the LER). None of the personality scales was significantly associated with the drinking variables or the LER, suggesting that personality trait markers were not effective in ameliorating negative environmental events. It should be noted that although some significant differences on some scales were noted, none of the scale scores for either group on the CPI approached abnormality. Published norms are not available for the TPQ, so it could not be determined if scale score evaluations were abnormal; however, given the relatively minor differences between the FH+ and FH− groups, it is doubtful that either of the groups was clinically remarkable. Because the personality scales that differentiated the two groups on both instruments were consistent with an antisocial type cluster, it will be interesting to see if individuals who exhibit varying degrees of this type will be overrepresented as alcohol abusers in the future.

This research has several implications for family studies of alcoholism. First, factors, associated with being a child of an alcoholic are most likely complex, multivariate, and interact in ways that are not yet well specified. Second, environmental factors that are potentially significant in the development of alcoholism are currently understudied and consequently not well understood. There is little research on which aspects of the environment provided by alcoholic or otherwise dysfunctional parents may play an etiologic role (e.g., modeling and strength of identification with the affected parent). Third, further work is necessary to specify those nonshared factors that may directly or indirectly contribute to alcohol abuse. One that has already been identified as peer group influence.[25] It should be clear that this type of research must include multiple members of the same family. One approach is to study siblings who are discordant for alcohol abuse.

Finally, a more complex etiological model that takes into account both genetic vulnerability and environmental reactivity is probably closer to the actual state of nature than are models positing simple effects of genes and environments. Individuals are not merely buffeted about as a consequence of their genetic predisposition nor do they engage in random environmental interactions. They actively select specific environments in which to operate, based partially on genetic propensities. Thus, an individual who inherited the trait of introversion is less likely to attend social events than is one who is genetically more of an extrovert. This concept, called gene-environment correlation, was recently proposed as a general model of development that allows for the simultaneous consideration of both genetic and environmental effects.[26,27] Research employing this concept which takes into account the dynamics of development by considering the impact of the environment on individuals as well as the effects of genetics on the environment is likely to yield extremely interesting and clinically meaningful results.

REFERENCES

1. GOODWIN, D. W., F. SCHULSINGER, L. HERMANSEN, S. B. GUZE & G. WINOKUR. 1973. Alcohol problems in adoptees raised apart from biological parents. Arch. Gen. Psychiatry **28:** 238–243.
2. CADORET, R. J., E. TROUGHTON & T. W. OGORMAN. 1987. Genetic and environmental factors in alcohol abuse and antisocial personality. J. Stud. Alcohol **48:** 1–8.
3. CLONINGER, C. R., M. BOHMAN & S. SIGVARDSSON. 1981. Inheritance of alcohol abuse: Cross-fostering analysis of adopted men. Arch. Gen. Psychiatry **38:** 861–868.
4. RUSSELL, M., M. L. COOPER & M. R. FRONE. 1990. The influence of sociodemographic characteristics on familial alcohol problems: Data from a community sample. Alcoholism: Clin. Exp. Res. **14:** 221–226.
5. MURRAY, R. M., C. A. CLIFFORD & H. M. D. GURLING. 1983. Twin and adoption studies: How good is the evidence for a genetic role? *In* Recent Developments in Alcoholism. M. Galanter, ed. Vol. 1: 25–48. Plenum Press. New York, NY.
6. BOHMAN, M., S. SIGVARDSSON & C. R. CLONINGER. 1981. Maternal inheritance of alcohol abuse: Cross-fostering analysis of adopted women. Arch. Gen. Psychiatry **38:** 965–969.
7. VON KNORRING, A.-L., C. R. CLONINGER, M. BOHMAN & S. SIGVARDSSON. 1983. An adoption study of depressive disorders and substance abuse. Arch. Gen. Psychiatry **40:** 943–950.
8. CADORET, T. J., T. W. OGORMAN, E. TROUGHTON & R. HEYWOOD. 1985. Alcoholism and antisocial personality: Interrelationships, genetic and environmental factors. Arch. Gen. Psychiatry **42:** 161–167.
9. LESTER, D. 1988. Genetic theory: An assessment of the heritability of alcoholism. *In* Theories on Alcoholism. C. D. Chaudron & D. A. Wilkinson, eds. Addiction Research Foundation. Toronto.
10. LITTRELL, J. 1990. The Swedish studies of the adopted children of alcoholics. J. Stud. Alcohol **49:** 491–499.
11. PEELE, S. 1986. The implications and limitations of genetic models of alcoholism and other addictions. J. Stud. Alcohol **47:** 63–71.
12. PEELE, S. 1990. The Diseasing of America. Lexington Books. Lexington, MA.
13. SEARLES, J. S. 1988. The role of genetics in the pathogenesis of alcoholism. J. Abnorm. Psychol. **97:** 153–167.
14. SEARLES, J. S. 1990. Behavior genetic research and risk for alcoholism among children of alcoholics. *In* Children of Alcoholics: Critical Perspectives. M. Windle & J. S. Searles, eds. Guilford Press. New York.
15. SEARLES, J. S. 1990. The contribution of genetic factors to the development of alcoholism: A critical review. *In* Alcoholism and the Family. R. L. Collins, K. E. Leonard & J. S. Searles, eds. Guilford Press. New York.
16. SEARLES, J. S. 1990. Methodological limitations of research on the genetics of alcoholism. *In* Genetics and Biology of Alcoholism. C. R. Cloninger & H. Begleiter, eds. Cold Spring Harbor Press. Cold Spring Harbor, NY.
17. WEST, M. O. & R. J. PRINZ. 1987. Parental alcoholism and childhood psychopthology. Psychol. Bull. **102:** 204–218.
18. RICHARDSON, J. L., K. DWYER, K. MCGUIGAN, W. B. HANSEN, C. DENT, C. A. JOHNSON, S. Y. SUSSMAN, B. BRANNON & B. FLAY. 1989. Substance use among eighth-grade students who take care of themselves after school. Pediatrics **84:** 556–566.
19. HOLMES, S. J. & L. N. ROBINS. 1988. The role of parental disciplinary practices on the development of depression and alcoholism. Psychiatry **51:** 24–36.
20. PICKENS, R. W., D. S. SVIKIS, M. MCGUE, D. T. LYKKEN, L. HESTON & P. J. CLAYTON. 1991. Heterogeneity in the inheritance of alcoholism: A study of male and female twins. Arch. Gen. Psychiatry **48:** 19–28.
21. MCGUE, M., R. W. PICKENS & D. S. SVIKIS. 1992. Sex and age effects on the inheritance of alcohol problems: A twin study. J. Abnorm. Psychol. **101:** 3–17.

22. POKORNY, A. D., B. A. MILLER & H. B. KAPLAN. 1972. A shortened version of the Michigan Alcoholism Screening Test. Am. J. Psychiatry **129:** 342–345.
23. HOLMES, T. H. & R. H. RAHE. 1967. The social readjustment rating scale. J. Psychosom. Res. **11:** 213–218.
24. CLONINGER, C. R. 1986. A unified biosocial theory of personality and its role in the development of anxiety states. Psychiatric Dev. **3:** 167–226.
25. KANDEL, D. B. 1985. On processes of peer influence in adolescent drug use: A developmental perspective. Adv. Alcohol Subst. Abuse **4:** 139–163.
26. PLOMIN, R., J. C. DEFRIES & J. C. LOEHLIN. 1977. Genotype-environment interaction and correlation in the analysis of human behavior. Psychological Bull. **84:** 309–322.
27. SCARR, S. & K. MCCARTNEY. 1983. How people make their own environments: A theory of genotype → environment effects. Child Dev. **54:** 424–435.

Coexisting Problems and Alcoholic Family Risk among Adolescents[a]

MICHAEL WINDLE

Research Institute on Addictions
1021 Main Street
Buffalo, New York 14203

The investigation and utility of coexisting disorders and alcohol typologies have become a prominent focus in adult studies of alcoholism.[1–4] Much less research has focused on adolescent alcohol abusers and coexisting disordered conditions even though early onset problem behaviors are frequently symptomatic of some adult alcoholic subtypes (e.g., early onset, antisocial personality alcoholics). The limited number of adolescent studies focused on the coexistence of alcohol or substance abuse disorders with other psychiatric disorders has indicated a high prevalence of coexisting conditions. For example, Winters[5] reported that 74% of a drug clinic sample of adolescents manifested high levels of coexisting symptoms for other mental disorders. Neighbors *et al.*[6] studied an incarcerated juvenile delinquent sample, and their findings indicated significant comorbidity for substance abuse disorders with disorders of conduct, anxiety, and depression. (For a more extensive review, see Bukstein *et al.*[7])

Whereas studies of adolescents in chemical dependency treatment settings or penal institutions are supportive of high rates of coexisting psychiatric problems with alcohol or substance abuse disorders, the generalizability to nonclinical samples is largely untested and the limited available data on the issue are mixed. On the one hand, the problem behavior theory of Jessor and colleagues[8] suggests that a generalized construct of "deviance proneness" accounts for the observed pattern of covariation among a range of adolescent indicators of maladjustment (e.g., alcohol use, poor school performance, and delinquent behaviors). On the other hand, on the basis of data collected in the National Youth Survey, Elliott *et al.*[9] concluded that the evidence was weak for a singular, multiple problem syndrome. Rather, Elliott *et al.*[9] proposed three distinct etiological patterns for delinquency, mental health, and substance use, respectively.

The data presented here, derived from a nonclinical adolescent sample, focuses on three objectives related to maladjustment problems and coexisting conditions. First, the prevalence of three indicators of maladjustment are provided separately for boys and girls. The three indicators selected are problem drinking, depressive symptoms, and high levels of delinquency (or antisocial behaviors). Second, the prevalence of the coexistence of depressive symptoms, high levels of delinquency, and joint depression/delinquency with problem drinking is presented. Third, the association between family history of alcoholism and problem drinking is examined.

---

[a]This research was supported in part by NIAAA Grant No. 07861.

157

METHOD

Subjects

The data used in this paper are part of a four-wave longitudinal research design pertaining to vulnerability factors and adolescent substance use. The study is referred to by the acronym MAVS, which stands for Middle Adolescent Vulnerability Study. The principal objective of the MAVS is to assess the initiation, maintenance, and continuation (or termination) of substance use during the high school years in relation to a range of vulnerability, or high risk, factors. These risk factors include family history of alcoholism, childhood behavior problems, temperament, depression, stressful life events, family conflict, and peer functioning. The focus of the current paper is restricted to a single occasion of measurement (Time 4) and to a selected set of measures.

The demographic characteristics of the sample are presented in TABLE 1. The total sample size is 1,174 subjects with an approximately equal distribution of boys and girls. The adolescents participating in this study were selected from three suburban senior high schools in Western New York. The sample is best characterized as primarily white and middle to middle-upper class. Most of the adolescents came from maritally intact families.

Measures

A large sample school survey methodology was used in the study, and the following measures were used to derive the key substantive variables reported on in this article.

TABLE 1. Demographic Characteristics of Sample

Characteristics		Percent
Gender	Males ($n = 570$)	48.6
	Females ($n = 604$)	51.4
Grade level	Juniors	53.3
	Seniors	46.7
Age (yr)		$\bar{x} = 16.96$
Race	White	98.3
	Non-white	1.7
Religion	Catholic	74
	Protestant	16
	Other	10
Parental marital status	Currently married	86.7
Family income	<$12,000	2.2
	$12,000–$22,999	10.4
	$23,000–$39,999	36.4
	>$40,000	51.0

A *problem drinking categorization* was used that was designed to be consistent with DSM-III-R criteria[10] and was directed toward the notion of progressive levels of alcohol involvement. The criteria used for the categorization are listed and were derived on the basis of three indexes: a standard quantity/frequency index reflecting the average number of ounces of ethanol consumed per day, the number of heavy drinking episodes (i.e., 6 or more drinks in one setting), and the number of alcohol problems (principally adverse social consequences associated with alcohol use).

1. *Abstainers*: Have not consumed an alcoholic beverage in the last 30 days.

2. *Light-to-Moderate Drinkers*: Have consumed more than 1 but less than 45 drinks in the last 30 days.

3. *Heavy Drinkers*: Have consumed 45 drinks or more in the last 30 days *or* 6 or more drinks on at least 3 occasions in the last 30 days.

4. *Problem Drinkers*: Have consumed 45 drinks or more in the last 30 days *or* 6 drinks or more on at least 3 occasions in the last 30 days *and* have reported 5 or more alcohol problems (e.g., missing school because of drinking, "passing out") in the last 6 months.

TABLE 2 provides a listing and the frequency of endorsement of the alcohol problems index. In comparison with many other drinking categorizations, the "light" and "moderate" drinking categories were combined and a clear distinction was made between "heavy" and "problem" drinkers. Data are presented subsequently regarding the usefulness of these distinctions.

Depressive symptoms were assessed by the CES-D scale[11] and a criterion of ≥ 23

TABLE 2. Alcohol Problems Index: Frequency of Item Endorsement among Those Who Drink[a]

Item	Total (n)	Percentage Endorsing Affirmative Response[a]		
		Boys	Girls	χ^2 (1 df)
1. Drank before or during school	12.1	16.6	8.5	12.61**
2. Missed school because of drinking	8.9	11.2	7.1	4.05*
3. Had a fight with parents about your drinking	16.1	19.3	13.4	5.36*
4. Did things while you were drinking that you regretted the next day	47.4	49.9	45.4	1.59
5. Thought about cutting down on your drinking	24.0	30.2	18.9	14.16**
6. Got drunk or high from alcohol several days in a row	28.3	36.8	21.4	23.42**
7. Passed out from drinking	29.1	33.1	25.8	5.10*
8. Had a fight with your girlfriend or boyfriend about your drinking	15.9	16.1	15.8	0.01
9. Got into a fight or heated argument with someone you didn't know while drinking	16.1	25.1	8.7	40.21**
10. Got into trouble with the law (other than driving-related) while drinking	6.4	11.2	2.4	25.75**
11. Drank alone	21.6	25.3	18.5	5.61*
12. Drank alcohol to get rid of a hangover	7.2	8.7	6.0	2.20
13. Drank to forget your troubles	36.9	34.9	38.5	1.16

[a]Abstainers were excluded from these analyses. Time referent for alcohol problems was the last 6 months.
*$p <0.05$; **$p <0.001$.

was used as a cut-off indicative of problematic depressive symptoms (a cut-off of 16 is typically used with adults).[12] For convenience, those who met or exceeded the criterion were regarded as depressed. However, this shorthand for problematic depressive symptoms should not be misconstrued as isomorphic with clinical depression.[12,13]

Delinquent activity was assessed by 30 items gleaned from and representative of those used in prior delinquency research.[9] Subjects responded to the frequency of engaging in each of the 30 items during the preceding 6-month interval. Items ranged in severity of delinquency/criminality and included activities such as skipped school, hit teacher or parent, stole something valued at $20 or less, stole something valued at more than $20, beat up someone, and was suspended from school. A categorical index reflecting a "non-high" *versus* "high" level of delinquency was created by summing the number of delinquent acts *for the respective gender groups* and assigning a score of 1 to those adolescents who scored in the upper 20% of the distribution, indicative of higher levels of delinquent activity, and a score of 0 to the remaining subjects.

Family history of alcoholism was determined by having the primary caregivers of the participating adolescents complete a modified self-report form of the Family History-Research Diagnostic Criteria.[14] Primary caregivers (in most instances, mothers) completed the family history form for the adolescents' first-degree relatives. A diagnosis of alcoholism was assigned to those members who met the criteria of at least two affirmative responses to the following items associated with having a serious drinking problem: had legal problems associated with alcohol use (e.g., losing driver's license), had health problems associated with alcohol use (e.g., delirium tremens), was treated for alcoholism, alcohol problems created problems in marriage or with the family, and alcohol use created problems at work (e.g., job loss due to alcoholism). (See Domenico and Windle[15] for additional information on the family history data used in this study.) In this study, a positive family history of alcoholism was indicated if either parent met the criteria. In over 90% of the cases, paternal alcoholism was the referent.

RESULTS

TABLE 3 is a summary of the prevalence of drinker status, depressive symptoms, and delinquent activity for boys, girls, and the total sample. Similar to prior studies, there were no marked gender differences in the percentage of abstainers and drinkers; however, boys were more highly represented in heavy and problem-drinking groups.[16,17] A gender by drinker status comparison indicated significant group differences (χ^2 with 3 df = 48.49, p <0.001). The reverse pattern was evident with respect to depressive symptoms, as more girls than boys reported problematic levels of depression (χ^2 with 1 df = 8.39, p <0.01). Again, this finding is consistent with research literature.[12,18,19] The prevalence of delinquent activity was based on statistically normative considerations (i.e., 80th percentile cut-points), and this cut-point for boys corresponded to the commission of 9.53 delinquent acts in the last 6 months, whereas the cut-point for girls corresponded to 8.61 delinquent acts.

In TABLE 4, the prevalence of depression, high delinquency, and joint depression and high delinquency is presented for the drinking status categories. Separate contingency table analyses were conducted for boys and girls for each of the three problem domains. Depression for girls was relatively high and constant across abstainers, light-to-moderate, and heavy drinkers, but increased somewhat for prob-

TABLE 3. Gender Distribution of Drinker Status, Depression, and Delinquent Activity[a]

	Males		Females		Total	
	n	%	n	%	n	%
Drinker Status						
Abstainer	169	34.5	164	30.5	334	32.4
Light-to-moderate	171	34.9	291	54.1	462	44.9
Heavy	65	13.3	38	7.1	103	10.0
Problem	85	17.3	45	8.4	130	12.7
Depression Status						
Low	418	84.3	415	77.1	833	80.6
High	78	15.7	123	22.9	201	19.4
Delinquent Activity						
Low	386	79.9	416	80.2	802	80.0
High	97	20.1	103	19.8	200	20.0

[a]"High" delinquent activity was determined on the basis of a cut-point at the 80th percentile for the respective distributions of the gender groups.

lem drinkers (although a nonsignificant association was indicated [χ^2 with 3 df = 1.04, p >0.05]). A different pattern emerged for boys, as heavy drinkers reported the lowest levels of depression and problem drinkers reported rates of depression similar to those of female problem drinkers. This association was statistically significant for boys (χ^2 with 3 df = 9.17, p <0.05). A strong association between problem drinking and high delinquent activity was noted for both boys (χ^2 with 3 df = 27.77, p <0.001) and girls (χ^2 with 3 df = 53.85, p <0.001), and heavy drinking was also associated with high delinquent activity for girls. The joint problem pattern indicated that the highest prevalence was among problem drinkers, both boys and girls. The associated contingency table statistics for boys was a χ^2 with 3 df = 7.79, p <0.05 and for girls was a χ^2 with 3 df = 13.29, p <0.01. The joint problem pattern was more than twice as frequent for the problem drinker category as for other drinker status categories.

The cross-tabulation of drinker status by family history risk status is provided in TABLE 5. The χ^2 test statistic was significant (χ^2 with 3 df = 10.49, p <0.05), supporting an association between the two variables. Examination of the percentages in TABLE 5 indicates that the family history positive group was (relatively) over-represented in both the abstainer and the problem drinker categories. Two other

TABLE 4. Gender Distribution of Coexisting Problems

Drinker Status	% Depressed		% High Delinquency		% Both Depressed and High Delinquency	
	Males	Females	Males	Females	Males	Females
Abstainer	14.8	22.7	12.4	7.0	3.6	2.4
Light-to-moderate	14.0	22.3	17.0	19.4	5.3	7.6
Heavy	9.2	21.1	25.4	40.5	6.2	5.3
Problem	25.9	28.9	40.0	51.1	11.8	17.8

TABLE 5. Drinker Status and Family History of Alcoholism

Drinker Status	Family History Negative (86.8%) (n = 844)	Family History Positive (13.2%) (n = 128)
Abstainer	30.3	39.1
Light-to-moderate	47.5	34.4
Heavy	10.3	8.6
Problem	11.8	18.0

cross-tabulations of family history positive by Depression and Delinquency, respectively, yielded statistically nonsignificant χ^2 values.

DISCUSSION

The findings of this study with a nonclinical adolescent sample are generally consistent with those reported with chemically dependent and incarcerated delinquent youth.[7] That is, although the prevalence of problems was not as great as those reported in treatment samples, a significant subset of adolescents manifested serious problems with regard to drinking behavior, depression, and/or delinquent activity. Furthermore, some gender specificity with regard to prevalence of problems was noted, with boys having more alcohol-related problems and girls having more problems related to depression.

Three findings were of particular significance with regard to coexisting problems and family history of alcoholism associated with drinker status. First, the proposed adolescent drinker typology was supported in that problem drinkers consistently differed from heavy drinkers (as well as abstainers and light-to-moderate drinkers) with regard to the prevalence of the coexisting problems of depression, high delinquency, and joint depression/high delinquency. Problem drinkers were also more highly represented in the high risk family history subgroup. Future research will need to carefully examine biological, neuropsychological, and psychosocial (e.g., temperament, peer, and family) factors that differentiate problem drinkers from other drinker status members, especially for comparisons with heavy drinkers. In addition, with the MAVS longitudinal data, the study of stability and change of problem drinkers across time will be investigated, as will the study of family history status as an even more potent variable in predicting future drinking status as adolescents make transitions into adulthood.

A second important finding was the gender differences in the pattern of coexisting problems with drinker status. Specifically, among the males, heavy drinkers reported the lowest levels of depression. For female drinkers, however, depressive symptoms were approximately equal across abstainers, light-to-moderate, and heavy drinkers. This gender difference may arise from differences in sex role socialization and group drinking practices in that heavy drinking behavior may be perceived as sex role appropriate for boys but deviant for girls. The socially unsanctioned heavy drinking behavior for girls may be associated with higher levels of negative affect (e.g., due to guilt) or may be symptomatic of problems at school, with peers, or with family members. For boys, the lower prevalence of depression among heavy drinkers may reflect a stage in development preceding a progression to problem drinking or it may reflect distinct characteristics that are associated with a heavy drinking, socially nonproblematic drinking style. Such styles have been reported in the adult alcoholism typology literature.[4] Longitudinal research will facilitate efforts to determine if this is a high risk transitional group that will eventually progress to a problem drinking group.

A third finding was the significant association between family history of alcoholism and drinker status. The research literature, at best, has been mixed with regard to family history group differences for alcohol use and alcohol problems with adolescent samples.[20–22] The findings of this study indicate that high risk status was associated with a larger percentage of both problem drinkers and abstainers. Thus, with this nonclinical adolescent sample, it appeared that high risk status was associated with more extreme drinking patterns at both ends of the spectrum. It is important to note, however, that a drinker status group effect was not found in relation to alcohol consumption. Many family history studies of children and adolescents have focused exclusively on alcohol consumption; the inclusion of alcohol problems or a problem drinker type may facilitate a more comprehensive assessment of drinking behaviors and perhaps contribute to more consistent findings in the literature.

SUMMARY

A substantial subset of nonclinical adolescents reported single or multiple problems with regard to drinking behavior, depressive symptoms, and delinquency. Similar to previous research,[17,19] boys reported higher levels of heavy and problematic drinking, whereas girls reported higher levels of depressive symptoms. Higher levels of alcohol involvement were associated with higher levels of delinquency for both boys and girls. Family history of alcoholism was associated with an overrepresentation of problem drinkers (as well as abstainers). These findings are fairly consistent with the adult alcohol typology literature in identifying coexisting problem patterns for the triumvirate of problem drinking, depressive symptoms, and antisocial behavior.[1,3,4] Furthermore, these data support the manifestation of these patterns *prior to* the onset of alcohol dependence (and maybe alcohol abuse). Future longitudinal research needs to focus on the short- and long-term patterns of problem behaviors among adolescents and to identify common and unique precursors, correlates, and consequences associated with varying levels of alcohol involvement.

REFERENCES

1. BABOR, T. F., Z. S. DOLINSKY, R. E. MEYER, M. HESSELBROCK, M. HOFFMAN & H. TENNEN. 1992. Types of alcoholics: Concurrent and predictive validity of some common classification schemes. Br. J. Addict. **87:** 1415–1431.
2. ALTERMAN, A. I. & R. E. TARTER. 1986. An examination of selected typologies: Hyperactivity, familial, and antisocial alcoholism. *In* Recent Developments in Alcoholism. M. Galanter, ed. IV: 169–189. Plenum Press. New York, NY.
3. PENICK, E. C., B. J. POWELL, E. OTHMER, S. F. BINGHAM, A. S. RICE & B. S. LIESE. 1984. Subtyping alcoholics by coexisting psychiatric syndromes: Course, family history, outcome. *In* Longitudinal Research in Alcoholism. D. W. Goodwin & S. A. Mednick, eds. : 167–196. Kluwer-Nijhoff Publishing Co. Hingham, MA.
4. ZUCKER, R. 1987. The four alcoholisms: A developmental account of the etiologic process. *In* Nebraska Symposium on Motivation. : 27–84. University of Nebraska Press. Lincoln, NE.

5. WINTERS, K. 1990. Clinical considerations in the assessment of adolescent chemical dependency. J. Adol. Chem. Depend. **1:** 31–52.

6. NEIGHBORS, B., T. KEMPTON & R. FOREHAND. 1992. Co-occurrence of substance abuse with conduct, anxiety, and depression disorders in juvenile delinquents. Addict. Behav. **17:** 379–386.

7. BUKSTEIN, O. G., D. A. BRENT & Y. KAMINER. 1989. Comorbidity of substance abuse and other psychiatric disorders in adolescents. Am. J. Psychiatry **146:** 1139–1141.

8. JESSOR, R., J. E. DONOVAN & F. M. COSTA. 1991. Beyond Adolescence: Problem Behavior and Young Adult Development. Cambridge University Press. Cambridge, MA.

9. ELLIOTT, D. S., D. HUIZINGA & S. MENARD. 1989. Multiple Problem Youth: Delinquency, Substance Use, and Mental Health Problems. Springer-Verlag. New York, NY.

10. American Psychiatric Association. 1987. Diagnostic and Statistical Manual of Mental Disorders (DSM-III-R). 3rd Ed. Revised. Washington, DC.

11. RADLOFF, L. S. 1977. The CES-D scale: A self-report depression scale for research in the general population. Appl. Psychol. Meas. **1:** 385–401.

12. ROBERTS, R. E., J. A. ANDREWS, P. M. LEWINSOHN & H. HOPS. 1990. Assessment of depression in adolescents using the Center for Epidemiologic Studies Depression Scale. Psychol. Assess. **2:** 122–128.

13. MYERS, J. K. & M. M. WEISSMAN. 1980. Use of a self-report symptom scale to detect depression in a community sample. Am. J. Psychiatry **137:** 1081–1084.

14. ANDREASEN, N. C., J. ENDICOTT, R. L. SPITZER & G. WINOKUR. 1977. The family history method using diagnostic criteria. Arch. Gen. Psychiatry **34:** 1229–1235.

15. DOMENICO, D. & M. WINDLE. 1993. Intra- and interpersonal functioning among middle-aged female adult children of alcoholics (ACOAs). J. Cons. Clin. **61:** 659–666.

16. JOHNSTON, L. D., P. M. O'MALLEY & J. G. BACHMAN. 1986. Drug Use Among American High School Students, College Students, and Other Young Adults: National Trends Through 1985. National Institute on Drug Abuse. Rockville, MD.

17. WINDLE, M. 1991. Alcohol use and abuse: Some findings from the National Adolescent Student Health Survey. Alcohol Health Res. World **15:** 5–10.

18. ANGOLD, A. & M. RUTTER. 1992. Effects of age and pubertal status on depression in a large clinical sample. Dev. Psychopathol. **4:** 5–28.

19. HORWITZ, A. V. & H. R. WHITE. 1987. Gender role orientations and styles of pathology among adolescents. J. Health Soc. Behav. **28:** 158–170.

20. CLAIR, D. & M. GENEST. 1987. Variables associated with the adjustment of offspring of alcoholic fathers. J. Stud. Alcohol **48:** 345–355.

21. HARBURG, E., D. R. DAVIS & R. CAPLAN. 1982. Parent and offspring alcohol use: Imitative and aversive transmission. J. Stud. Alcohol **43:** 497–516.

22. PANDINA, R. J. & V. JOHNSON. 1990. Serious alcohol and drug problems among adolescents with a family history of alcoholism. J. Stud. Alcohol. **51:** 278–282.

Psychological Expectancy as Mediator of Vulnerability to Alcoholism

GREGORY T. SMITH

Department of Psychology
University of Kentucky
Lexington, Kentucky 40506–0044

Psychological expectancy theory[1] is a basic learning theory, that is, a theory of how new behaviors are acquired. Its aim is to identify the mechanism by which early learning experiences come to influence later behavioral choices. Applied to alcohol misuse, the notion is that individuals who learn, or perceive, repeatedly that consumption is associated with reinforcing outcomes will store these learned associations in memory in the form of expectancies for if-then relations between drinking and its consequences.[2,3] These memories then form the basis for decisions made at future choice points; if one has learned to anticipate valued reinforcers from drinking, one is more likely to drink in the future. Thus, early learning experiences are seen as the *original causes* of later behavior; their influence is transmitted forward in time by the mechanism of stored memories concerning the behavior or expectancies. Expectancy, then, can be seen as the *proximal cause* that *mediates* the influence of original causes on current behavior. In this sense, alcohol-related expectancy is viewed as the final common pathway by which a host of historical events exert their influence on current drinking behavior.[2,4,5]

Both biological (e.g., temperament, reactivity to ethanol) and psychological (e.g., parental and peer modeling) models have been proposed to explain alcoholism and problem drinking. None of these models describes how the causal factors themselves automatically "commandeer" drinking behavior. Rather, the proposed factors are causal in that they increase the probability that learning events (original causes) will occur to then increase the likelihood and rate of future consumption. There still must be some controlling mechanism that transcends the time lag between early learning and later drinking and that connects these early experiences with later drinking patterns.[2]

Alcohol expectancies are proposed to be this mechanism. As memories relating alcohol consumption to anticipated consequences (e.g., social enhancement, relaxation), they constitute a kind of summary representation or distillation of a myriad of relevant learning experiences.[2,3] They are what is available to the drinker in the present from his or her learning history. It is in this sense that alcohol expectancy is thought to be a *mediator* of the influence of early learning (original causes) on later drinking.[5] Thus, for example, a child whose parents model excessive consumption and a dearth of alternative coping skills is more likely to learn to rely on alcohol to obtain desired outcomes than is a child exposed to different modeling. Perhaps a sensation-seeking adolescent is more likely to both (1) experiment with alcohol and (2) find the stimulating behaviors disinhibition brings reinforcing (including the experience of violating social constraints by drinking heavily) and thus acquire stronger expectancies for reinforcement from alcohol. This paper reviews three recent studies supporting this expectancy model and relating expectancy to other, dispositional risk factors.

STUDY 1. EXPECTANCY PREDICTS FUTURE ONSET
OF TEEN DRINKING

Method

The method is described in detail by Smith *et al.*[6] Briefly, 353 originally non-drinking seventh graders taken from a multiyear study of teenage drinking were assessed thrice annually over a 24-month period. Subjects came from diverse ethnic and national backgrounds, and their parents' drinking behavior appeared to approximate previously reported figures for the US population. Analyses of dropouts and of new cohorts added to the study during the longitudinal period indicated minor biasing effects due to attrition.

Subjects completed self-report measures of drinking behavior (validated by collateral report) and of the alcohol expectancy most predictive of adolescent drinking, the expectancy for enhanced social experience from drinking. They were tested in their regular classrooms without teachers present and were exposed to extensive reassurances regarding confidentiality.

Results and Discussion

Covariance structure analysis was applied to test the model of expectancy-drinking behavior relations depicted in FIGURE 1. The horizontal arrows in FIGURE 1 depict the simple notion that early expectancies influence later expectancies and Year 2 drinking influences Year 3 drinking. The diagonal arrows reflect the hypothesis that expectancies held at one time point (and at Year 1, before drinking experience) influence drinking behavior reported in subsequent years. The vertical arrows reflect the notion that typical drinking behavior, reported retrospectively during the second and third annual data collections, influences expectancy endorsement at the time of those data collections.

This model fits the data quite well, producing a comparative fit index of .999. Equally important as a test of expectancy theory, the model fit significantly better than did a comparison model that eliminated the two diagonal arrows. In other words, a model in which expectancy did not cause future drinking and was merely an artifact of drinking experience proved inferior to the expectancy-as-cause model (chi-square [df = 2] = 46.372, p <0.001). Additional analyses supported the expectancy-as-mediator of drinking experience notion; expectancies appeared to mediate the influence of early drinking on later drinking in addition to predicting the onset of drinking. In fact, a kind of vicious cycle was observed; originally high-expectancy teens later drank more, which in turn led to still more positive expectancies, which in turn led to still more drinking. In contrast, originally low-expectancy teens remained low in both expectancy and drinking behavior across the 24 months.

Expectancies predate drinking experience, predict subsequent drinking onset, and mediate the influence of early drinking experience on later drinking increases.

STUDY 2. EXPECTANCY MEDIATES
FAMILY DRINKING INFLUENCES

Method

Smith and Goldman[7] describe this study in more depth. Briefly, subjects were 608 high school juniors and seniors from a range of socioeconomic backgrounds. Their

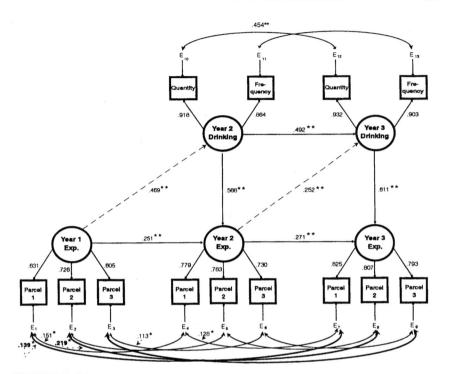

FIGURE 1. Diagram of cognitive up-regulation model of reciprocal expectancy-drinking behavior influence. *Straight lines* represent hypothesized causal influence. *Dotted lines* represent paths tested by a comparison model. *Curved lines* represent associations of undetermined causality. At bottom of figure, *heavier curved lines* represent correlations between Year 1 and Year 3 error terms; *lighter lines* relate error terms from adjacent years. For clarity of presentation, only significant correlations among error terms are presented. Coefficients are generalized least-squares estimates. * p <0.05; ** p <0.001.

parents' drinking behavior appeared broadly representative of the US population as a whole. Following extensive assurances of confidentiality, subjects completed self-report measures of family drinking and of alcohol expectancy. A composite index of family drinking variables was developed to represent parents' likely influence on their teens' consumption. The index included (1) history of alcoholism in first-, second-, and third-degree relatives; (2) history of problems associated with drinking in the same relatives; (3) parents' experience of a set of life problems associated with drinking, including hangovers, blackouts, family problems, work problems, legal difficulties, fighting, early morning drinking, and drinking alone; (4) frequency of both father and mother's drinking; and (5) each parent's attitude toward adult drinking. The expectancy measure was again the expectancy for social enhancement scale taken from the Alcohol Expectancy Questionnaire-Adolescent Form (AEQ-A).[8]

Using a test of mediation described by Baron and Kenny,[9] Smith and Goldman[7] found that alcohol expectancies mediated the impact of the family influence composite on teenage drinking; the decrease in the relation between family influence and teen drinking when expectancy was included in the model was statistically significant. Furthermore, they found that expectancies appear to mediate other, as yet unspecified original causes as well, because expectancy accounted for over 10 times the variance in teen drinking than did the family influence composite. Interestingly, expectancy's mediational influence was partial; with expectancies included, families' independent influence on teen drinking, although smaller (accounting for 1% of the variance), remained statistically significant. If this finding is replicated, it may suggest some small contributors to adolescent drinking that do not operate through the expectancy-learning process. These findings are depicted in FIGURE 2. (One other recent study provided additional support for this mediational hypothesis. In an extensive study of children of alcoholics, Sher et al.[4] found evidence that expectancy mediates the influence of personality variables such as negative affectivity and behavioral undercontrol on alcohol involvement.)

FIGURE 3 is a schematic model that summarizes the findings of these two studies. The model identifies original causes (including family factors) that lead to the proximal cause (expectancy), that in turn initiates a vicious cycle of progressively more drinking and more positive expectancy for drinking's consequences.

STUDY 3. EXPECTANCY, ANTISOCIAL PROPENSITIES, AND SOCIAL ANXIETY

This study[10] proposed a distinction between general, dispositional risk factors, that is, those that might predispose one to risk for a variety of disorders, and alcohol-specific risk factors, or those involving learning about alcohol and its effects.

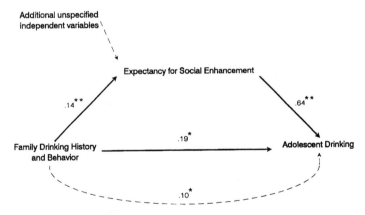

FIGURE 2. A mediational model of alcohol expectancy. *Solid arrows* indicate direct paths, and *curved, dotted arrow* indicates the residual influence of family drinking on adolescent drinking, when family drinking is controlled for the expectancy mediator. * p <0.01; ** p <0.001.

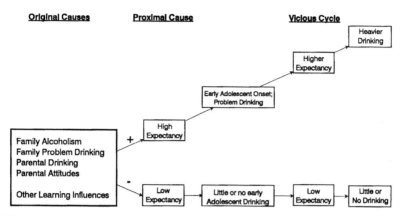

FIGURE 3. Depiction of a Model of Expectancy as (1) mediating early learning influences on later drinking and (2) leading to a vicious cycle of increased drinking during the adolescent years.

Method

Miller and Smith[10] provide a detailed description of the method. Briefly, subjects were 365 undergraduates, ranging in background from upper middle class professional to working class without a family history of higher education. Approximately 70% were white and 30% black. Subjects completed reliable and well-validated measures of three classes of variables: general, dispositional risk was assessed by measures of antisocial propensities and social anxiety; alcohol-specific risk was assessed by family drinking descriptions and by alcohol expectancy for social facilitation; and drinking behavior was assessed by a self-report measure of problem drinking validated by Smith *et al.*[11]

Results and Discussion

Hierarchical multiple regression and other correlational analyses produced the following findings. First, both the general and specific risk factors contributed significantly to the prediction of problem drinking; they combined additively to enhance prediction, producing a multiple correlation of $R = 0.64$, $p < 0.001$. Second, the alcohol-specific risk factors of family modeling and alcohol expectancy accounted for substantially more variance in problem-drinking behavior than did the general risk factors, that is, 27% of the variance *versus* 15% for men and 40% *versus* 14% for women. Third, women demonstrated a significant interaction between the two sets of risk factors ($p < 0.001$). For women in the upper quartile on either social anxiety or antisocial propensities, the correlation between expectancy and problem drinking was significantly higher than that for women in the lower quartile (0.68 *vs* 0.47 for social anxiety; 0.68 *vs* 0.46 for antisocial propensities).

Alcohol-specific learning, acquired through parental modeling and measured by alcohol expectancies, adds to general, dispositional personality risk factors to predict problem drinking in young adults. For women, these two classes of risk factors interact; expectancies for social facilitation are more likely to lead to drinking for women high in personality risk than for women low in personality risk.

CONCLUSIONS

As a currently active, proximal cause of adolescent drinking choices, expectancy is hypothesized to mediate a whole range of potential early causal influences (original causes). Studies 1 and 2 provide clear support for this model. Expectancy mediates family influences on adolescent drinking, mediates the influence of early drinking experience on increased subsequent drinking, and leads to a vicious cycle of increases in expectancy and alcohol consumption during the teenage years. Study 3 begins to explore the relation between expectancy and other risk factors. Expectancy may be an important determinant of whether acting-out propensities or other personality risk factors get expressed as alcohol misuse rather than as some other form of pathology.

SUMMARY

Alcohol expectancy has proven to be a powerful predictor of drinking behavior, including alcoholism, in a wide range of groups. Three recent studies that begin to address expectancy's relation to other alcoholism vulnerability factors are reviewed. Results indicate that: (1) expectancies for reinforcement from alcohol predate teens' first drinking experiences; (2) expectancies predict subsequent drinking onset and problem drinking; (3) high initial expectancies lead to a vicious cycle of progressively more drinking and more positive expectancies during the adolescent years; (4) expectancy mediates the influence of family drinking history on adolescent drinking; and (5) as an alcohol-specific risk factor, expectancy adds to and (in women) interacts with more general, dispositional (personality) risk factors to predict problem drinking in young adults. These findings support the model of expectancy as a mediator of the original causal influences of earlier learning experiences.

REFERENCES

1. TOLMAN, E. G. 1932. Purposive Behavior in Animals and Man. Appleton-Century-Crofts. New York.
2. GOLDMAN, M. S., S. A. BROWN, B. A. CHRISTIANSEN & G. T. SMITH. 1991. Alcoholism etiology and memory: Broadening the scope of alcohol expectancy research. Psychol. Bull. **110:** 137–146.
3. SMITH, G. T. 1989. Expectancy theory and alcohol: The situational insensitivity hypothesis. Psychol. Addict. Behav. **2:** 108–115.
4. SHER, K. J., K. S. WALITZER, P. K. WOOD & E. E. BRENT. 1991. Characteristics of children of alcoholics: Putative risk factors, substance use and abuse, and psychopathology. J. Abnorm. Psychol. **100:** 427–448.
5. SMITH, G. T. & M. S. GOLDMAN. 1993. Alcohol expectancy theory and the identification of high-risk adolescents. J. Res. Adolescence, in press.
6. SMITH, G. T., M. S. GOLDMAN & B. A. CHRISTIANSEN. 1992. The divergent paths of high-expectancy and low-expectancy adolescents. Submitted for publication.
7. SMITH, G. T. & M. S. GOLDMAN. 1990. Toward a mediational model of alcohol expectancy. Paper presented at the annual meeting of the American Psychological Association. Boston, MA.

8. BROWN, S. A., B. A. CHRISTIANSEN & M. S. GOLDMAN. 1987. The alcohol expectancy questionnaire: An instrument for the assessment of adolescent and adult expectancies. J. Stud. Alcohol **48:** 483–491.

9. BARON, R. M. & D. D. KENNY. 1986. The moderator-mediator distinction in social psychological research: Conceptual, strategic, and statistical considerations. J. Personality Soc. Psychol. **27:** 41–57.

10. MILLER, P. M. & G. T. SMITH. 1990. Dispositional and specific learning history risk factors for alcohol abuse. Paper presented at the annual meeting of the American Psychological Association. Boston, MA.

11. SMITH, G. T., M. S. GOLDMAN & B. A. CHRISTIANSEN. 1989. The Drinking Styles Questionnaire: Adolescent drinking self-report. Presented at the annual meeting of the American Psychological Association. New Orleans, LA.

Heterogeneity of Adolescent Alcoholism[a]

RALPH TARTER,[b] LEVENT KIRISCI, ANDREA HEGEDUS,
ADA MEZZICH, AND MICHAEL VANYUKOV

Department of Psychiatry
University of Pittsburgh Medical School
Pittsburgh, Pennsylvania 15213

Evidence marshalled from prospective, adoption, and retrospective investigations indicates that childhood hyperactivity and attentional deficits are associated with a heightened risk for alcoholism.[1] In view of the early age expression of attention deficit disorder and its frequent coexistence with conduct problems, it has been hypothesized that attention/hyperactivity problems are especially salient in cases of alcoholism that have an early age onset.[2]

Complementing the clinical literature, it is noteworthy that behavior activity level and attention/persistence are two well-established temperament traits.[3] High activity level and low attentional capacity have tentatively been observed to comprise temperament features of adolescent alcohol and drug abusers.[4] Direct measures of motor activity level also reveal higher activity under conditions that require behavioral inhibition and sustained vigilance in high risk prepubertal youth compared to youth at average risk.[5]

The early expression (first few weeks or months of life) of temperament traits allows them to be considered from an epigenetic perspective as the foundation of developing behavior. In this perspective, one hypothesized pathway to alcoholism has as its outset position deviations on the temperament traits of activity level and attention span-persistence which, when extreme and occurring in an unfavorable environmental context, culminate in a suprathreshold attention deficit disorder that in turn presages a conduct disorder.[6] Alcohol and other drug initiation is therefore the product of behavioral deviancy which has its roots in the child's temperament makeup.

One constellation of temperament traits that is attracting increased interest comprises the difficult temperament. Individuals presenting with a difficult temperament demonstrate high behavioral activity, low flexibility, mood instability, dysrhythmia, low task orientation, and social withdrawal.[7] This pattern of temperament characteristics in early childhood has been associated with an increased risk for both internalizing and externalizing psychiatric disorders.[8,9] Recent findings accrued from studies of substance abuse families also suggest that a difficult temperament makeup in childhood is associated with parent-child mutual dissatisfaction as well as poor parental disciplinary practices.[10] Furthermore, this constellation of temperament traits appears to mediate to some extent the effects of family substance abuse history on the child's deviant behavior.[11] Finally, a difficult temperament in childhood is associated with an increased risk for an early age involvement with alcohol and tobacco.[12]

[a]This work was supported by Grant AA08746 from the National Institute on Alcohol Abuse and Alcoholism.

[b]Address reprint requests to: Ralph Tarter, Western Psychiatric Institute and Clinic, 3811 O'Hara Street, Pittsburgh, PA 15213.

Systematic inquiry linking childhood temperament disposition to an alcohol and drug abuse outcome has not yet been undertaken. Within the epigenetic perspective, the alcohol or drug abuse outcome is conceptualized as the culmination of a succession of intermediary phenotypic behavioral deviations.[6] Phenotypes that have been shown in numerous studies to precede the expression of alcohol or drug abuse as suprathreshold intermediary outcomes include conduct disorder, anxiety disorder, and affective disorder. The sequential ordering of these endophenotypes, considered within an epigenetic framework, are hypothesized to comprise the myriad developmental trajectories that often culminate in an alcohol or drug abuse disorder.

FIGURE 1 conceptualizes the etiology of alcoholism from the epigenetic perspective. Temperament traits (t), measurable soon after birth, are normally distributed in the population. Through interaction with the environment, complex behaviors are

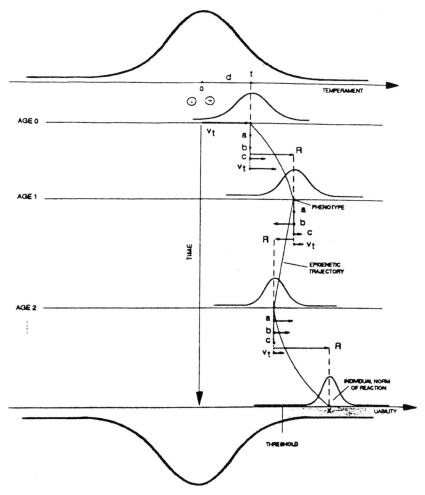

FIGURE 1. Epigenetic perspective of alcoholism etiology.

established. Each behavioral endophenotype influences the development of subsequent behavior. Thus, behavior can be viewed as having the properties of a vector (V_i) in that it is a quantity having both force and direction. Through reciprocal transactions with the environment, behavior is directed towards or away from an alcoholism outcome. Alcoholism, depicted in the shaded area at the bottom of the figure, encompasses individuals in the population who surpass the threshold and hence are defined as affected according to specific diagnostic criteria. As can be seen in FIGURE 1, the developmental trajectory from temperament to alcoholism is not straight. Depending on environmental factors, the trajectory can be oriented towards or away from an alcoholism outcome at different times during the lifespan. However, for purposes of this discussion, the main point to be made is that successive behaviors, considered as an epigenetic process, determine the eventual development of alcoholism. (For a detailed discussion of the developmental behavior genetic perspective of etiology of the alcoholism, the reader is referred to Tarter et al.[6])

STUDY 1

The first study was conducted to determine if adolescent alcoholics can be classified into subgroups according to the internalizing-externalizing behavior dimension. Difficult temperament disposition was hypothesized to load on an externalizing factor, because the component trait deviations are characteristic of children with externalizing disorder.

Subjects. An inpatient sample of 43 adolescents who qualified for a DSM-III-R diagnosis of alcohol abuse was investigated. The characteristics of the sample are summarized in TABLE 1.

Methods. The following variables were employed to classify the subjects along the internalizing-externalizing dimension: (1) Multidimensional Personality Questionnaire subscale scores of positive affectivity, negative affectivity, and constraint[13]; (2) Youth Self-Report externalizing and internalizing scales[14]; (3) Trait Anxiety Scale[15]; (4) Fear Survey[16]; and (5) the difficult temperament index derived from the revised Dimensions of Temperament Scale.[17] The Drug Use Screening Inventory

TABLE 1. Demographic Characteristics of Alcoholic Adolescents ($n = 43$)

	M	(s)
Age (yr)	15.74	(1.47)
Grade level	9.12	(1.72)
Socioeconomic status	3.47	(1.32)

	N	(%)
Ethnicity		
White	32	74.4
African-American	10	23.3
Other	1	2.3
Gender	17	39.5
Male		
Female	26	60.5

(DUSI)[18] was the dependent measure employed to determine if the obtained subtypes could be discriminated on substance use and related disorders. The DUSI is a self-report 149-item questionnaire that quantifies problem severity in 10 domains: (1) substance use; (2) psychiatric disorder; (3) behavior patterns; (4) health status; (5) social competence; (6) work; (7) school; (8) family; (9) peer relationships; and (10) leisure and recreation.

Results. A principal components factor analysis with varimax rotation was first conducted to extract factors in order to simplify and identify the structures of the multivariate observations. These factor scores are presented in TABLE 2. Four factors were identified: Factor I (Fear); Factor II (Negative Affect); Factor III (Behavior Dyscontrol); and Factor IV (Hypophoria). The four factors explained 38.7, 13.5, 10.7, and 8.2% of the total variance.

Ward's[19] minimum variance hierarchical agglomerative method with squared euclidian distance was employed to determine if the subjects comprised one or more subtypes of adolescent alcoholism. This analysis yielded two clusters. Cluster 1 contained 10 subjects (23.3% of sample), whereas cluster 2 contained 33 subjects (76.7% of the sample). As can be seen in TABLE 3, an opposite pattern of scores across the four factors is observed for clusters 1 and 2 (Pillais trace statistic = 1.83, $p <0.001$). Cluster 1 subjects are high in fear and negative affect, whereas cluster 2 is featured by dyscontrol and hypophoria.

Cluster 1 subjects compared to cluster 2 subjects had a higher mean age of onset of first substance use and older age at which they qualified for a diagnosis of substance abuse. The two clusters did not differ with respect to age of first psychiatric diagnosis, gender, or family history of alcoholism. These results are summarized in TABLE 4.

FIGURE 2 shows that on the Drug Use Screening Inventory, cluster 1 subjects had less severe involvement with alcohol and drugs (F = 14.79, $p <0.001$), less severe behavior disorder (F = 18.85, $p <0.001$), and less severe psychiatric disorder (F = 7.80, $p <0.005$). The overall problem density score, reflecting the mean score across all 10 domains, also discriminated the two clusters (F = 10.02, $p <0.002$).

TABLE 2. Factor Structure and Loadings of the Alcoholic Subjects

	Factor I: Fear	Factor II: Negative Affect	Factor III: Dyscontrol	Factor IV: Hypophoria
Positive affectivity	−.12851	.07195	.26723	−.69894
Negative affectivity	.19259	.75656	.28505	−.15727
Constraint	.21684	.19120	.83756	−.21553
State anxiety	.16570	.76728	.12550	.13030
Internalizing	.29502	.74862	−.18326	.00387
Externalizing	.17601	.55941	−.66311	−.12495
Fear/failure/criticism	.87857	.16591	.13783	.07924
Fear/unknown	.89446	.17685	−.00900	.90856
Fear/injury/animals	.91657	.21238	.07449	.08240
Fear/danger/death	.83750	.29235	.16353	.03493
Medical fears	.81548	.23069	.05846	−.07780
Difficult temperament (Index)	.29258	.68173	−.06662	.22901
Illegal drug	−.12212	−.02106	−.58682	−.05287
Suicide attempt	−.00426	.19352	.21654	.74412

TABLE 3. Mean and Standard Deviation Scores of Factor Loadings for Each Cluster

	Factor I: Fear		Factor II: Negative Affect		Factor III: Dyscontrol		Factor IV: Hypophoria	
	M	(s)	M	(s)	M	(s)	M	(s)
Cluster 1	.90882	(1.450)	.33048	(.882)	−.95846	(1.099)	−.53480	(.739)
Cluster 2	−.35616	(.560)	−.08682	(1.095)	.28057	(.863)	.12428	(1.018)

Discussion. The main finding to emerge from this study is that based on the internalizing and externalizing behavior dimension, adolescent alcoholics can be classified into two general groups that differ according to age of first use of substances, age of first diagnosis, and severity of problems. This result indicates that alcohol abuse of early age onset is a heterogeneous disorder. The majority of subjects are features by behavior dysregulation and a minority featured primarily by negative affect. Surprisingly, the difficult temperament index loaded on the negative affect factor and not, as hypothesized, on the behavior dyscontrol factor. The next study determined the extent to which difficult temperament preceded the onset of psychiatric disorders in adolescent alcoholics.

STUDY 2

Subjects. The subjects were 48 adolescents who qualified for a DSM-III-R diagnosis of either alcohol or drug abuse. Diagnoses were made by a board-certified

TABLE 4. Differences between Two Adolescent Alcoholic Clusters According to Age of Onset of Substance Use and Age to First Diagnosis for Psychoactive Substance Use Disorder

	Cluster 1: Anxious/Negative Affect ($n = 10$)		Cluster 2: Disinhibited/Hypophoric ($n = 33$)			
	M	(s)	M	(s)	F	p
Age of first substance use diagnosis	13.43	(2.16)	11.85	(1.68)	5.54	.02
Age of first substance use	12.03	(2.63)	10.00	(2.86)	5.14	.03
Age of first psychiatric diagnosis	9.45	(3.94)	8.08	(3.77)	1.11	ns
	N	%	N	%	Chi^2	p
Female gender	3	30.0	14	42.4	.11	NS
Alcoholic parents (mother and father or both)	5	50.0	8	24.2	.17	NS

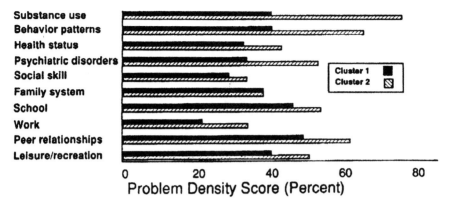

FIGURE 2. Differences between anxious/negative (cluster 1) and disinhibited/hypophoric (cluster 2) adolescent alcoholics on the Drug Use Screening Inventory.

psychiatrist employing an expanded version of the Kiddie Schedule for Affective Disorders and Schizophrenia.[20] All of the subjects were in treatment for a substance abuse disorder at Western Psychiatric Institute and Clinic. The research protocol was initiated only after detoxification was completed. TABLE 5 summarizes the personal and demographic characteristics of the sample.

Method. The subjects completed the Dimensions of Temperament Scale—Revised (DOTS-R)[13] from which the difficult temperament index was derived. This index aggregates the temperament traits of flexibility/rigidity, activity level, mood, rhythmicity, task persistence, and social approach/withdrawal.

Results. FIGURE 3 summarizes the results of an analysis of the conditional probabilities. From among the total sample of adolescent alcoholics, 35% were classified as having a difficult temperament. The probability of developing an attention deficit disorder with hyperactivity (ADHD) if the child had a difficult temperament was 0.22. For conduct disorder, anxiety, and depression, the probability values were 0.22, 0.28, and 0.11, respectively. As can be seen in FIGURE 3, the probability of manifest-

TABLE 5. Characteristics of Alcoholic Adolescents

Characteristics	M	(s)
Age (yr)	15.69	(1.55)
Grade level	9.08	(1.71)
Socioeconomic Status	3.47	(1.32)

	N	(%)
Ethnicity		
White	36	(7.5)
African-American	12	(2.5)
Sex		
Female	22	(45.8)
Male	26	(54.2)

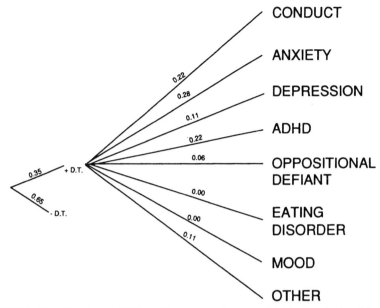

FIGURE 3. Conditional probabilities of temperament and psychiatric disorder in adolescent alcoholics.

ing other disorders as the first diagnosis is either zero or very low; however, these results must be interpreted tentatively in light of the small sample size and the low prevalence of these conditions.

Discussion. Two findings emerged from this analysis. First, a difficult temperament is present in only a minority of cases of adolescent substance abusers. Second, among those having this temperament disposition, there is a variable pattern of first psychiatric diagnosis; conduct disorder, anxiety disorder, and attention deficit disorder each occurs with approximately the same probability.

DISCUSSION

The results from these two studies underscore the substantial developmental heterogeneity of the adolescent substance abuse population. Caution needs to be exercised, therefore, before presuming that early age onset alcoholism invariably has an antisocial presentation. Whereas this may be the most common end product of an externalizing behavior disposition, the present results also point to an internalizing variant that expresses primarily as a disorder of negative affect.

Approximately 35% of adolescent substance abusers were classified as having a difficult temperament in early childhood; this prevalence level is consistent with prospective studies of psychopathology in children who have a difficult temperament.[9] Moreover, this temperament disposition did not predictably lead to a specific type of psychiatric disorder. Significantly, the difficult temperament index loaded most highly on a negative affect factor. However, inasmuch as negative affect is comorbid to a number of different psychiatric diagnoses, it cannot be concluded that this temperament disposition predisposes to one specific psychopathological disorder or one particular variant of alcohol abuse. Hence, it appears that there are many developmental pathways to a substance abuse outcome when there is the disposition of a difficult temperament.

It is salient to note that a difficult temperament predates the onset of substance abuse in approximately 35% of cases. This observation should not be interpreted to mean that temperament is unimportant in the remaining 65% of cases. Rather, this finding indicates only that specifically a difficult temperament is not present; other temperament traits or configurations of traits may potentially be important antecedents to substance abuse.

In conclusion, referring to FIGURE 1, it can be seen that the pathways to early age onset alcohol abuse are multiply determined and varied. A difficult temperament disposition presages a number of different psychiatric diagnoses which, in turn, culminate in alcohol abuse. Whether the difficult temperament disposition has greater specificity in identifying a particular subset of the adult alcoholic population who eventually qualify for a diagnosis of dependence remains unknown. It is possible that this temperament configuration in childhood predisposes to adult alcoholism with antisocial propensities; however, prospective research is required to address this issue. The present data indicate that the difficult temperament constellation does not lead to one particular variant of adolescent alcohol abuse even though, as shown in study 1, this temperament makeup is most closely associated with negative affectivity.

SUMMARY

Adolescent alcoholics in treatment were classified into two broad clusters. The larger group consisted of youth who demonstrated behavioral dyscontrol and hypophoria, whereas the smaller group consisted of individuals with primarily negative affect. These two clusters differed with respect to age of first drug use, age of first substance abuse diagnosis, severity of substance use, behavior disturbance, and psychiatric disorder. Difficult temperament loaded on the negative affect factor.

In a second study, it was observed that difficult childhood temperament led to a number of different first diagnoses in the adolescents. No specific psychiatric outcome followed a difficult temperament.

These results illustrate the heterogeneity of the adolescent alcoholic population and suggest that there are many developmental pathways to this outcome where a predisposing risk factor is a difficult temperament makeup.

REFERENCES

1. TARTER, R. A., A. ALTERMAN & K. EDWARDS. 1985. Vulnerability to alcoholism in men: A behavior genetic perspective. J. Stud. Alcohol **46:** 329–356.
2. ALTERMAN, A., R. TARTER, T. BAUGHNAN, B. BOBER & S. FABIAN. 1985. Differentiation of alcoholics high and low in childhood hyperactivity. Drug Alcohol Depend. **15:** 111–121.
3. BUSS, A. & R. PLOMIN. 1975. A Temperament Theory of Personality Development. Wiley. New York.
4. TARTER, R., S. LAIRD, M. KABENE, O. BUKSTEIN & Y. KAMINER. 1990. Drug abuse severity in adolescents is associated with magnitude of deviation in temperament traits. Br. J. Addict. **85:** 1501–1504.

5. Moss, H., T. Blackson, C. Martin & R. Tarter. 1992. Heightened motor activity level in male offspring of substance abusing fathers. Biol. Psychiat. **32:** 1135–1147.
6. Tarter, R., H. Moss & M. Vanyukov. 1994. Behavior genetics and the etiology of alcoholism. *In* Alcohol and Alcoholism. H. Begleiter & H. Kissin, eds. Vol. 1. Oxford University Press. New York. In press.
7. Thomas, A. & S. Chess. 1984. Genesis and evolution of behavioral disorders. From infancy to early adult life. Am. J. Psychiatry. **140:** 1–8.
8. Maziade, M., C. Caron, R. Cote, C. Merrette, H. Bernier, B. LaPlante, P. Boutin & J. Thiverge. 1990. Psychiatric status of adolescents who had extreme temperaments at age 7. Am. J. Psych. **147:** 1531–1537.
9. Maziade, M., C. Caron, C. Cote, P. Boutin & J. Thivierge. 1990. Extreme temperament and diagnosis. A study in a psychiatric sample of consecutive children. Arch. Gen. Psych. **47:** 477–484.
10. Tarter, R., T. Blackson, C. Martin, R. Loeber & H. Moss. 1993. Characteristics and correlates of child discipline practices in substance abuse and normal families. Am. J. Addict. **2:** 18–25.
11. Blackson, T., R. Tarter, C. Martin & H. Moss. 1993. Temperament mediates the effects of family history of substance abuse on externalizing and internalizing child behavior. Am. J. Addict., in press.
12. Lerner, J. & J. Vicary. 1984. Difficult temperament and drug use: Analysis from the New York Longitudinal Study. J. Drug Educ. **14:** 1–8.
13. Tellegen, A. 1982. Manual for the Multidimensional Personality Questionnaire. Department of Psychiatry, University of Minnesota. Minneapolis, MN. Unpublished manuscript.
14. Achenbach, T. 1991. Manual for the Youth Self-Report. Department of Psychiatry, University of Vermont. Burlington, VT.
15. Spielberger, C. 1973. Preliminary test manual for the State-Trait Anxiety Inventory for Children. Consulting Psychologists Press. Palo Alto, CA.
16. Ollendick, T. 1983. Reliability and validity of the revised fear survey schedule for children (FSSC-R). Behav. Res. Ther. **6:** 685–692.
17. Windle, M., K. Hucker, K. Lernerz, P. East, J. Lerner & R. Lerner. 1986. Temperament, perceived competence, and depression in early and late adolescence. Dev. Psychol. **22:** 384–392.
18. Tarter, R. 1990. Evaluation and treatment of adolescent substance abuse. A decision tree method. Am. J. Drug Alcohol Abuse **16:** 1–46.
19. Everitt, B. 1974. Cluster Analysis. Wiley. New York.

Anxiety and Conduct Disorders in Early Onset Alcoholism[a]

DUNCAN B. CLARK,[b] ROLF G. JACOB,

AND ADA MEZZICH

Pittsburgh Adolescent Alcohol Research Center
Department of Psychiatry
University of Pittsburgh
Pittsburgh, Pennsylvania 15213

Epidemiologic and clinical studies show that alcohol disorders,[1] antisocial behavioral disorders,[2,3] and anxiety disorders[4] often have an onset during or before adolescence. Although the study of adolescents may reveal important insights concerning the development of these disorders, there is little published data from adolescents themselves on the relationships between these disorders.

Antisocial characteristics and anxiety have been thought to contribute independently to alcohol disorders.[5] For example, Cloninger[6] and Cloninger *et al.*[7] hypothesize two subtypes of alcoholics: Type 1 alcoholics are described as having anxious personality traits with high harm avoidance, high reward dependence, and low novelty seeking, and Type 2 alcoholics are described as having antisocial personality traits with low harm avoidance, low reward dependence, and high novelty seeking. Type 2 alcoholism is considered to be of earlier onset than Type 1. Findings from psychophysiological studies are consistent with the validity of this classification system with respect to harm avoidance, as sociopaths show diminished anticipatory fear responses and individuals with anxiety disorders have enhanced responses.[8] Such models imply that indicators of antisocial behaviors and anxiety would be inversely correlated and suggest that anxiety would be a less common factor in early onset alcoholism.

Gray[9,10] proposed that conduct disorder may be viewed as the result of (1) excessive activation of a behavioral activation system, mediating consummatory and related aggressive behavior through the medial forebrain bundle and the lateral hypothalamus, and (2) deficient activity of a behavioral inhibition system, mediating the inhibition of behavior in the presence of cues signaling punishment through the septal-hippocampal system, the Papez circuit, locus coeruleus, raphe nuclei, and prefrontal cortex. Increased activity in the behavioral inhibition system is hypothesized to mediate anxiety. With reference to this model, Walker and colleagues[10] reasoned that children who exhibit conduct disorder with significant anxiety symptoms would engage in less antisocial behavior than would those without significant anxiety. In their study of boys aged 7–12 years, subjects with conduct disorder without anxiety disorders had more police contacts and more school suspensions than did those with conduct disorder and separation anxiety or overanxious disorder.

The present study had three purposes: (1) to determine the frequency of conduct disorder, anxiety disorders, and their co-occurrence in adolescents with alcohol abuse or dependence; (2) to determine the severity of anxiety symptoms, depression symptoms, history of suicidal behavior, and psychosocial impairment associated with these disorders; and (3) to test the hypothesis that conduct disorder cases with

[a]This work was supported by grants from the NIAAA (AA08746) and NIMH (MH19816).
[b]Address for correspondence: 3811 O'Hara Street, Pittsburgh, PA 15213.

181

anxiety disorders have less psychosocial impairment than do conduct disorder cases without anxiety disorders.

METHODS

The subjects were 48 adolescents between 12 and 18 years old identified primarily by participation in inpatient treatment programs as having a history of alcohol dependence or abuse, and 40 adolescents without psychiatric disorders recruited as volunteers from the community. Informed consent was obtained. In addition to describing diagnoses in the cases with a lifetime history of alcohol abuse or dependence, we also compared three groups with respect to psychiatric symptoms and psychosocial impairment: (1) current alcohol abuse or dependence, conduct disorder, and anxiety disorder (CD + AD: n = 18); (2) current alcohol abuse or dependence and conduct disorder without anxiety disorder (CD − AD: n = 21); and (3) normal controls without current psychiatric disorders (NC: n = 40).

The primary assessment instrument for psychiatric diagnoses was a modified version of the Schedule for Affective Disorders and Schizophrenia for School Age Children (K-SADS), utilizing aspects of several versions and adding additional items[11–13] and the Lifetime History of Alcohol and Drug Use Questionnaire.[14] The K-SADS interview was expanded to include lifetime and current assessment of all DSM III-R child and adult anxiety disorder diagnoses. Current diagnoses referred to the last 6 months. The K-SADS interview was administered to both the adolescent and to one parent with reference to the child by a trained clinical interviewer. Evaluation was performed a minimum of 2 weeks after the last use of alcohol and in the absence of withdrawal symptoms. Interviews were reviewed and psychiatric diagnoses were confirmed by clinically experienced faculty members of the Department of Psychiatry.

An interview rating of anxiety was performed using the Hamilton Anxiety Rating Scale.[15] The Beck Depression Inventory[16] was used as a continuous measure of depression. Suicidal ideas and attempts were measured by a screening interview. Information on areas of psychosocial impairment was obtained by the Drug Use Screening Inventory (DUSI).[17] This comprehensive self-report assessment was designed to determine the functioning of adolescents in 10 domains: (1) substance use; (2) behavior patterns; (3) health status; (4) psychiatric disorder; (5) social skills; (6) family system; (7) school adjustment; (8) work; (9) peer relationships; and (10) recreation. Each domain consists of 10–20 true-false items. Three types of indices are obtained: (1) Problem Density in each domain; (2) Relative Problem Density; and (3) Summary Problem Index. The Problem Density, which consists of the number of "yes" endorsements in each problem domain divided by the number of items and multiplied by 100, will be used for this analysis. When there were directional hypotheses, one-tailed significance tests were used for paired comparisons.

RESULTS

From the initial sample of 48 adolescents with a lifetime history of alcohol abuse or dependence, 39 (81%) had a lifetime history of conduct disorder and 24 (50%) had anxiety disorder diagnoses. The most common anxiety disorder diagnosis was Post-traumatic Stress Disorder (PTSD: 12 cases, 25%), with abuse by family members being the most frequent trauma. Other common anxiety disorders were Social Phobia (*n* = 10, 21%), Separation Anxiety Disorder (*n* = 9, 19%), Avoidant Disorder of Childhood or Adolescence (*n* = 8, 17%), and Agoraphobia without a history of panic disorder (*n* = 7, 15%). In most of these cases, the onset of the anxiety disorder preceded alcohol abuse or dependence. Of those cases with Conduct Disorder (CD), 18 of 39 (46%) had a diagnosis of anxiety disorder.

The three groups compared (CD+AD, CD−AD, and NC) differed in age (16.0 ± 1.5, 15.5 ± 1.5, and 14.6 ± 1.8 years, respectively; F = 4.2, p <0.01) and socioeconomic status (SES: 3.6 ± 1.3, 3.3 ± 1.6, and 2.0 ± 1.1; F = 11.6, p <0.001), so ANOVA analyses were done with age and SES as covariates.

As expected, these groups were significantly different on the Hamilton Anxiety Rating Scale (12.3 ± 8.4, 7.4 ± 6.5, and 1.9 ± 2.2; F = 9.1, p <0.001). Pairwise comparisons showed the CD+AD and CD−AD had higher anxiety scores than did the NC group (p <0.005, one-tailed tests), and CD+AD had higher anxiety scores than did CD−AD (p <0.05, one-tailed test). These groups were also significantly different on the Beck Depression Inventory (11.9 ± 9.8, 11.8 ± 10.2, and 3.3 ± 3.9; F = 12.4 p <0.001). Pairwise comparisons showed that CD+AD and CD−AD had higher scores than did NC (p <0.001, one-tailed tests), and CD+AD and CD−AD were not significantly different. With regard to suicidal ideas and attempts, the groups were significantly different on all variables, including a history of a wish to die (72%, 48%, 2%; chi-square = 32.6, p <0.001), thoughts to kill self (67%, 43%, 10%; chi-square = 20.1, p <0.001), plan to kill self (56%, 33%, 0%; chi-square = 25.0, p <0.001), and suicide attempts (50%, 38%, 0%; chi-square = 23.0, p <0.001). Pairwise comparisons showed that CD+AD and CD−AD cases more frequently reported all categories of suicidal thoughts and behavior than did NC cases (p <0.01). The CD+AD and CD−AD groups were not significantly different in any suicide variables.

Analyses of the 10 DUSI impairment subscales showed significant differences between groups on all scales (TABLE 1). Pairwise comparisons showed that impairment in the CD+AD and CD−AD groups was greater than that in the NC group on every scale. Consistent with our prediction, the CD+AD group was significantly less impaired than the CD−AD group on several subscales, including substance abuse behavior, school problems, peer relationships, and overall problem density. On the other hand, these groups were not significantly different in other problem areas, including behavior patterns, health status, psychiatric symptoms, social skills, family system, work, and recreation.

DISCUSSION

The study found that anxiety disorders are common in adolescents with early onset alcoholism. Nearly half the adolescents with alcohol abuse or dependence and conduct disorder also had an anxiety disorder diagnosis. In addition, adolescents with alcohol abuse or dependence and conduct disorder with no anxiety disorder diagnosis had relatively high anxiety compared with a control group. These results suggest that anxiety symptoms are common in conduct disorder adolescents with early onset alcoholism.

Other studies similarly found anxiety and antisocial characteristics to be associated. In a study of adult Type I and Type II alcoholics, von Knorring and colleagues[18] found that both groups had high psychic and somatic anxiety. In fact, Type II alcoholics had significantly higher scores than did Type I alcoholics on somatic anxiety. Aggression was found to be associated with anxiety in children and ado-

TABLE 1. Drug Use Screening Inventory (DUSI) Subscales by Diagnostic Group

DUSI Variable	CD+AD	CD–AD	NC	Overall F	Pairwise, One-Tailed
1. Substance use behavior	41.9 ± 6.7	57.4 ± 5.4	3.7 ± 4.3	27.4***	1>3***; 2>3***; 1<2*
2. Behavior patterns	42.8 ± 5.9	51.5 ± 4.8	18.3 ± 3.8	13.5***	1>3***; 2>3***
3. Health status	36.0 ± 4.5	32.8 ± 3.6	14.6 ± 2.9	9.2**	1>3***; 2>3***
4. Psychiatric disorders	37.3 ± 5.6	42.8 ± 4.5	14.0 ± 3.6	11.5***	1>3***; 2>3***
5. Social skills	42.8 ± 5.9	51.5 ± 4.8	18.3 ± 3.8	13.5***	1>3***; 2>3***
6. Family system	32.6 ± 5.4	41.9 ± 4.4	13.3 ± 3.5	11.8***	1>3***; 2>3***
7. School problems	38.9 ± 5.4	51.2 ± 4.4	13.9 ± 3.5	19.8***	1>3***; 2>3***; 1<2*
8. Work problems	24.5 ± 4.4	26.9 ± 3.6	4.2 ± 2.8	11.9***	1>3***; 2>3***
9. Peer relationships	40.5 ± 6.2	62.8 ± 5.0	19.5 ± 4.0	20.6***	1>3*; 2>3***; 1<2**
10. Recreation	37.3 ± 5.4	43.1 ± 4.4	18.8 ± 3.5	8.8***	1>3**; 2>3***
Overall problem density	37.1 ± 3.5	45.1 ± 2.8	13.6 ± 2.2	34.5***	1>3***; 2>3***; 1<2*

NOTE: DUSI subscale scores are given, with ANCoVA result and pairwise comparisons, for subjects with Alcohol Abuse or Dependence, Conduct Disorder and Anxiety Disorders (CD+AD), subjects with Alcohol Abuse or Dependence, Conduct Disorder without Anxiety Disorders (CD–AD), and normal control subjects (NC). *$p <0.05$; **$p <0.01$; ***$p <0.001$.

lescents. In a study of 8-, 12-, and 17-year-olds,[19] high ratings of physical aggression were associated with high anxiety ratings.

Morbidity in these adolescents is also illustrated by relatively high levels of depression and suicidal behaviors. Of the adolescents with alcohol abuse or dependence, conduct disorder, and anxiety disorders, 72% reported a wish to die, 67% reported suicidal thoughts, and 50% reported a suicide attempt. As most were recruited from inpatient services, these subjects may have had a higher rate of suicide attempts than would other subgroups of adolescents with alcohol abuse or dependence. This group also had a high prevalence of PTSD, which was present in 25% of cases.

There was some evidence to support the hypothesis that anxiety disorders are associated with less social impairment in cases of conduct disorder. Cases of anxiety disorders reported less problems with substance use behavior, school adjustment, and peer relationships as well as lower overall problem density on the DUSI. These findings are consistent with those of Walker and colleagues.[10] However, the statistical significance of the results was marginal in some cases, and these findings need confirmation with a larger sample.

In summary, adolescents with alcohol abuse or dependence and simultaneous conduct disorder frequently had anxiety disorders and relatively high levels of anxiety symptoms. The combination of alcohol abuse or dependence, conduct disorder, and anxiety disorders was associated with significant morbidity. However, the presence of anxiety disorders had a positive moderating influence in some areas.

REFERENCES

1. HELZER, J. E., A. BURNAM & L. T. McEVOY. 1991. Alcohol abuse and dependence. *In* Psychiatric Disorders in America: The Epidemiologic Catchment Area Study. L. N. Robins & D. A. Regier, eds. Free Press. New York.
2. ROBINS, L. N., J. TIPP & T. PRZYBECK. 1991. Antisocial personality. *In* Psychiatric Disorders in America: The Epidemiologic Catchment Area Study. L. N. Robins & D. A. Regier, eds. Free Press. New York.
3. KELLER, M. B., P. W. LAVORI, W. R. BEARDSLEE, J. WUNDER, C. E. SCHWARTZ, J. ROTH & J. BIEDERMAN. 1992. The disruptive behavioral disorder in children and adolescents: Comorbidity and clinical course. J. Am. Acad. Child Adolesc. Psychiatry **31:** 204–209.
4. BERNSTEIN, G. A. & C. M. BORCHARDT. 1991. Anxiety disorders of childhood and adolescents. A critical review. J. Am. Acad. Child Adolesc. Psychiatry **30:** 519–532.
5. CLARK, D. B. & M. SAYETTE. 1993. Anxiety and the development of alcoholism: Clinical and scientific issues. Am. J. Addict. **2:** 59–76.
6. CLONINGER, C. R. 1987. Neurogenetic adaptive mechanisms in alcoholism. Science **236:** 410–416.
7. CLONINGER, C. R., S. H. DINWIDDIE & T. REICH. 1989. Epidemiology and genetics of alcoholism. *In* Review of Psychiatry. A. Tasman, R. E. Hales & A. J. Frances, eds. Vol. **8:** 293–308. American Psychiatric Press. Washington, DC.
8. JACOB, R. G., C. BRODBECK & D. B. CLARK. 1992. Physiological and behavioral assessment. *In* Research in Psychiatry: Issues, Strategies and Methods. G. L. Hsu & M. Hersen, eds. Plenum Press. New York.
9. GRAY, J. 1987. The Psychology of Fear and Stress. Cambridge University Press. Cambridge, MA.
10. WALKER, J. L., B. B. LAHEY, M. F. RUSSO, P. J. FRICK, M. A. CHRIST, K. McBURNETT, R. LOEBER, M. STOUTHAMER-LOEBER & S. M. GREEN. 1991. Anxiety, inhibition, and conduct disorder in children. I. Relations to social impairment. J. Am. Acad. Child Adolesc. Psychiatry **30:** 187–191.
11. PUIG-ANTICH, J., H. ORVASCHEL, M. A. TABRIXI & W. J. CHAMBERG. 1981. The Schedule

for Affective Disorders and Schizophrenia for School Age Children, Epidemiologic Version (Kiddie-SADS-E), 3rd Ed. Department of Child and Adolescent Psychiatry, New York State Psychiatric Institute. New York.

12. LAST, C. G. 1986. Modification of the K-SADS-P for use with anxiety disordered children and adolescents. Unpublished manuscript.

13. ORVASCHEL, H., J. PUIG-ANTICH, W. CHAMBERS, M. A. TABRIZI & R. JOHNSON. 1982. Retrospective assessment of prepubertal Major Depression with the Kiddie-SADS-E. J. Am. Acad. Child Adolesc. Psychiatry 21: 392–397.

14. SKINNER, H. A. 1982. Development and validation of a lifetime alcohol consumption assessment procedure. Final report, Health and Welfare Grant #6606-1693-42.

15. HAMILTON, M. 1959. The assessment of anxiety states by rating. Br. J. Med. Psychol. 32: 50–55.

16. BECK, A. T., R. A. STEER & M. S. GARBIN. 1988. Psychometric properties of the Beck Depression Inventory: Twenty-five years of evaluation. Clin. Psychol. Rev. 8: 77–100.

17. TARTER, R. E. 1990. Evaluation and treatment of adolescent substance abuse: A decision tree model. J. Drug Alcohol Abuse 16: 1–46.

18. VON KNORRING, L., Å. VON KNORRING, L. SMIGAN, U. LINDBERG & M. EDHOLM. 1987. Personality traits in subtypes of alcoholics. J. Stud. Alcohol 48: 523–527.

19. KASHANI, J. H., W. DEUSER & J. C. REID. 1991. Aggression and anxiety: A new look at an old notion. J. Am. Acad. Child Adolesc. Psychiatry 30: 218–223.

Alcohol and Drug Abusers Subtyped by Antisocial Personality and Primary or Secondary Depressive Disorder[a,b,c]

ELIZABETH E. EPSTEIN,[d] BENSON E. GINSBURG,[e,f,g]
VICTOR M. HESSELBROCK,[f] AND J. CONRAD SCHWARZ[g]

[d]Center of Alcohol Studies
Rutgers—The State University of New Jersey
Piscataway, New Jersey 08855

[e]Biobehavioral Sciences
University of Connecticut
Storrs, Connecticut 06269

[f]Department of Psychiatry
University of Connecticut Health Center
Farmington, Connecticut 06030

[g]Department of Psychology
University of Connecticut
Storrs, Connecticut 06269

Both antisocial personality disorder (ASP) and depression can be viewed as disorders of affect. Antisocials have been shown to exhibit and perhaps experience limited affect or arousal,[1,2] whereas individuals with depression seem to have difficulty controlling their experience of negative affect such as anxiety and sadness. The relationship between ASP and depression, both of which are overrepresented among substance abusers,[3,4] is an intriguing research question in relation to substance abuse subtypes.

Unidimensional typological models based on psychiatric disorders that are co-morbid with substance use disorders have typically focused on ASP or depression as monothetic subtypes.[5,6] Overlap between the two disorders in substance-abusing populations, and affective profiles that may underlie these psychiatric disorders have not been studied extensively. It may be that people with ASP and people with depression each represent genetically and phenotypically distinct substance abuse subtypes. However, some studies have combined depressed and nondepressed ASP subjects into a single group[7,8] or have been unclear about the method used to classify subjects vis-à-vis multiple comorbid disorders. Only a small number of unidimensional studies[9–11] have addressed the issue of co-occurrence, despite data suggesting that ASP and depressive symptoms can aggregate in the same individual[7,12–14] and despite the theoretical distinctions between the two disorders.

The more complex multidimensional typologies that have been proposed[15–17]

[a]This study was funded by grants from the University of Connecticut Research Foundation and by a Grant-in-Aid of Research from Sigma Xi and an APA Dissertation Research Award to the first author.
[b]This manuscript is based on dissertation research of the first author at the University of Connecticut Department of Psychology.
[c]Second, third, and fourth authors are listed in alphabetic order.

typically include some combination of comorbid psychopathology, family history, patterns of drinking, and premorbid vulnerability factors. The co-occurrence of ASP and depression is directly addressed in only one of three of the most prominent multidimensional typologies. Read et al.'s[18] bidimensional model is based on family history of psychopathology and psychiatric comorbidity in adulthood. In some studies[8] this research group has attempted to separate subjects with ASP from depressed subjects. Cloninger's Type I/Type II schema has been roughly equated to an ASP (Type II) versus a non-ASP (Type I) distinction.[19] Type I alcoholics are hypothesized to score higher on measures of traits such as harm avoidance and anxiety, which are related to depression, and Type II alcoholics are considered to be more antisocial and to score higher on measures of novelty-seeking traits. However, the relationship between ASP and depression is not addressed in Cloninger's typology. For example, would it be possible for an individual with ASP and a history of depression to have elevated scores on harm avoidance, and if so, how would this person fit into the Type I/Type II schema? Babor et al.'s[15] Type A/Type B multidimensional typology describes subjects with elevated rates of psychopathology, including ASP, depression, and anxiety, as Type B. The relationship among these three disorders is not addressed. However, Babor et al.[15] suggest that in future research, variants of ASPs (such as ASPs with no history of depression) within the Type B classification be examined.

A careful look at the relationship between ASP and depression in substance abuse subtypes requires attention to the primary/secondary distinction in diagnosing psychopathology. Schuckit[20] and others[21,22] have stressed the need to differentiate between primary and secondary disorders in establishing subtypes, but their methods have generally not yet been incorporated into research on multidimensional typologies.

The current study was designed to test if two unidimensional subtyping schemas based on (1) ASP and (2) depressive disorder could be used to differentiate a substance-abusing population according to a "low" (or novelty seeking, as used here) versus "high" (anxious/depressed) arousal personality. We reasoned that low arousal personality would be correlated with a dearth of affectivity that would manifest itself in novelty seeking and reckless behavior. A high arousal personality would be characterized by anxiety, emotionality, and sadness. We then tested two different multidimensional (in the sense of multisymptomatic) schemas based on combinations of (1) ASP and depressive disorder, and (2) ASP and more refined diagnoses of primary versus secondary depressive disorder. This schema incorporates theoretical considerations suggested by Tarter et al.[23] and Read et al.,[18] on underlying affect-related personality characteristics, and by Guze[21] and others,[20,24] on the importance of distinguishing between primary and secondary disorders. To this end, we employed a multivariate research strategy to test whether a multisymptomatic (based on co-occurrence of ASP and primary or secondary depression) typological schema identifies groups of substance abusers with unique personality and affective profiles.

We predicted that personality characteristics in the affective domain for ASP alcohol/substance abusers would be high sensation seeking, high novelty seeking, low anxiety, low depression, and low reward dependence. Compared with the depressive substance abusers, these ASP substance abusers were also predicted to have a higher prevalence of polysubstance abuse and to view danger as exciting and violence as appealing. High sensation seeking and novelty seeking are assumed to be associated with a "low arousal" affective profile.

We hypothesized that affect-related personality correlates for depressed substance abusers would be a "high arousal" pattern: high trait anxiety, high harm avoidance, and low sensation seeking (people who are already anxious would be less likely, according to this view, to seek experiences that would heighten already uncomfortable levels of emotionality).

We further predicted that the "ASP only" subgroup would be differentiated from an ASP/secondary depressed group, which would in turn look different from an ASP/primary depressed group, on measures of affect-related personality characteristics.

METHOD

Subjects

Data were collected from 83 male inpatients from various alcohol and drug rehabilitation programs and dual-diagnosis units in Connecticut and New Jersey. Thirty-six subjects (43.5%) were from three state-subsidized alcoholism/drug abuse halfway houses. Twenty-nine (34.9%) subjects were from shorter-term alcohol and drug abuse units of hospitals (one state hospital alcohol unit and two private alcohol and drug rehabilitation programs) and 18 (21.7%) were from dual-diagnosis units at two private psychiatric hospitals. A multisite approach was used to obtain a heterogeneous sample in terms of socioeconomic status. Halfway houses were chosen as recruitment sites to increase the probability of finding individuals with ASP, because many patients in these programs are court-referred in lieu of a prison sentence. Criteria for subjects' participation in the study were: age 18 to 45, male, high school reading level of English, a history of alcohol and/or drug abuse, and no history of psychosis.

Instruments

Assessment of Psychopathology

The Family History Research Diagnostic Criteria interview (FHRDC)[25,26] was used to ascertain diagnoses of antisocial personality disorder, alcoholism, drug abuse, and depressive disorder. The FHRDC interview can be given to index subjects about themselves or their relatives. This interview, which was administered to all 83 subjects, is somewhat less conservative than some other diagnostic measures. To check its reliability, we administered the National Institute of Mental Health Diagnostic Interview Schedule (NIMH-DIS)[27,28] antisocial section to 54 of the 83 subjects. We also used Research Diagnostic Criteria (RDC)[29] to recode 83 subjects regarding alcohol abuse, to recode 50 subjects regarding ASP, and to recode all 83 subjects regarding primary *versus* secondary major depression. DSM-III[30] criteria were used to diagnose dysthymia, because the FHRDC does not provide diagnostic criteria for this disorder. Chi-square values were calculated among the various measures for each disorder to determine concordance of diagnoses among these different instruments. For antisocial personality disorder, FHRDC overlapped adequately with NIMH-DIS (chi-square value = 41.09, p <0.000, n = 55) and with RDC (chi-square value = 43.31, p <0.000, n = 51). For alcohol abuse, concordance between FHRDC and RDC (chi-square value = 62.47, p <0.000, n = 83) and between FHRDC and DSM-III-based diagnoses (chi-square value = 66.67, p <0.000, n = 83) was high. As these chi-square analyses indicated that the overlap between FHRDC and other diagnostic measures was adequate, we used the measure that had the least number of missing values, the FHRDC, to represent the diagnoses for ASP, alcohol abuse, and drug use disorder. FHRDC criteria were modified for major depression so that we could use the RDC distinction between primary and secondary major depression.

Personality Measures Used to Assess Psychopathology

Three subscales of the Million Clinical Multiaxial Inventory (MCMI)[31] were administered for a self-report assessment of personality disorder, psychopathy, and depression: MCMI scales 4 (gregarious/histrionic), 6 (antisocial), and D (neurotic depression). Scales 4 and 6 are reported to tap more enduring personality traits, whereas scale D focuses on both "state" and "trait" depression.

The Socialization (So) scale of the California Personality Inventory (CPI)[32] was given. The So scale has been used extensively to obtain a continuous measure of sociopathy.[33] This scale is scored so that a high score indicates a high degree of socialization so that it is typically inversely correlated with measures of antisocial personality.

Additional Personality Assessment

The Sensation Seeking Scale V (SSS)[34] is a 40-item questionnaire designed to assess (1) thrill and adventure seeking behavior (TAS scale), (2) experience seeking behavior (ES scale), (3) disinhibition (Dis scale), and (4) boredom susceptibility (BS scale). The Thrill and Adventure Seeking scale measures stated attraction to risky, adventurous activities such as parachute jumping and motorcycle riding. The Experience Seeking scale taps one's "seeking . . . arousal through the mind and senses in a nonconforming life-style." Individuals who score high on this subscale seek external stimulation in music, art, or travel or internal stimulation in drugs. The Disinhibition scale reflects a need to "seek release and social disinhibition through drinking, partying, gambling, and sex" and rebellion against codes of acceptable social behavior. Finally, the Boredom Susceptibility scale reflects, in males, an aversion to repetitive experience.

The Trait Anxiety Subscale of the State-Trait Anxiety Inventory[35] is designed to measure an enduring quality of anxiety.

The Tridimensional Personality Questionnaire (TPQ, version 1) is an 80-item inventory designed by Cloninger[36] to assess novelty seeking (intensity of response to excitement), harm avoidance (intensity of response to harm), and reward dependence (intensity of response to aversive stimuli). These three dimensions can be measured on a continuum or can be divided by a median split to form eight categories. Internal consistencies with the Horst formula with correction for item difficulty for the three TPQ scales for a normative sample of 326 white males are as follows: .77 for Novelty Seeking, .87 for Harm Avoidance, and .70 for Reward Dependence (Cloninger, personal communication, 1989).

The Hypermasculinity Inventory (HI) developed by Mosher and Sirkin[37] assesses a "macho personality constellation." It is a forced-choice scale with three subscales: calloused sex attitudes toward women (.79), violence as manly (.79), and danger as exciting (.71). The values in parentheses are Cronbach alphas for a sample of 135 college males. The Cronbach alpha for the total "macho" score was .89.[37]

Demographic Information and Substance Use History

Demographic information was obtained. An alcohol and drug use survey was administered to assess the quantity and frequency of drug use and to ascertain the drug of choice. Questions on motivation for using alcohol and drugs were included, and the responses fell into three categories: to alleviate depression, to alleviate boredom, and addiction. The Michigan Alcoholism Screening Test (MAST)[38] was administered to ascertain severity of the alcohol problem.

Procedure

Patients were recruited for participation through an announcement at a group meeting at their treatment site. Patients were told that "people who had a history of trouble with the law and/or depression were needed." There was no further screening except in the dual diagnosis units, where hospital records were reviewed to exclude patients with psychotic symptoms. Each subject participated voluntarily in the study and was compensated with $10.00.

Each subject gave informed consent and was individually administered a structured interview by the first author (E.E.) to ascertain history of antisocial personality disorder, alcoholism, drug abuse, and depressive disorder.

Each subject then filled out the battery of personality inventories.

Subject Classification

All diagnoses were made by the first author (E.E.). Ambiguous cases were resolved in collaboration with the third author (V.H.).

Each subject was classified into one of six groups: (1) ASP, Primary Depressive Disorder (*n*) = 15; (2) ASP, Secondary Depressive Disorder (*n* = 7); (3) ASP, no history of depressive disorder (*n* = 23) (also called "Pure ASPs"); (4) Primary Depressive Disorder, no ASP (*n* = 18) (also called "Pure Depressive"); (5) Secondary Depressive Disorder, no ASP (*n* = 9); and (6) no history of depressive disorder, no ASP (*n* = 11).

For this study, individuals who abused alcohol, drugs, or both were included in the sample. "Substance abusers" refers in this paper to the entire sample of alcohol and/or drug abusers. In the sample, 21 (25%) subjects used alcohol only, 19 (23%) used drugs only, and 43 (52%) used both alcohol and drugs. Twenty-eight (34%) subjects reported that alcohol was their substance of choice. The rest (55 subjects, 66%) cited a drug as their substance of choice (30, or 36%, said cocaine was the drug of choice; 10, or 12%, said marijuana; 13, or 16%, said narcotics; and 2, or 2%, reported hallucinogens as their drug of choice).

Depressive and antisocial symptoms that were associated with alcoholism and drug abuse were not counted as positive towards depressive and ASP diagnoses. For questions on ASP, for instance, if the subject reported having had a poor employment history, he was asked, "Were the days missed related at all to alcohol or drug use? How so?" Similar follow-up questions were asked of any symptoms positive for depression or antisocial personality disorder that would also have been positive for substance abuse.

Diagnoses of a history of depression primary to substance abuse were assigned on the basis of age of onset, severity of disorder, and, where possible, symptoms in the absence of substance abuse. For instance, if a subject reported having suffered symptoms of major depression, he was asked, "Were these symptoms (or, Was this depression . . .) related at all to alcohol or drug use?" or "Were you depressed when you weren't using drugs or alcohol?" Thus, primary major depression was diagnosed if the subject reported having had the symptoms prior to or in the absence of drug or alcohol use. If the depressive episodes occurred after or were related to substance abuse, a diagnosis of secondary depression was made. To summarize, for this classification system, a subject was considered to have a history of primary major depression if he had a lifetime history of dysthymia and/or a major depressive episode that pre-dated or was unrelated to substance abuse. Secondary depression refers to at least one major depressive episode that occurred after the onset of substance abuse or was directly related to substance abuse. (Subjects who had a history of both primary and secondary depression were classified in the primary depression group.) The primary depression/secondary depression distinction is used here only in relation to substance use and not other disorders such as ASP. Differential diagnosis has been cited as a significant problem in studies of this type.[39–41]

RESULTS

Demographics

TABLE 1 presents a demographic description of the sample. Socioeconomic status was significantly higher in the non-ASP group than in the ASP group. Subjects with any form of depression (either "Pure Depressives," those with secondary depression and no ASP, or secondary depression and ASP) had a significantly higher socioeconomic status than did nondepressed subjects. Socioeconomic status was used as a covariate in subsequent analyses.

History of Substance Use

Analysis of variance (ANOVA) was used to test differences among groups. TABLE 2 shows that subjects with ASP had a more severe drinking and drug history (i.e., began drinking earlier [$F = 5.98$; $df = 5,76$; $p < 0.017$], had alcohol-related problems earlier [$F = 7.77$; $df = 5,64$; $p < 0.007$], had drug-related problems earlier [$F = 5.44$; $df = 5,61$; $p < 0.023$], and used more drugs [$F = 7.31$; $df = 5,79$; $p < 0.009$]) than did non-ASPs.

Personality

Monothetic subtypes were compared using multivariate analysis of variance (MANOVA) on continuous personality measures. Polythetic subtypes were compared using ANOVA to minimize the chance for Type 2 (false-negative) errors, because cell sizes were small. Post hoc Tukey Studentized Range tests were used to assess significance (at the 0.05 level) of differences among groups.

Unidimensional Subtypes: ASP

Main effects for ASP on individual personality measures are shown in TABLE 3. The Socialization scale of the California Personality Inventory, the Aggressive-

TABLE 1. Demographic Characteristics of Entire Sample of 83 Male Substance Abusers

Demographic Characteristic	Mean	Range	%	n
Age 28.06	18–45	—	83	—
Marital Status				
Married	—	—	13.9	11
Divorced or separated	—	—	26.6	21
Single	—	—	59.5	47
Education				
Jr. high school or partial HS	—	—	34.2	28
High school diploma	—	—	37.3	31
Partial college or technical school	—	—	19.3	16
College diploma	—	—	9.6	8
Income				
0–10,000	—	—	27.6	21
11,000–20,000	—	—	30.3	23
21,000–40,000	—	—	31.6	24
41,000+	—	—	7.9	6
Socioeconomic status (SES)[a]	5.91	2–11	—	83
Ethnicity				
Afro-American	—	—	17.3	14
Caucasian	—	—	79.0	64
Hispanic	—	—	3.7	3
Religion				
Catholic	—	—	52.6	40
Protestant	—	—	22.4	17
Jewish	—	—	2.6	2
Baptist	—	—	11.8	9
Other	—	—	9.2	7

[a]SES is a composite of father's education, father's occupation, family of origin's income, subject's education, subject's occupation, and subject's income.

TABLE 2. Substance Use History in Relation to a History of Antisocial Personality Disorder (ASP)

Substance Use History		Antisocial Personality	Non-antisocial Personality	DF	F	p
Age started drinking	Mean (SD)	13.46 (1.93)	14.67 (1.90)	5,76	5.98	0.017*
	n	44	33			
Age alcohol-related problems began	Mean (SD)	17.87 (5.72)	21.70 (5.80)	5,64	7.77	0.007**
	n	38	27			
Age started using drugs	Mean (SD)	14.36 (3.57)	15.53 (3.45)	5,78	1.60	0.21
	n	45	34			
Age drug-related problems began	Mean (SD)	16.58 (3.64)	20.19 (5.62)	5,61	5.44	0.023*
	n	36	26			
Michigan Alcohol Screening Test (MAST)	Mean (SD)	36.77 (21.34)	30.56 (19.74)	5,78	1.51	0.22
	n	43	36			
Number drugs used in life	Mean (SD)	7.71 (3.90)	5.44 (3.44)	5,79	7.31	0.009**
	n	44	36			

*$p < 0.05$; **$p < 0.01$.

TABLE 3. Comparison of Personality Measures in Relation to a History of Antisocial Personality Disorder (ASP)

Personality Measures	Antisocial			Non-antisocial			DF	F	p
	Mean	(SD)	n	Mean	(SD)	n			
Sensation Seeking Scale									
Total Score	22.41	(6.41)	40	20.50	(5.66)	34	1,73	1.82	0.18
Tridimensional Personality Questionnaire									
Novelty seeking	17.03	(4.17)	39	14.95	(4.77)	32	1,70	3.81	0.055
Harm avoidance	9.68	(6.52)	42	10.26	(6.88)	35	1,76	.14	0.71
Reward dependence	13.43	(2.99)	42	13.34	(2.55)	35	1,76	.02	0.889
Trait Anxiety Scale	45.29	(11.41)	42	44.22	(9.41)	35	1,76	.20	0.66
Socialization scale of CPI	29.01	(4.66)	39	26.66	(5.96)	32	1,70	8.36	0.005**
Gregarious/histrionic	17.85	(4.40)	39	16.85	(4.89)	32	1,70	.81	0.371
Antisocial (MCMI)	20.16	(3.61)	39	17.39	(5.60)	32	1,70	6.32	0.014*
Neurotic depression	11.02	(7.08)	42	9.57	(7.86)	35	1,76	.72	0.40
Hypermasculinity Inventory									
Total score	14.08	(6.21)	40	10.93	(6.75)	34	1,73	4.38	0.039*
Callous sex attitudes	4.07	(2.71)	39	2.60	(1.89)	32	1,70	6.63	0.012*
Violence as manly	5.01	(2.73)	39	3.94	(3.21)	32	1,70	2.30	0.134
Danger as exciting	5.19	(2.48)	39	3.82	(2.37)	32	1,70	5.59	0.021*

*p <0.05; **p <0.01.

Antisocial Scale of the MCMI, the total Hypermasculinity Inventory (HMI) score, and the "Callous Sex Attitudes Towards Women" and "Violence as Manly" subscales of the HMI were found to differentiate antisocial from nonantisocial subjects. The TPQ Novelty Seeking score differentiated the two groups at a p <0.055 level.

Unidimensional Subtypes: Depressive Disorder

Personality differences due to type of depression are shown in TABLE 4. The substance abuser with a history of primary depression as compared to those with depression secondary to substance abuse or no history of depression at all was: more harm avoidant ($F = 7.58$; $df = 2,76$; p <0.001), more anxious ($F = 6.73$; $df = 2,76$; p <0.002), less gregarious ($F = 3.46$; $df = 2,70$; p <0.037), more depressed ($F = 6.98$; $df = 2,76$; p <0.002), and less likely to view danger as exciting ($F = 3.90$; $df = 2,70$; p <0.025). They also tended to be less antisocial ($F = 2.47$; $df = 2,70$; p <0.09) and less sensation seeking ($F = 2.92$; $df = 2,73$; p <0.061). Subjects with secondary depression were less anxious ($F = 6.73$; $df = 2,76$; p <0.002), reported less trait depression ($F = 6.98$; $df = 2,76$; p <0.002), and were more likely to see danger as exciting ($F = 3.90$; $df = 2,70$; p <0.025) than subjects with primary depression. Subjects with substance-related depression tended to be more sensation seeking ($F = 2.92$; $df = 2,73$; p <0.061) than subjects with primary depression and no depression.

Multidimensional Subtypes: ASP × Primary and Secondary Depression

TABLE 5 presents the six-cell design and additional information about the nondepressive group, because it separates the subjects who have reported no history of depression from those who have a history of depression secondary to substance abuse. The Pure Antisocial group (i.e., those with ASP and no history of depression) differed from the Pure Depressive group (i.e., non-ASP subjects with a history of primary depression) in expected directions for all personality variables. The Antisocial Secondary Affected group was lowest on depression ($F = 3.28$; $df = 5,77$; p <0.010), lowest on anxiety ($F = 3.05$; $df = 5,79$; p <0.015), highest on the gregarious/histrionic scale ($F = 3.0$; $df = 5,76$; p <0.016), and highest on the "danger as exciting" subscale ($F = 2.42$; $df = 5,73$; p <0.044) of hypermasculinity index compared to the ASP primary depressed and Pure ASP groups. This group was higher on socialization ($F = 2.88$; $df = 5,78$; p <0.020) than the other two Antisocial groups.

Comparison of Unidimensional and Multidimensional Subtypes

There was no main effect of ASP for the gregarious/histrionic scale of the MCMI. TABLE 5 shows that the ASP secondary depressed group, not the ASP primary depressed group, scored highest on this measure. Actually, the ASP primary depressed group's score on gregarious/histrionic scale was significantly lower than the score for the ASP secondary depressed group ($F = 3.00$; $df = 5,76$; p <0.016) and the non-ASP groups on this measure.

When ASP was examined without regard to comorbid depression, the ASP *versus* non-ASP groups were similar on trait depression ($F = .72$; $df = 1,76$; p <0.40). This is a surprising finding, given the ASP literature about the dearth of affectivity in ASPs. However, when subgroups of ASP were examined in conjunction with comorbid depression, the ASP/primary depressed subgroup scored the highest. Fur-

TABLE 4. Comparison of Personality Measures for Subjects with History of Primary Major Depression or Dysthymia, Major Depression Associated with Substance Abuse, and No History of Depression

Personality Measures	Primary Depression			Substance-Related Depression			No Depression			DF	F	p
	Mean	(SD)	n	Mean	(SD)	n	Mean	(SD)	n			
SSS (total)	20.19	(5.59)	30	24.71	(6.13)	15	21.28	(6.24)	29	2,73	2.92	0.061
TPQ												
Novelty Seeking	15.17	(4.83)	30	18.09	(4.72)	12	16.22	(4.00)	29	2,70	1.84	0.17
Harm Avoidance	13.34	(6.82)	30	8.28	(5.97)	15	7.55	(5.51)	32	2,76	7.58	0.001***
Reward Dependence	13.47	(2.65)	30	14.53	(2.62)	15	12.79	(2.88)	32	2,76	2.09	0.131
MCMI												
Gregarious/histrionic	15.79	(5.06)	30	19.00	(4.43)	12	18.40	(3.76)	29	2,70	3.46	0.037*
Neurotic depression	14.03	(7.70)	30	8.47	(7.17)	15	7.81	(5.93)	32	2,76	6.98	0.002**
Antisocial	17.85	(5.16)	30	17.97	(5.41)	12	20.40	(3.79)	29	2,70	2.47	0.09
HMI												
Total score	11.30	(6.02)	30	14.28	(7.77)	15	13.17	(6.52)	29	2,73	1.18	0.313
Callous sex attitudes	3.24	(2.63)	30	2.85	(2.71)	12	3.81	(2.21)	29	2,70	.75	0.478
Violence as manly	4.37	(2.88)	30	5.33	(3.21)	12	4.36	(3.01)	29	2,70	.52	0.598
Danger as exciting	3.69~	(1.95)	30	5.75~	(2.93)	12	5.00	(2.60)	29	2,70	3.90	0.025*
Trait Anxiety Scale	49.91	(10.85)	30	42.09	(7.55)	15	41.29	(9.62)	32	2,76	6.73	0.002**
CPI So Scale	24.73	(6.37)	30	25.95	(5.34)	12	24.04	(4.77)	29	2,70	.50	0.607

ABBREVIATIONS: SSSV = Sensation Seeking Scale V; TPQ = Tridimensional Personality Questionnaire; MCMI = Millon Clinical Multiaxial Inventory; HMI = Hypermasculinity Inventory; CPI So = California Personality Inventory Socialization Scale.
~Post hoc Tukey Studentized Range Test significant at the 0.05 level.
*p <0.05; **p <0.01; ***p <0.001.

TABLE 5. Means on Selected Personality Measures in Relation to a History of Antisocial Personality Disorder (ASP) and/or Primary vs Secondary Depressive Disorder

Personality Measures	Antisocial Depressive Mean (SD) / n	Antisocial Secondary Affected Mean (SD) / n	Pure Antisocial Mean (SD) / n	Pure Depressive Mean (SD) / n	Non-ASP Secondary Affected Mean (SD) / n	Non-ASP No Depression Mean (SD) / n	DF	F	P
Sensation Seeking	21.42 (5.69) / 15	24.18 (3.34) / 7	22.98 (6.10) / 23	19.40 (5.51) / 16	25.05 (5.0) / 9	18.92 (4.3) / 11	5,80	2.16	0.067
Novelty seeking	16.37 (4.68) / 15	19.39 (3.96) / 7	16.98 (3.59) / 23	14.42 (4.99) / 16	15.42 (4.20) / 9	15.24 (4.45) / 11	5,80	1.63	0.16
Harm avoidance	13.47^ (6.36) / 15	6.60 (4.95) / 7	8.41 (5.75) / 23	13.19 (7.23) / 16	9.64 (6.29) / 9	6.36^ (4.52) / 11	5,80	3.44	0.008**
Reward dependence	13.38 (3.23) / 15	15.57 (2.64) / 7	12.89 (2.60) / 23	13.71 (2.04) / 16	13.11 (2.73) / 9	12.61 (3.31) / 11	5,80	1.27	0.287
Trait anxiety	51.23^ (11.76) / 14	41.86 (9.26) / 7	42.88 (10.28) / 23	48.75 (10.24) / 16	41.15 (6.86) / 9	39.03^ (7.09) / 11	5,79	3.05	0.015**
Socialization	22.79 (5.53) / 15	25.78 (3.22) / 7	22.50 (4.09) / 22	26.50 (6.55) / 16	28.25 (7.09) / 8	27.64 (4.97) / 11	5,78	2.88	0.020
Gregarious/histrionic	14.93^ ~(4.38) / 15	21.67^ (3.45) / 6	19.29~ (3.43) / 21	16.79 (5.53) / 16	17.00 (3.46) / 8	17.51 (4.46) / 11	5,76	3.00	0.016**
Antisocial	19.64 (4.02) / 15	20.17 (5.46) / 6	20.64 (2.53) / 21	16.49 (5.73) / 16	16.21 (4.53) / 8	20.37 (5.59) / 11	5,76	2.49	0.039*
Neurotic depression	15.73^ ~(7.55) / 15	7.71 (5.41) / 7	8.56^ (5.29) / 21	11.99 (7.56) / 16	9.13 (8.76) / 8	6.37~ (7.05) / 11	5,77	3.28	0.010**
HMI - Total[a]	14.54 (5.51) / 14	14.59 (7.78) / 7	13.56 (6.39) / 19	8.46 (5.02) / 16	14.01 (8.29) / 8	12.41 (7.05) / 10	5,73	1.89	0.108
Callous sex attitudes	4.43 (2.95) / 14	2.73 (3.08) / 7	4.08 (2.48) / 19	2.20 (1.83) / 16	3.64 (3.01) / 8	3.30 (1.57) / 10	5,73	1.63	0.165
Violence as manly	5.72 (2.42) / 14	5.71 (2.81) / 7	4.12 (2.78) / 19	3.19 (2.79) / 16	5.25 (3.54) / 8	4.81 (3.51) / 10	5,73	1.56	0.182
Danger as exciting	4.39 (2.02) / 14	6.14 (2.67) / 7	5.36 (2.62) / 19	3.07 (1.72) / 16	5.13 (3.00) / 8	4.30 (2.58) / 10	5,73	2.42	0.044*

[a]HMI = Hypermasculinity Inventory.
^, ~, # Pairs of means sharing the same symbol differ at the 0.05 level on the post hoc Tukey Studentized Range Test.
*p <0.05; **p <0.01; ***p <0.001.

ther differentiation among ASP subgroups indicated that the ASP/secondary depressed group was lower in depression than even the ASP/no depression group ($F = 3.28$; $df = 5,77$; $p <0.010$). The ASP/depressed subgroup elevated the ASPs' depression score; the ASP/secondary depressed and the ASP/nondepressed groups were lower than or in the same general range as the non-ASP depressed groups.

The antisocial group's score on the trait anxiety scale was not significantly different from that of the non-ASP group. This is contrary to expected results. ASPs with secondary depression reported a level of anxiety much more similar to that of "pure ASPs" than to that of depressed antisocials ($F = 3.05$; $df = 5,79$; $p <0.015$). Thus, the ASP/depressed subgroup of the ASP group elevated the anxiety score of the ASPs. This information would have been lost if ASPs had been examined as a monosymptomatic rather than a polysymptomatic group.

These data highlight the importance of differentiating the ASP subtype by the type of depressive disorder. It is important to differentiate primary from secondary disorders and to view the ASP/secondary affected group as a subtype of substance abuser with unique characteristics.

DISCUSSION

Our findings suggest that the unidimensional typologies did identify low *versus* high arousal profiles. Differences due to antisocial personality disorder and depressive disorder on several personality measures were found. A personality profile of antisocial personality disorder (*versus* non-antisocial personality disorder) included higher sensation seeking, novelty seeking, and hypermasculine traits, lower harm avoidance, and lower socialization. Antisocials were also higher on trait anxiety and depression.

Differences were found due to depressive disorder as a unidimensional typological schema; subjects with primary depression had unique affective profiles vis-à-vis nondepressed subjects and, most interestingly, vis-à-vis subjects who had major depressions that were secondary to substance abuse. Subjects with primary depressions were generally less sensation seeking, more harm avoidant, less reward dependent, more anxious, less gregarious/histrionic, more characteristically depressed, and less "macho" than were subjects with substance-related major depressions. These results support findings of Guze,[21] Winokur,[22] and Yerevanian and Akiskal[42] and suggest that in research on comorbid major depressive disorder, major depression should be broken down into primary and secondary major depression. Care should be taken to report separately on subjects with each type of major depression.

Our data suggest that precise diagnoses that take into account comorbid disorders yield more information about each subtype. On the basis of our findings, it may not be useful to identify affective profiles of ASP *versus* non-ASP substance abusers. When these subtypes were further divided by presence/absence of primary or secondary depressive disorder, the affective and personality profiles differed. Thus, our data are in accord with Gerstley *et al.*'s[12] speculations about diagnostic heterogeneity among ASPs.

It seems necessary to test theories about underlying affective or temperament profiles of subtypes of substance abusers. For instance, there appear to be at least three different subtypes of antisocial substance abusers in terms of their affective profiles. Antisocial subjects who have depression before the onset of alcohol or drug abuse are an interesting, probably understudied group of people who exhibit elevated levels of trait depression, anxiety, and antisocial, hypermasculine traits, and undersocialization and low sociability. Antisocial subjects who have a history of substance-related major depression stand out as another intriguing and even more understudied group. They seem to have a "high arousal" pattern of personality characteristics along with low depression and low anxiety and a socialized but "macho" stance. Pure antisocial subjects (those with no history of any kind of depression) are differentiated from depressed non-ASP substance abusers, but seem to be less disturbed in terms of anxiety, depression, and undersocialization than either ASPs with either primary or secondary depression.

Our ability to compare results from this study to others in the literature is somewhat limited by two factors: (1) We combined alcohol and drug abusers in the same sample; most subtype studies require subjects to meet criteria for either alcohol or drug abuse. However, most studies on "alcoholics," for instance, do not exclude subjects who also have a history of drug abuse or dependence. Because the majority of alcoholics in any recent treatment population also have a history of other drug abuse,[43] our sample should be fairly representative of the typical substance abuse non-V.A. inpatient treatment population. (2) We included dysthymia and major depressive episodes that predated alcohol abuse in one group called primary depression. This classification scheme is also atypical of similar studies reported thus far.

Despite the relatively small sample sizes in this study, the findings were consistently in directions predicted by our subtype hypotheses based on comorbid psychopathology and affect.

SUMMARY

Our data show that when substance abusers are subtyped simultaneously by antisocial personality disorder and the onset of depression relative to alcohol or drug abuse, groups of people with unique personality and affective profiles are identified. The profiles are represented by measures of affect-related personality variables such as trait anxiety, trait depression, histrionic traits, sensation seeking, and novelty seeking. These measures were chosen in an attempt to show that a "low arousal" personality type may be associated with antisocial personality and may thus indirectly be linked to a certain type (i.e., ASP/nondepressed) of substance abuser. By using a multisymptomatic typological schema (i.e., a constellation of diagnostic categories rather than just one), we can show that different personality or affective profiles are indeed associated with certain subtypes of substance abusers and that depressed people who use drugs or alcohol are different affectively from antisocial types. We also show that the relationship between "low" and "high" arousal personality profile and subtypes based on comorbid psychopathology is highlighted even more when we take into account the onset of dysthymia or depression that is primary *versus* secondary to substance abuse. Our findings are in accord with others' descriptions of the "affective arousal" dimensions of personality and are the first to link these dimensions with subtypes based on ASP and depression.

REFERENCES

1. CLECKLEY, H. 1976. The Mask of Sanity. 5th Ed. C.V. Mosby. St. Louis, Missouri.
2. HARE, R. D. & D. N. COX. 1978. Psychophysical research on psychopathy. *In* The Psychopath: A Comprehensive Study of Antisocial Disorders and Behaviors. W. H. Reid, ed. :209–233. Brunner-Mazel. New York.

3. Helzer, J. E. & T. R. Pryzbeck. 1988. The co-occurrence of alcoholism with other psychiatric disorders in the general population and its impact on treatment. J. Stud. Alcohol **49:** 219–224.
4. Hesselbrock, M., R. E. Meyer & J. J. Keener. 1985. Psychopathology in hospitalized alcoholics. Arch. Gen. Psychiatry **42:** 1050–1055.
5. Babor, T. F., Z. S. Dolinsky, R. E. Meyer, M. Hesselbrock, M. Hofmann & H. Tennen. 1992. Types of alcoholics: Concurrent and predictive validity of some common classification schemes. Br. J. Addict. **87:** 23–40.
6. Hesselbrock, M. 1986. Alcoholic typologies: A review of empirical evaluations of common classification schemes. Recent Dev. Alcohol. **4:** 191–206.
7. Rounsaville, B. J., T. Kosten, M. M. Weissman, B. Prusoff, D. Pauls, S. F. Anton & K. Merikangas. 1991. Psychiatric disorders in relatives of probands with opiate addiction. Arch. Gen. Psychiatry **48:** 33–42.
8. Penick, E. C., B. J. Powell, E. Othmer, S. F. Bingham, A. S. Rice & B. S. Liese. 1984. Subtyping alcoholics by coexisting psychiatric syndromes: Course, family history, outcome. *In* Longitudinal Research in Alcoholism. D. W. Goodwin, N. T. Van Dusen & S. A. Mednick, eds. Klufer-Nijhoff. Boston.
9. Hesselbrock, V. M., M. N. Hesselbrock & K. L. Workman-Daniels. 1986. Effect of major depression and antisocial personality on alcoholism: Course and motivational patterns. J. Stud. Alcohol **47:** 207–212.
10. Lewis, C. E., J. Rice, N. Andreasen, P. Clayton & J. Endicott. 1985. Alcoholism in antisocial and nonantisocial men with unipolar major depression. J. Affect. Dis. **9:** 253–263.
11. Lewis, C. E., J. Rice, N. Andreasen, J. Endicott & A. Hartman. 1986. Clinical and familial correlates of alcoholism in men with unipolar major depression. Alcoholism: Clin. & Exp. Res. **10:** 657–662.
12. Gerstley, L. J., A. I. Alterman, A. T. McLellan & G. E. Woody. 1990. Antisocial personality in patients with substance abuse disorders: A problematic diagnosis? Am. J. Psychiatry **147:** 173–178.
13. Liskow, B., B. J. Powell, E. Nickel & E. Penick. 1991. Antisocial alcoholics: Are there clinically significant diagnostic subtypes? J. Stud. Alcohol **52:** 62–69.
14. Schuckit, M. A. 1986. Genetic and clinical implications of alcoholism and affective disorder. Am. J. Psychiatry **143:** 141–147.
15. Babor, T. F., M. Hoffmann, F. K. Delboca, V. M. Hesselbrock, R. E. Meyer, Z. S. Dolinsky & B. Rounsaville. 1992. Types of alcoholics: I. Evidence for an empirically-derived typology based on indicators of vulnerability and severity. Arch. Gen. Psychiatry **49:** 599–608.
16. Cloninger, C. R., M. Bohman & S. Sigvardsson. 1981. Inheritance of alcohol abuse: Cross-fostering analysis of adopted men. Arch. Gen. Psychiatry **38:** 861–868.
17. Penick, E. C., E. J. Nickel, B. J. Powell, S. F. Bingham & B. I. Liskow. 1990. A comparison of familial and nonfamilial male alcoholic patients without a coexisting psychiatric disorder. J. Stud. Alcohol **51:** 443–447.
18. Read, M. R., E. C. Penick, B. J. Powell, E. J. Nickel, S. F. Bingham & J. Campbell. 1990. Subtyping male alcoholics by family history of alcohol abuse and co-occurring psychiatric disorder: A bi-dimensional model. Br. J. Addict. **85:** 367–378.
19. Parsian, A. P., R. D. Todd, E. J. Devor, K. L. Omalley, B. K. Suarez, T. Reich & C. R. Cloninger. 1991. Alcoholism and alleles of the Human D2 dopamine receptor locus. Arch. Gen. Psychiatry **48:** 655–663.
20. Schuckit, M. A. 1985. The clinical implications of primary diagnostic groups among alcoholics. Arch. Gen. Psychiatry **42:** 1043–1049.
21. Guze, S. B. 1990. Secondary depression: Observations in alcoholism, briquet's syndrome, anxiety disorder, schizophrenia, and antisocial personality. A new form of comorbidity? Psychiatric Clin. North Am. **13:** 651–659.
22. Winokur, G. 1990. The concept of secondary depression and its relationship to comorbidity. Psychiatric Clin. North Am. **13:** 567–583.
23. Tarter, R. E., A. I. Alterman & K. L. Edwards. 1985. The vulnerability to alcoholism in men: A behavior-genetic perspective. J. Stud. Alcohol **46:** 329–356.

24. CORYELL, W., G. WINOKUR, M. KELLER, W. SCHEFTNER & J. ENDICOTT. 1992. Alcoholism and primary major depression: A family study approach to co-existing disorders. J. Affective Dis. **24:** 93–99.
25. ANDREASEN, N. C., J. ENDICOTT, R. SPITZER & G. WINOKUR. 1977. The family history method using diagnostic criteria. Arch. Gen. Psychiatry **34:** 1229–1235.
26. ANDREASEN, N. C., J. R. RICE, J. ENDICOTT, T. REICH & W. CORYELL. 1986. The family history approach to diagnosis. Arch. Gen. Psychiatry **43:** 421–429.
27. ROBINS, L. N., J. E. HELZER, J. CROUGHAN, J. B. W. WILLIAMS & R. L. SPITZER. 1981. NIMH Diagnostic Interview Schedule: Version III. National Institute of Mental Health. Rockville, MD.
28. BURKE, J. D. 1986. Diagnostic categorization by the Diagnostic Interview Schedule: A comparison with other methods of assessment. *In* Mental Disorders in the Community. J. Barrett & R. Rose, eds. : 225–285. Guilford Press. New York.
29. SPITZER, R. L., J. ENDICOTT & E. ROBINS. 1978. Research Diagnostic Criteria: Rationale and reliability. Arch. Gen. Psychiatry **35:** 773–782.
30. Diagnostic and Statistical Manual of Mental Disorders. Third Edition. 1980. American Psychiatric Association. Washington, DC.
31. MILLON, T. 1983. Millon Clinical Multiaxial Inventory (3rd Ed.) Interpretive Scoring Systems.
32. GOUGH, H. G. 1969. Manual for the California Psychological Inventory. Consulting Psychologists Press. California.
33. WIDOM, C. S. 1976. Interpersonal conflict and cooperation in psychopaths. J. Abnorm. Psychol. **85:** 330–334.
34. ZUCKERMAN, M. 1979. Sensation Seeking: Beyond the Optimal Level of Arousal. Lawrence Erlbaum Associates. Hillsdale, NJ.
35. SPIELBERGER, C. D. 1975. The measurement of state and trait anxiety: Conceptual and methodological issues. *In* Emotions—Their Parameters and Measurement. L. Levi, ed. :713–725. Raven Press. New York.
36. CLONINGER, C. R. 1987. Neurogenetic adaptive mechanisms in alcoholism. Science **236:** 410–416.
37. MOSHER, D. L. & M. SIRKIN. 1984. Measuring a macho personality constellation. J. Res. Personality **18:** 150–163.
38. SELZER, M. L. 1971. The Michigan Alcoholism Screening Test: The quest for a new diagnostic instrument. Am. J. Psychiatry **127:** 1653–1658.
39. HESSELBROCK, M. N., V. M. HESSELBROCK, H. TENNEN, R. E. MEYER & K. L. WORKMAN. 1983. Methodological considerations in the assessment of depression in alcoholics. J. Consulting Clin. Psychol. **5:** 399–405.
40. SCHUCKIT, M. A. 1973. Alcoholism and sociopathy—diagnostic confusion. Q. J. Stud. Alcohol **34:** 157–164.
41. SCHUCKIT, M. A. 1983. Alcoholic patients with secondary depression. Am. J. Psychiatry **140:** 711–714.
42. YEREVANIAN, B. I. & H. S. AKISKAL. 1979. "Neurotic," characterological, and dysthymic depressions. Psychiatric Clin. North Am. **2(3):** 595–617.
43. CARROLL, J. F. 1986. Treating multiple substance abuse clients. Recent Dev. Alcohol. **4:** 85–103.

Familial Alcoholism and Personality-Environment Fit

A Developmental Study of Risk in Adolescents[a]

MARSHA E. BATES AND ERICH W. LABOUVIE

Center of Alcohol Studies
Rutgers University
Piscataway, New Jersey 08854

Multiple lines of convergent evidence indicate that vulnerability to the negative consequences of alcohol and other drug use is determined only in part by consumption level.[1–4] These studies argue for the need to distinguish between risk factors that *directly* affect alcohol-problem vulnerability and those that affect the experience of use consequences *indirectly* through their influence on intensity of drinking behaviors. In a number of high-risk studies, youth with a familial risk for alcoholism (FH+) did not drink alcohol more frequently or in higher quantities than did others with no such family history (FH−).[5–8] Yet, some of these FH+ samples still showed heightened vulnerability to the negative consequences of use.[7,8] This pattern of results suggests that a conceptual and empirical distinction between high risk for heavy alcohol use and high risk for use consequences may be of particular relevance to understanding the developmental vulnerability of FH+ offspring. The distinction may be especially useful during adolescence when use behaviors are often initiated, use patterns are initially formed, and the risk of use consequences begins to accrue.

Both personality and social support constructs have figured prominently in theoretical accounts of increased vulnerability to psychopathology in general and to alcoholism in familial risk populations in particular. However, research has been hindered by the failure to consider the mutual embeddedness and reciprocity of these constructs.[9–11] An interactionist approach to the study of use behaviors and associated outcomes suggests that knowledge of the *relation* between person and environment may provide more information about the roles of personality risk and the social support systems that directly affect or moderate risk than either of these elements viewed separately.[9,10] Our conception of risk is based on the notion of goodness-of-fit or "match"[9,12] between personality needs and the individual's perception of opportunities for meeting these needs within his/her environment. Note that this approach is different from those that attach a unilateral positive or negative valence to individual person or environmental characteristics. Rather, adaptive behavioral development is viewed as a function of the congruence between the individual's dispositional needs and the social context in which they occur. Needs and contexts are dynamic and mutually transform one another throughout development.[9,12]

We specifically test the mismatch between adolescents' social personality needs and their perceptions of the responsivity of the social environment, because a lack of match in this area may be central to the emergence of many deviant behaviors including alcohol abuse.[13] We focus on the chronic persistence of mismatch across a substantial proportion of adolescence. Sustained movement within a high-risk

[a]This work was supported by grant DA/AA 03395 from the National Institute of Drug Abuse, Washington, DC, grant AA 08747-02 from the National Institute of Alcohol Abuse and Alcoholism, Washington, DC, and the Alcoholic Beverage Medical Research Foundation, Baltimore, Maryland.

developmental pathway may better portend continuity of intensive use or use consequences into early adulthood than would the static measurement of personality-environmental match at one point in time. Finally, we operationalize familial and peer support systems in terms of *perceived* environmental support. Psychological distress and problem behavior often appear to be most proximally connected to and invariant with the symbolic, subjectively meaningful environment.[4,15]

According to this perspective, personality needs for social support and approval as well as social support in the perceived environment will be differentially related to developmental outcome depending on the functional meaning of each, which derives from their unique interaction across time. For example, disrupted social support systems are thought to characterize the family relations of many FH+ youth. However, this may only increase vulnerability to relatively intensive alcohol use or other externalizing behaviors in those adolescents who maintain high needs for succorance and approval from others. Alternately, the continuing perception of inadequate social supports by youth with substantial needs in this area may affect increased vulnerability to negative social consequences of use, even when use behaviors do not appear excessive during adolescence.

The specific aims of this study are twofold: (1) to determine if the developmental patterns of use intensity *versus* use consequences systematically vary on the basis of a history of familial alcoholism alone or in interaction with gender or age and (2) to test the main and interactive effects of family history and one specific aspect of personality-environment mismatch in contributing to variability in developmental patterns of use intensity and use consequences. These relations are studied across two age sequences, from 12 to 15 to 18 years and from 15 to 18 to 21 years.

Differences in extent and rate of developmental change as well as in level of use intensity and use consequences will be examined. These analyses will provide a strong test of the hypothesis that a history of familial alcoholism, in combination with one aspect of personality/environment mismatch, promotes the development of patterns of use behaviors or related problems that are unique to this subtype of adolescents and that may affect the salience of risk factors at different points in time.

METHOD

Subjects

A sample of eligible adolescents stratified by age and gender was recruited through random telephone calls covering all but five counties of New Jersey. Taking into account unlisted numbers and variations in population density, this procedure is estimated to reach 95% of all households in the specified area. An initial anonymous telephone interview was used to identify households with eligible adolescents and to obtain basic demographic information. Eligibility was based on birth year, the absence of language barriers, and the absence of severe mental or physical handicaps. Comparisons of nonparticipants and participants indicated that higher levels of family income and of parental education are slightly overrepresented in the study sample.

Samples of 12-year-olds (birth years 1967–1969), 15-year-olds (birth years 1964–1966), and 18-year-olds (birth years 1961–1963) were first tested in 1979–1981 (T1), retested in 1982–1984 (T2) at ages 15, 18, and 21, and again retested in 1985–1987 (T3) at ages 18, 21, and 24, respectively, yielding a 92% retest rate. Analyses and results reported here are based on test-retest data of the two youngest age groups (217 males and 213 females tested at 12, 15, and 18 years of age; 217 males and 223 females tested at 15, 18, and 21 years of age). The third group was excluded for several reasons: (1) the legal status of alcohol use changes at age 21; (2) patterns of alcohol use are generally well established by age 21; (3) after leaving the parental home, young adults are more likely to actively search for social environments that can match their social needs; and (4) one of the variables used to operationalize person-environment fit (i.e., parental warmth) was not assessed beyond age 18.

Demographic characteristics of the sample of 870 adolescents at T1 were generally comparable to those of the State of New Jersey[16] and were as follows: (a) 31% were the oldest, 28% middle-born, 37% the youngest, and 4% only children; (b) 22% came from families with a yearly income of $20,000 or less, 30% from families with an income of $20,000–30,000, and 48% from families with an income of over $30,000; (c) 50% were Catholic, 32% were Protestant, 10% Jewish, and 8% had another or no religion; (d) 79% lived with both biological parents, 10% with a single parent, and the remaining 11% with other parental arrangements (e.g., foster parents, stepparents, adoptive parents); (e) 89% were white, 8% black, and the remaining 3% of another race.

Each subject was assigned a trained interviewer upon arrival at the laboratory. All self-report questionnaires were administered individually, and subjects were repeatedly assured of the anonymity and confidentiality of their responses. Similar batteries of measures were administered at all three test occasions.

Measures

Included in the present analyses are the following self-report data:

Use consequences. Adolescents were presented a list of 53 problems associated with alcohol use and asked to indicate how often or how many times each problem had happened ("ever" at T1 *versus* "within the last 3 years" at T2 and T3). Five-point response scales ranged either from 1 (never) to 5 (always or almost always) or from 1 (never) to 5 (more than 10 times). Previous factor analyses yielded the Rutgers Alcohol Problem Index (RAPI), a 23-item scale with an internal consistency of .92.[17] Negative physical, psychological, and social consequences of use are represented. The most commonly endorsed (by approximately 20–35% of users) consequences included getting into fights, neglecting responsibilities, feeling the need for more alcohol to get the same effect, memory loss, noticing a change in personality, interference with homework or studying, and causing shame or embarrassment. At T1, 77% of the sample (ages 12 and 15) reported the experience of one or no use-related consequences. This proportion dropped to 54% at T2 (ages 15 and 18) and to 43% at T3 (ages 18 and 21). At each age, only 3% of the sample scored at or above the mean of a matched clinical substance abuse sample.[18]

Use intensity. All measures combined information on the use of beer, wine, and distilled liquor and included: (a) frequency of alcohol use in the last year, (b) typical quantity of alcohol consumed per drinking occasion, (c) maximum quantity of alcohol consumed within the last 3 years, (d) frequency with which that quantity was consumed in the last year, (e) number of times getting "drunk" in the last 3 years, and (f) relative frequency of getting drunk when drinking in the last year. Data on the prevalence of alcohol use in the Health and Human Development Project sample at T1 and T2 are comparable to national survey data for same-age subjects in this geographic region.[19,20] Statistics descriptive of subjects' relatively more intensive drinking behaviors are presented in TABLE 1 for each age/sex group at T1, T2, and T3.

TABLE 1. Use Intensity by Age and Gender

Gender		Age (yr)			
		12	15	18	21
	Maximum Quantity Consumed during the Last 3 Years				
Females	0–1 units[a]	88.5[b]	44.2	13.5	
	2–3 units	7.2	16.8	18.3	
	4–5 units	2.8	12.5	22.1	
	6+ units	1.5	26.5	46.2	
	0–1 units		40.1	12.6	10.4
	2–3 units		13.5	12.7	10.4
	4–5 units		15.3	20.3	22.5
	6+ units		31.1	54.5	56.7
Males	0–1 units	79.9	39.3	14.5	
	2–3 units	11.6	10.7	7.5	
	4–5 units	6.1	10.3	7.5	
	6+ units	2.4	39.7	70.6	
	0–1 units		27.3	9.3	6.0
	2–3 units		15.3	4.2	3.3
	4–5 units		13.5	7.9	8.8
	6+ units		43.9	78.7	81.9
	Frequency of Consuming Maximum Quantity during the Last Year				
Females	0–2 times	93.3	79.3	61.5	
	3–9 times	4.3	14.4	24.0	
	10+ times	2.4	6.2	14.4	
	0–2 times		72.5	61.7	56.8
	3–9 times		17.6	22.5	25.7
	10+ times		9.1	15.8	17.6
Males	0–2 times	91.6	77.6	47.7	
	3–9 times	6.1	14.5	20.1	
	10+ times	2.3	7.9	32.2	
	0–2 times		75.5	51.4	42.1
	3–9 times		12.5	22.7	22.7
	10+ times		12.0	25.9	35.2
	Number of Times Drunk during the Last Year				
Females	0 times	95.7	57.2	25.0	
	1–2 times	1.4	11.5	9.1	
	3–5 times	2.0	5.3	11.1	
	6–19 times	.5	19.3	31.2	
	20+ times	.5	6.7	23.6	
	0 times		47.7	23.4	18.5
	1–2 times		11.3	9.0	12.2
	3–5 times		7.2	7.3	11.3
	6–19 times		24.8	32.0	33.8
	20+ times		9.0	28.4	24.3

continued

TABLE 1. *continued*

Gender		Age (yr)			
		12	15	18	21
	Number of Times Drunk during the Last Year				
Males	0 times	93.5	54.2	21.5	
	1–2 times	2.3	7.9	11.7	
	3–5 times	2.8	6.5	4.2	
	6–19 times	1.4	21.5	24.7	
	20+ times	.0	9.8	37.9	
			46.3	18.5	12.0
	1–2 times		11.6	4.2	6.9
	3–5 times		6.9	6.9	5.1
	6–19 times		20.8	31.5	28.2
	20+ times		14.4	38.9	47.7

[a]One unit = one 12 oz. regular beer = 5 oz. wine = 1.5 oz. 80 proof distilled liquor.
[b]Percentage of the total age/sex group.

Principal components analyses of the six use measures were carried out separately by occasion, with gender and age of subjects partialed out. In each case, eigenvalues indicated the extraction of a single component accounting for between 69 and 75% of the common variance. Variable loadings ranged between .69 and .92. Individual scores for use intensity were obtained as the sum of the six use measures.

Family history of alcoholism. Subjects and their parents completed a family medical history questionnaire about their own diseases, disorders, illnesses, and treatment histories as well as those of the subjects' siblings and grandparents. Subjects and their parents were also asked to indicate parents' typical drinking patterns and whether alcoholism was a factor in parental separation or divorce. Information was updated at each test time. Subjects were classified in terms of (1) presence of biological parental alcoholism ($n = 127$) and (2) presence of biological grandparental alcoholism ($n = 132$). Criteria for parental alcoholism included alcoholism treatment, health problems due to alcohol use, alcoholism as a cause of death, divorce or separation attributed to alcoholism, or parental or subject reports of parental drinking history consistent with alcoholism. Classification of grandparental alcoholism was based on the same variables except for the last two criteria, because this information was not available. These classifications were entered into the analyses as two dichotomous variables (0 = absence; 1 = presence of alcoholism).

Social mismatch. Measurement of social mismatch was based on three components: (1) social need, (2) perceived parental warmth, and (3) perceived social status/integration. Each component measure was assessed at T1 and T2. With a shortened version of Jackson's[21] Personality Research Form E (PRF E), individual scores on social need were obtained as the sum of six scale scores including need for affiliation, need for exhibition, need for nurturance, need for social recognition, need for succorance, and (low) need for autonomy. Three-year stability coefficients ranged between .37 and .57. Perceived parental warmth was assessed by an 18-item scale (e.g., how often do your parents praise you; comfort you when you are afraid; make you feel wanted; enjoy doing things with you; smile at you) with an internal consistency of .94[22] and 3-year stabilities ranging between .54 and .60. To assess perceived social status/integration, adolescents were presented eight items and asked to indicate if the item was true for them and (if yes) how much they were bothered by the described condition (e.g., friends don't respect my opinions; I don't get along with parent(s); classmates don't like me; I get into trouble in school; I am not popular with the opposite sex) on 4-point scales ranging from 1 (not at all) to 4 (very much). A *high* score on this scale reflects a subjectively unsatisfactory social status and poor social integration. Three-year stability coefficients ranged between .23 and .43.

Social mismatch was defined as a pattern of high scores on the social needs scale, high scores on the social status scale, and low scores on the parental warmth scale. Furthermore, individual differences in *chronic* social mismatch were measured by adding social mismatch scores at T1 and T2 with resulting high scores implying chronically high levels of social mismatch over a 3-year time span. (See also Labouvie *et al.*[23])

Analyses

Two-step hierarchical regression analyses were computed to examine developmental patterns of change in use intensity and use consequences, with step 1 yielding statistical tests for all main effects of interest and step 2 providing statistical tests for all two-way interactions. The modeling of individual growth curves was based on the method of orthogonal polynomials.[24] Specifically, individual patterns of intraindividual change in both use intensity and use consequences were characterized in terms of three scores: (1) level (T1 + T2 + T3) representing an individual's overall level on each variable across the full time interval, (2) a linear trend (T3 − T1) representing an individual's net increase or decrease in each variable from beginning to end of the specified time span, and (3) a quadratic trend [(T3 − T2) − (T2 − T1)] representing deceleration on acceleration in change in each variable during the 6-year time interval.

In the first set of hierarchical regression analyses, level, linear trend, and quadratic trend of use intensity were the dependent variables and examined in relation to a predictor set including gender (0 = female, 1 = male), age sequence (12–15–18 years *versus* 15–18–21 years), parental alcoholism, grandparental alcoholism, and chronic social mismatch. In the second set of hierarchical regression analyses, level, linear trend, and quadratic trend of use consequences served as dependent variables. To the predictors used in the first set of analyses were added (1) level of use intensity as a predictor of level of use consequences, (2) level and linear trend of use intensity as predictors of differential linear trends in use consequences, and (3) level, linear trend, and quadratic trend of use intensity as predictors of differential quadratic trends in use consequences.

RESULTS

Descriptive Statistics

Correlations between the independent variables (i.e., parental alcoholism, grandparental alcoholism, gender, age, and chronic mismatch) were all small, ranging between −.11 and +.15 and indicating a high degree of independence among the predictors. Correlations between the dependent variables are shown in TABLE 2. As can be seen, the highest correlations were found between corresponding trends in use intensity and in use consequences, suggesting temporal synchrony between both use

TABLE 2. Correlations between Dependent Variables

| | Variable | | | | |
	1	2	3	4	5
Use Intensity					
1. Level	—				
2. Linear Trend	−.02	—			
3. Quadratic Trend	−.30*	−.01	—		
Use Consequences					
4. Level	.58*	−.07	−.17*	—	
5. Linear Trend	.08	.41*	−.12*	.17*	—
6. Quadratic Trend	−.10*	.03	.32*	−.19*	−.01

*$p < 0.01$.

characteristics. Also note the fair degree of independence between the three curve components of each dependent variable.

Average longitudinal patterns for use intensity and use consequences are shown in FIGURE 1a and b. A steady linear increase in use intensity from 12 to 18 is followed by a subsequent leveling-off or deceleration from 18 to 21. As far as use consequences are concerned, a pattern of decelerating increases (i.e., negative quadratic trend) is evident in both longitudinal sequences, but the pattern is more pronounced for the age span from 15 to 21.

Regression Analyses

Of six incremental F tests assessing the significance of a combined total of 90 two-way interactions (i.e., step 2), only one reached significance ($p < 0.01$). Spec-

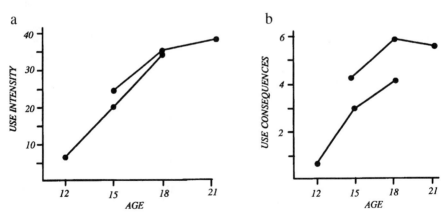

FIGURE 1. (a) Longitudinal patterns of alcohol use intensity across six years for two adolescent age sequences. (b) Longitudinal patterns of alcohol use consequences across six years for two adolescent age sequences.

ifically, the level of use intensity by social mismatch and the level of use intensity by age interactions were significant for level of use consequences as the dependent variable ($p < 0.05$ for the corresponding t tests). However, both interactions accounted collectively for less than 2% of the variance. In comparison, all six F tests used to assess the significance of a combined total of 36 main effects (i.e., step 1) were significant ($p < 0.01$). The proportions of variance (R^2) accounted for were as follows: 26.6% for level of use intensity, 15.4% for linear trend in use intensity, 4.2% for quadratic trend in use intensity; 36.6% for level of use consequences, 17.8% for linear trend in use consequences; and 11.3% for quadratic trend in use consequences. All statistically significant main effects with corresponding unstandardized regression weights are shown in FIGURES 2 through 4.

Developmental synchrony between use intensity and use consequences is evident in the fact that (1) differences in level of use intensity predict differences in level of use consequences, (2) differential linear trends in use intensity predict differential linear trends in use consequences, and (3) differential quadratic trends in use intensity predict differential quadratic trends in use consequences. The corresponding regression weights are all statistically significant and positive. In addition, a positive main effect of level of use intensity on linear trends in use consequences suggests the presence of a cumulative effect, with chronically higher levels of use intensity fostering stronger increases in use consequences over time.

Main effects of gender indicate that males exhibit higher overall levels of use intensity and use consequences as well as stronger linear trends in use intensity. Main effects of age are consistent with the developmental patterns of use intensity presented in FIGURE 1. They indicate that the pattern of use intensity from 15 to 21 years compared to that for the 12–18 age period exhibits a higher overall level, a weaker linear trend, and a stronger deceleration (i.e., stronger negative quadratic trend). Furthermore, the absence of any direct effects of age on use consequences indicates that age-related differences in the patterns of use consequences, as shown

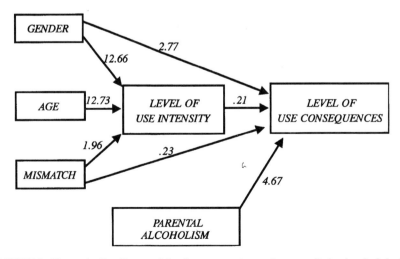

FIGURE 2. Unstandardized beta weights from regression analyses predicting level of alcohol use intensity and consequences.

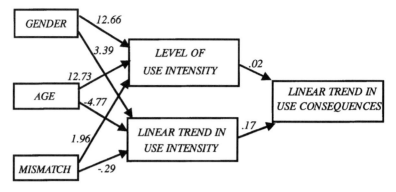

FIGURE 3. Unstandardized beta weights from regression analyses predicting linear trends in alcohol use intensity and consequences.

in FIGURE 2, are mediated through age-related differences in the patterns of use intensity.

Main effects for chronic social mismatch indicate that higher levels of mismatch are predictive of patterns of use intensity that exhibit higher overall levels but weaker linear trends. In addition, mismatch also has a direct effect on use consequences, suggesting that higher levels of chronic social mismatch are predictive of higher overall levels of use consequences independent of the overall level of use intensity.

As far as family history is concerned, the present findings indicate that the presence of grandparental alcoholism does not predict any differences in the developmental patterns of use intensity or use consequences. This interpretation is supported by the low correlation between parental and grandparental alcoholism ($r = 0.15$) and the lack of significant interaction effects between these two factors in the regression analyses. In comparison, the presence of parental alcoholism is predictive of (1) patterns of use intensity that exhibit stronger deceleration (i.e., stronger negative quadratic trends) and (2) patterns of use consequences that display higher overall levels and stronger deceleration. It is worth noting that the effects of parental alcoholism on use consequences are, for the most part, direct rather than mediated through use intensity.

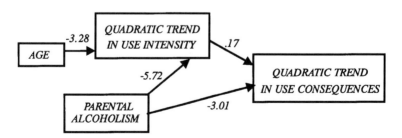

FIGURE 4. Unstandardized beta weights from regression analyses predicting quadratic trends in alcohol use intensity and consequences.

DISCUSSION

As far as the generalizability of our findings is concerned, note that the sample is most representative of working-and middle-class white adolescents. The expected main effects of age and gender were found. Males exhibited higher levels of, as well as greater increases in, use intensity and consequences across time than did females. Large increases in use intensity and consequences were observed from age 12–15–18, with a deceleration in both of these aspects of use from 18–21. The inclusion of representative numbers of female subjects is noteworthy given their exclusion from many high risk studies. Finally, the relative equivalence of the effects of chronic mismatch across the two age sequences studied suggests that adolescence appears to be a rather uniformly sensitive period with respect to the negative influence of mismatch on use behaviors and consequences.

Although differences in alcohol use between males and females were normative, we found no evidence of gender-linked differences in the relation between family history or social mismatch and use intensity or use consequences. In comparison, Windle[25] recently found that low familial support predicted problem behaviors of female, but not male, adolescents in a short-term longitudinal study of social support and life stressors. The present results suggest that social support may play a role in the development of adaptive and nonadaptive behaviors in both sexes. However, it may be necessary to study this aspect of the environmental context in relation to differences in individuals' needs for such support. That is, we have shown that the discrepancy between social personality needs and perceived social support in both males and females tends to exacerbate negative consequences of use both directly and through effects mediated by higher levels of consumption.

It is sometimes assumed that the heightened vulnerability of FH+ youth is linked to risk factors that are unique or that operate through developmental processes that are different in timing or nature from those that increase risk in the general population. Within the confines of our measures, the results of this 6-year longitudinal study do not support this notion. Chronic social personality-environment mismatch was a significant risk factor for adolescents of both sexes regardless of whether their family history was positive or negative for alcoholism on the part of first-degree relatives or grandparents. That is, discrepancies between chronically maintained social needs and perceptions of familial and peer support affected the intensity of alcohol use behaviors over time in both FH+ and FH– adolescents. Regardless of family history, chronic social mismatch also increased the level of negative consequences of use experienced during adolescence, indirectly through its effect on increased levels of use as well as through a direct pathway independent of level of use. Finally, FH+ and FH– adolescents did not report different levels of chronic mismatch. This finding is consistent with the notion that both groups receive comparable emotional support.[26] Overall, the data clearly suggest that the effects of family history on use consequences is not mediated through social mismatch.

As found at earlier test occasions with this sample[7] and in some other samples,[8] parental alcoholism directly influenced the level of negative consequences of use, but not use intensity. FH+ adolescents of both sexes reported experiencing higher levels of negative consequences of use across time. The effect of parental alcoholism on quadratic trends in both use intensity and consequences suggests that the use behaviors of FH+ adolescents accelerate faster during early and mid-adolescence relative to FH– adolescents. However, their use curves tend also to level off or decelerate somewhat faster than do those of FH– youth during later adolescence. Given the high degree of developmental synchrony between use intensity and use consequences, it is noteworthy that the effect of parental alcoholism on use consequences was not mediated by use intensity. Thus, it appears that independent of use intensity, FH+ youth may be more vulnerable than others to negative consequences of relatively low levels of alcohol use early in adolescence. It is possible that FH+ adolescents are more sensitized than others to behavioral cues that may be related to potentially problematic use. It is, of course, also possible that they are more likely to attribute behavioral changes or problems to alcohol use *per se* than are FH– youth. With respect to social consequences of use, normative levels of use in FH+ adolescents may also elicit responses from others aimed at limiting and/or restraining further increases in use intensity. The same levels of alcohol consumption in FH– adolescents may not elicit such responses. At the same time, grandparental alcoholism showed no reliable relations to either use intensity or consequences, underscoring the need to define and contrast the various typologies used to operationalize genetic pedigrees in high risk research.[11] Overall, the modest strength of familial alcoholism history as a unique predictor of negative use consequences emphasizes the heterogeneity of FH+ adolescents in this nonclinical sample. Our data support the idea that factors that increase vulnerability to the problematic use of alcohol may often operate similarly in both "high risk" and "low risk" populations. However, the mechanisms through which these and factors such as cognitive motivations for use and other aspects of personality directly increase use consequences are not well understood.[27]

REFERENCES

1. CLONINGER, C. R. 1987. Neurogenetic adaptive mechanisms in alcoholism. Science **236:** 410–416.
2. SADAVA, S. W. 1990. Problem drinking and alcohol problems: Widening the circle of covariation. *In* Recent Developments in Alcoholism. M. Galanter, ed. Vol. **8:** 173–201. Plenum Press. New York, NY.
3. WHITE, H. R. 1987. Longitudinal stability and dimensional structure of problem drinking in adolescence. J. Stud. Alcohol. **48:** 541–550.
4. WHITE, H. R. & M. E. BATES. 1993. Self-attributed consequences of cocaine use. Int. J. Addict. **28:** 187–209.
5. ALTERMAN, A. I., J. S. SEARLES & J. G. HALL. 1989. Failure to find differences in drinking behavior as a function of familial risk for alcoholism: A replication. J. Abnorm. Psychol. **98:** 50–53.
6. KNOP, J., T. W. TEASDALE, F. SCHULSINGER & D. W. GOODWIN. 1985. A prospective study of young men at high risk for alcoholism: School behavior and achievement. J. Stud. Alcohol. **46:** 273–278.
7. PANDINA, R. J. & V. JOHNSON. 1990. Familial drinking history as a predictor of alcohol and drug consumption among adolescent children. J. Stud. Alcohol **50:** 245–253.
8. SCHUCKIT, M. A. & S. SWEENEY. 1987. Substance use and mental health problems among sons of alcoholics and controls. J. Stud. Alcohol. **48:** 528–534.
9. LERNER, R. M. 1983. A "goodness of fit" model of person-context interaction. *In* Human Development: An Interactional Perspective. D. Magnusson & V. L. Allen, eds.: 279–294. Academic Press. New York, NY.
10. MONROE, S. M. & S. C. STEINER. 1986. Social support and psychopathology: Interrelations with preexisting disorder, stress and personality. J. Abnorm. Psychol. **95:** 29–39.
11. SHER, K. J. 1991. Children of Alcoholics: A Critical Appraisal of Theory and Research. University of Chicago Press. Chicago, IL.
12. CHESS, S. & A. THOMAS. 1984. Origins and Evolution of Behavior Disorders. Brunner/Mazel. New York, NY.
13. KAPLAN, H. B. 1980. Deviant Behavior in Defense of Self. Academic Press. New York, NY.

14. JESSOR, R., J. E. DONOVAN & F. M. COSTA. 1991. Beyond Adolescence: Problem Behavior and Young Adult Development. Cambridge University Press. New York, NY.
15. LIEBERMAN, M. A. 1982. The effects of social supports on responses to stress. *In* Handbook of Stress: Theoretical and Clinical Aspects. L. Goldberger & S. Breznitz, eds.: 764–783. Free Press. New York, NY.
16. BUREAU OF THE CENSUS. 1981. Current Population Survey: Money, Income and Poverty Status of Families and Persons in the United States, 1980 (current population reports, Series P-60, No. 127). US Government Printing Office. Washington, DC.
17. WHITE, H. R. & E. W. LABOUVIE. 1989. Towards the assessment of adolescent problem drinking. J. Stud. Alcohol **50:** 30–37.
18. WHITE, H. R., E. W. LABOUVIE, W. J. FILSTEAD, J. CONLIN, R. J. PANDINA & D. PARRELLA. 1988. Assessing alcohol problems in clinical and nonclinical adolescent populations. Alcoholism: Clin. Exp. Res. **12:** 328.
19. JOHNSTON, L. D., P. M. OMALLEY & J. G. BACHMAN. 1985. Use of Licit and Illicit Drugs by American High School Students, 1975–1984. National Institute on Drug Abuse. Rockville, MD.
20. MILLER, J. D., I. H. CISIN, H. GARDNER-KEATON, A. V. HARRELL, P. W. WERTZ, H. I. ABELSON & P. M. FISHBURNE. 1983. National Survey on Drug Abuse: Main Findings 1982. National Institute on Drug Abuse. Rockville, MD.
21. JACKSON, D. N. 1974. Personality Research Form Manual. Research Psychologists Press. Goshen, NY.
22. STREIT, F. 1978. Technical Manual: Youth Perception Inventory. Essence Publications. Highland Park, NJ.
23. LABOUVIE, E. W., R. J. PANDINA & V. JOHNSON. 1991. Developmental trajectories of substance use in adolescence: Differences and predictors. Int. J. Behav. Dev. **14:** 305–328.
24. COHEN, J. & P. COHEN. 1983. Applied Multiple Regression/Correlation Analysis for the Behavioral Sciences. 2nd Edit. Lawrence Erlbaum. Hillsdale, NJ.
25. WINDLE, M. 1992. A longitudinal study of stress buffering for adolescent problem behaviors. Dev. Psychol. **28:** 522–530.
26. CLAIR, D. & M. GENEST. 1987. Variables associated with the adjustment of offspring of alcoholic fathers. J. Stud. Alcohol **48:** 345–355.
27. BATES, M. E. 1993. Psychology. *In* Recent Developments in Alcoholism. M. Galanter, ed.: 45–72. Vol. **11.** Plenum Press. New York, NY.

Research Opportunities in Typology and Genetics of Alcoholism

ENOCH GORDIS AND JOHN P. ALLEN

National Institute on Alcohol Abuse and Alcoholism
5600 Fishers Lane
Rockville, Maryland 20857

The two themes addressed by this conference, alcoholism subtypes and genetic influences on alcoholism, are of considerable interest to the National Institute on Alcohol Abuse and Alcoholism (NIAAA), the alcoholism research and treatment communities, and the health care system at large. Despite noteworthy advances in understanding these aspects of alcoholism, fundamental questions remain about each and, perhaps most importantly, about the two topics when considered in light of each other.

In the sections that follow we address some issues that surround research in these areas. We first offer impressions on development of useful alcoholism typologies, then invite attention to salient topics in alcoholism genetic research, and finally pose research questions deriving from reflection on how genetics and typology research might contribute to each other.

TYPOLOGY OF ALCOHOLISM

Even cursory observation leads to the realization that alcoholics differ widely in drinking patterns, consequences they suffer as a result of drinking, age of onset, family history of alcoholism, personality variables, and a host of other variables. Appreciation of heterogeneity of alcoholics is hardly new, and as far back as the 19th century scholarly efforts were made to categorize alcoholics into meaningful subtypes.[1] Attempts at taxonomization, however, have recently grown in popularity, and today a rich variety of classificatory schemata have been tendered for consideration. Generally such systems focus on distinguishing alcoholics on the basis of a single dimension, most often personality or age of onset. Of greater heuristic value, "broad spectrum" taxonomies are also being proposed. These symptoms attempt to categorize alcoholics by simultaneously taking into account a range of conceptually diverse dimensions such as collateral psychopathology, treatment prognosis, and personality characteristics.

Ideally, a typology of alcoholism should identify a relatively small number of groupings that are at once mutually exclusive of each other, yet inclusive of all alcoholics. This typology should reflect meaningful similarities and differences in etiology, course of illness, choice of treatment, and prognosis. Statistical agglomerative techniques, such as cluster analysis or Q-type factor analysis, are often employed to yield subtypes; however, because these techniques can potentially group subjects according to a variety of inter-subject variables, it is important to assure that the basis of subject categorization is alcoholism specific. Moreover, as these sensitive, multivariate techniques run the risk of capturing study population-specific relationships, they should be replicated with new and diverse samples before they are legitimately accorded a high level of credence.

Despite the popularity of research on development of alcoholism typologies, unfortunately few investigations have explored the relationship of typologic membership to treatment response. Although one might argue that demonstration of a rich array of patient-treatment matching studies provides *a fortiori* evidence of the meaningfulness and applied value of alcoholism typologies, only two matching studies have actually included explicit alcoholism types.[2,3] Other patient-treatment matching studies have found a differential response to interventions as a function of a single, generally median-dichotomized, patient variable.

NIAAA's large-scale, randomized clinical trial of patient-treatment interactions, Project MATCH, should shed light on the utility of patient typologies in predicting differential response to three alternative interventions and may generate a typological scheme developed empirically from observed variation in responsiveness to treatment. The Institute's continuously open announcements on Patient-Treatment Matching and Treatment Assessment Research also solicit rigorous investigations of alcoholism subtypes.

Another important issue in alcoholism typologies is delineation of the role of patient demographic characteristics. Research by Cloninger, Del Boca, Allen, and their colleagues presented in this volume argues that gender may prove an important underlying basis for alcoholism typology. In Phase II of its collaborative investigation with the Russian Research Center of Addictions, NIAAA is further investigating the role of culture in alcoholism taxonomies, a topic heretofore unexplored. Investigations are also needed to determine how other sociodemographic variables such as ethnicity, social status, and educational achievement may relate to taxonomies. Importantly, research on the role of demographic factors in taxonomies should determine if demographic groups differ in distribution of cases across a common set of subtypes or if the demographic variables are themselves a foundation of the taxonomy.

Although research on typologies is heartily encouraged, investigators should continue to consider the possibility that the heterogeneity of alcoholism may well be more apparent than real. For example, prior to understanding the etiology and course of tuberculosis, its multi-organ manifestations could have been seen as separate and distinct illnesses. But while skepticism on the existence of alcoholic taxonomies continues to represent a healthy scientific attitude, it should also be borne in mind that heterogeneity of a disorder may be fundamental. Despite a common nomenclature, the two variants of diabetes differ dramatically in etiology, severity, prognosis, and treatment needs.

GENETICS OF ALCOHOLISM

Research on genetics of alcoholism is no less important than is research on subtypes. Identifying genes and their products associated with alcoholism would significantly advance our understanding of the origin and nature of alcoholism and perhaps its prevention and treatment as well. Investigations in this field already suggest that phenomena traditionally associated with the alcohol dependence syndrome are influenced by different genes. For example, in a dependent organism severity of reaction to abrupt withdrawal of alcohol is mediated by genes different from those associated with initial sensitivity to alcohol and at least some forms of tolerance.[4]

At a highly applied level, research on genetics may assist in the recognition of markers of risk for alcoholism. Hence, individuals at high risk for alcoholism could be identified before the problem becomes severe and consequently more difficult to treat. Recent studies on markers such as the P-300 brain waive and low voltage alpha brain activity suggest the potential of this line of research activity. Early, effective intervention targeted to those at high risk for alcoholism would decrease the financial cost of treating alcoholism and, no less importantly, would diminish the human suffering associated with alcoholism. When the range of genetic contributions to the development of alcoholism are ultimately identified, research might also lead to techniques to disrupt them, thereby directly and definitively interrupting the disease process. Finally, specification of the genetic bases for alcoholism should also improve our understanding of the contribution of environmental influences on emergence and progression of alcohol dependence.

The key focus of genetics research is on mapping the genetic determinants of alcoholism. This is a near-herculean task because a large variety of genes apparently contribute to the development of alcoholism. Fortunately, new techniques in quantitative trait locus mapping have greatly improved the prospects that this quest will be successful. The NIAAA-supported multi-site cooperative agreement on genetics will take full advantage of these exciting technological advances. NIAAA, through its ongoing RO1 program announcement on genetics, also specifically solicits further research in this regard.

As noted, several lines of research suggest that alcoholism is polygenic rather than simply caused by the actions of one or two genes. Although little doubt exists that genes enhance risk for alcoholism, research has yet to specify vulnerability factors that are inherited, such as susceptibility to ethanol effects, proclivity to engage in high risk situations, and sensitivity to the reinforcing properties of alcohol.

An emerging and intriguing research topic involves identification of genetic characteristics that may protect against the development of alcoholism. It has been argued, for example, that inheritance of the inactive allele of the gene producing aldehyde dehydrogenase, by making ethanol consumption aversive, may also decrease the risk of alcoholism. Clearly, research in our field has emphasized determinants of alcoholism. Asking the converse question, however, why do most people not become alcoholic despite ready exposure to alcohol?, might provide new insights on the issue.

Whereas research attention should be directed primarily toward understanding the genetics of alcoholism, research on the genetic bases for the consequences of alcoholism as well as for drinking-related phenomena also merits attention. These phenomena are important in their own right, inasmuch as consequences such as liability for cirrhosis may be partially genetically based.

RESEARCH ON RELATIONSHIPS BETWEEN ALCOHOLIC SUBTYPES AND GENETICS

Even though typically studied in isolation, the topics of alcoholic subtypes and genetics of alcoholism, in fact, have much in common. Carefully conceptualized, well-designed research on either topic should therefore inform the other.

For example, as the genetic basis of alcoholism and the interactive roles of genetics and environment are elucidated, more meaningful systems should emerge in which alcoholics are classified into subtypes based on the contribution of genetic, developmental, and other factors. Current typologies categorize phenotypes of alcoholism. Future research might also identify genotypes.

More precise description and quantification of alcohol dependence, alcohol-seeking behaviors, drinking-related consequences, and the like will help advance research on typology as well as on genetics because both depend on accurate definition of patient characteristics. So too, research on alcoholism genetics and typology in alcoholism may suggest critical variables in patients that must be more precisely measured.

Research on the course of alcoholism, with and without formal intervention, should also suggest promising directions for constructing meaningful subtypes of patients as well as predicting genetic bases of and pathways to alcoholism.

Finally, the development of more sophisticated statistical techniques should simultaneously facilitate interpretation of complex results of quantitative trait loci research and should aid in the derivation of more meaningful, multifaceted subtypes of alcoholism.

Exciting advances in the development of analytic and statistical techniques, more refined conceptualization of alcoholism, and enhancement in research design have rapidly occurred during the last decade. Incorporation of this knowledge base into investigations on subtypes and genetics of alcoholism will materially improve further efforts to understand, prevent, and treat alcoholism.

REFERENCES

1. BABOR, T. F. & Z. S. DOLINSKY. 1988. Alcoholic typologies: Historical evaluation and empirical evaluation of some common classification schemes. *In* Alcoholism: Origins and Outcome. R. M. Rose & J. E. Barrett, eds. :245–266. Raven Press. New York, NY.
2. OXFORD, J. E., E. OPENHEIMER & G. EDWARDS. 1976. Abstinence or control: The outcome for excessive drinkers two years after consultation. Behav. Res. Ther. **14:** 409–418.
3. LITT, M. D., T. F. BABOR, F. K. DEL BOCA, R. M. KADDEN & N. COONEY. 1992. Types of alcoholics. II. Application of an empirically-derived typology to treatment matching. Arch. Gen. Psychiatry **49:** 609–614.
4. AGARWAL, D. P. & H. W. GOEDDE. 1989. Human aldehyde dehydrogenases: Their role in alcoholism. Alcohol **6:** 517–523.

Matching Alcoholics to Coping Skills or Interactional Therapies

Role of Intervening Variables

RONALD M. KADDEN,[a] MARK D. LITT,[b]
AND NED L. COONEY[c]

[a]University of Connecticut School of Medicine
Farmington, Connecticut 06030

[b]University of Connecticut School of Dental Medicine
and School of Medicine
Farmington, Connecticut 06030

[c]West Haven Veterans Affairs Medical Center
West Haven, Connecticut 06516
and Yale University School of Medicine
New Haven, Connecticut 06510

In recent years there has been growing interest in aptitude-treatment interactions (ATI), involving "the degree to which alternative treatments have different effects as a function of person characteristics."[1] In the alcoholism field particularly, interest has been increasing in the possibility of identifying patient characteristics that would predict better (or worse) outcomes with various treatment approaches. We previously reported findings from an alcoholism treatment outcome study that suggest two patient characteristics as potential bases for patient-treatment matching.[2] The present paper explores several additional variables assessed in that study that are potential treatment-process variables or independent contributors to outcome variance. To provide a context for this work, we first provide a brief review of some pertinent conceptual issues related to patient-treatment matching for alcoholism.

MATCHING STRATEGY

In clinical settings, patient-treatment matching for alcoholism, to the extent it is done at all, is based on clinical judgment. Unfortunately, however, this approach is less effective than matching based on systematic assessment methods. In a prior report,[2] we noted that the clinicians treating those inpatients who were later to become subjects of our study were asked to choose the type of aftercare treatment that would be most effective for each of their patients. Patients who were then assigned, by chance, to the clinician-selected treatment had no better outcomes than did those assigned to the treatment not selected. Matching based on clinical judgment may suffer from the inaccuracy of clinical predictions in general[3] and with substance abusers in particular.[4]

In research settings, one strategy for studying matching is to conduct exploratory data analyses on the outcomes of clients assigned to different treatments without specific *a priori* matching hypotheses. For example, Smart[5] examined 186 possible interactions, but found that none was significant. Although exploratory data analyses could potentially yield useful findings, a consensus is developing that it is more productive to test *a priori* matching hypotheses that are based on theoretical considerations.[6,7]

PATIENT VARIABLES

Alcoholism treatment matching studies have focused mainly on patient variables, such as severity of alcohol dependence[8] or global psychopathology.[9] For the present study we chose three patient characteristics as matching variables that have been shown to indicate poor prognosis and to impede the treatment process: sociopathy, global psychopathology, and neuropsychological impairment.

Sociopathy. Sociopathy is often associated with alcoholism.[10,11] An early onset of excessive drinking and a more rapid progression to alcoholism have been found in those who exhibit antisocial personality traits,[12] consistent with Cloninger's[13] Type 2 and Babor *et al.*'s[14] Type B alcoholism. Mandell[11] documented a disproportionately high rate of dropout from alcoholism treatment by sociopathic alcoholics, and Woody *et al.*[15] showed that antisocial personality is a predictor of poor psychotherapy outcome in nondepressed opiate addicts.

Psychopathology. Psychopathology is also associated with alcoholism.[16,17] Woody *et al.*[18] found that treatment outcome was a function of an interaction between the severity of psychopathology and type of treatment; opiate abusers with low Addiction Severity Index (ASI) Psychiatric Severity scores[19] had equivalent outcomes with either professional psychotherapy (cognitive-behavioral or supportive-expressive psychotherapy) or paraprofessional drug counseling, whereas those subjects with middle and high range ASI Psychiatric Severity scores had better outcomes in psychotherapy than in drug counseling. Another study by McLellan *et al.*[20] found higher order interactions involving ASI Psychiatric Severity scores, social functioning variables, and type of substance abuse treatment program.

Neuropsychological Impairment. This patient characteristic is often observed in alcoholics.[21,22] Some studies have shown that the *process* of treatment is affected by neuropsychological impairment, whether in the form of resistance to treatment,[23] retention of treatment-relevant information,[24] or treatment dropouts.[25] Other studies have demonstrated a relation between neuropsychological deficits and *outcome*,[26,27] but some have found this relation to be weak or inconsistent.[28,29] Nevertheless, McCrady and Smith[30] recommended tailoring alcohol rehabilitation efforts to accommodate patients' neuropsychological deficits.

TREATMENT VARIABLES

Two distinct treatment modalities were selected for the present study based on prior demonstrations of their effectiveness with alcoholic clients and their widespread use in treatment settings.

Cognitive-Behavioral Coping Skills Training. Several controlled clinical trials have demonstrated that the addiction of social skills training to inpatient alcoholism treatment improved outcomes compared to standard inpatient treatment.[31–34] However, Jones *et al.*[35] failed to replicate Chaney *et al.*[31] using a higher socioeconomic sample from a private hospital.

Three studies evaluated outpatient skills training for alcoholics: Annis and Davis[36] found marked reductions in alcohol consumption and adverse consequences over a 6-month follow-up, but Ito *et al.*[37] and Sjoberg and Samsonowitz[38] found no difference between outpatient skills training and alternative treatment. In general, however, clinical outcome studies support the effectiveness of behavioral skills training, and elements of it have been increasingly incorporated in substance abuse programs.

Interactional Group Therapy. Evidence regarding the efficacy of interactional group therapy comes from uncontrolled outcome studies. For example, Yalom *et al.*[39] reported an evaluation of an interactional group approach with alcoholics: of those who entered therapy, 50% showed improvement on the drinking criteria, 20% deteriorated, and 30% remained about the same. Patients, therapists, and independent judges all gave high ratings on achievement of goals and reduction in distress, similar to a comparison sample of neurotic patients. Vanicelli[40] also reported success with the use of interactional group therapy with alcoholics.

POTENTIAL CONTRIBUTORS TO PATIENT-TREATMENT MATCHING EFFECT

Previous reports from our treatment matching study have focused on the interaction between types of patients (characterized by degrees of sociopathy, psychopathology, and cognitive impairment) and types of treatment (coping skills training *versus* interactional group therapy). Cross-sectional analyses at posttreatment and survival analyses over 2 years[2,41] both reveal that relapse rates were lower when high psychopathology clients received coping skills treatment and when low psychopathology clients received interactional treatment. Relapse rates were also lower when high sociopathy clients were treated in coping skills groups and low sociopathy clients were treated in interactional group therapy. Those clients classified as cognitively unimpaired had similar outcomes in both treatments, whereas those classified as cognitively impaired had better outcomes with interactional therapy and worse outcomes with coping skills treatment. Exploratory analyses of several other potential matching variables, such as age, gender, educational level, WAIS vocabulary score, income, socioeconomic status, family history of alcoholism, and severity of dependence on alcohol, revealed that none of them was a significant predictor of outcome, either alone or in an interaction with treatment type.

We also examined the ability of an alcoholism typology to serve as a basis for treatment matching. The typology classifies alcoholics on measures of vulnerability and severity, with type B alcoholics manifesting an earlier onset of problem drinking, more familial alcoholism, greater dependence on alcohol, and more symptoms of antisocial personality than type A alcoholics.[14] We found that the typology of Babor *et al.*[14] was valid in our sample and that it could be used as the basis for patient-treatment matching. The matching results were similar to those found with the unidimensional patient characterizations, psychopathology, or sociopathy.[42]

In the analyses reported thus far, we did little to explore possible variables that may moderate the impact of the patient-treatment interaction on treatment outcome. Some of the more likely candidates for this role include attendance at group sessions, group process variables, and client involvement in treatment modalities outside of, and in addition to, those employed in the study.

Attendance. Ito and Donovan[43] have described the effect of attendance in outpatient aftercare treatment, noting that increased attendance at sessions is associated with better treatment outcome. Sussman *et al.*[26] also reported that aftercare attendance was an independent predictor of relapse. In our study, we assessed the possibility that matched patients had better outcomes because they attended more therapy sessions.

Group Process Variables. A number of instruments have been developed for the assessment of group process variables and their impact on treatment outcome.[44] In the present study, the Group Environment Scale (GES)[45] was employed because it is easily completed by both therapists and clients, has norms for both, and contains a number of potentially useful subscales. It was anticipated that subjects in the coping skills groups in this study would rate their groups higher on the Task Orientation, Order/Organization, and Leader Control subscales and that interactional group subjects would provide higher ratings on the Expressiveness, Self-Discovery, and Anger/Aggression subscales. We also examined the relation between GES subscale scores and treatment outcome.

Ancillary Treatment. There is evidence that involvement in additional therapeutic activities, such as Alcoholics Anonymous (AA), can affect treatment outcome,[46,47] as can the use of Antabuse[48,49] or involvement in individual psychotherapy.[50] The present data set was therefore examined to determine if the use of other therapeutic modalities in addition to the treatment offered in the study influenced outcome, especially if they were differentially utilized by the clients assigned to the different treatment conditions.

The present paper assesses the contributions of attendance, group process, and ancillary treatments to the outcomes we described in our previously reported treatment matching study.[2,41]

METHOD

The following presents a brief summary of the methods; further details can be found in a report by Kadden *et al.*[2]

One hundred eighteen male (66%) and female (34%) patients were recruited from a 3-week inpatient treatment program and were assigned in cohorts, based on their discharge date, to either of two aftercare treatments. Ninety-six subjects who completed at least three aftercare group sessions were considered exposed to treatment and were included in the data analyses. Each aftercare group met for 26 weekly, 90-minute sessions. Five coping skills and five interactional groups were led by cotherapist pairs who were experienced with the therapy orientation to which they were assigned.

A treatment manual was formulated for the coping skills therapists that provided a highly structured group experience designed to enable group members to identify and cope with situations that may put them at risk for relapse.[51] The manual for conducting the interactional groups[52] was based on previous development of Yalom's interactional group therapy approach for use with alcoholics.[53,54] The goal of the interactional groups was to explore participants' interpersonal relationships and pathology as manifested in "here and now" interactions within the group.

We developed a rating scale to assess the distinctiveness of the treatments from one another.[55] Trained judges scored audiotaped segments from the coping skills and interactional groups, rating the frequency with which seven different therapeutic activities took place. Interrater reliabilities ranged from .83 to .97. The judges' ratings confirmed that the groups were conducted as specified in the treatment manuals and that the distinctiveness of the two treatments was maintained. None of the separately rated group activities was significantly related to posttreatment abstinence.[55]

Patient characteristics were assessed prior to treatment. Global psychopathology was measured using the Psychiatric Severity composite score of the ASI. Sociopathy was assessed using the Socialization Scale of the California Psychological Inventory

(CPI-So). Neuropsychological status was based on a composite score derived from the Wisconsin Card Sorting Test, the Trail Making Test, the Four-Word Short Term Memory Test, the Digit Symbol subtest of the Wechsler Adult Intelligence Scale, and the Face-Name Paired Associates test. (See Kadden et al.[2] for details.)

Treatment process was assessed using the Moos[45] Group Environment Scale. Clients were asked to complete this scale after 3, 12, and 23 group sessions (or at the next session, in case of absence), to assess group process at different stages in the life of the 26-session groups.

Outcome assessments included evaluations of psychopathology, employment, social behavior, and three measures of drinking-related behavior (number of days of abstinence and number of days of heavy drinking, both derived from the Time-Line Follow-Back procedure, and number of current alcohol-related problems, taken from the alcohol section of the Diagnostic Interview Schedule for DSM-III). Informants' reports and laboratory tests were used to verify subjects' self-reports of their drinking behavior.

RESULTS

Characteristics of the treatment sample as well as evidence that subjects in both types of treatment were comparable in pretreatment characteristics and in their exposure to treatment were reported previously.[2] No significant differences in outcome were attributable to therapist effects.

Patient self-reports are regarded as reliable, inasmuch as collaterals agreed with subjects' reports of abstinence from 65–76% of the time and with drinking reports from 89–92% of the time.[41]

The previously reported patient-characteristic × treatment-type interaction effects[2] that were sustained over a 2-year outcome period[41] provide very promising support for the concept of patient-treatment matching. The following analyses were conducted to determine if the matching effects were mediated by attendance at treatment sessions, group process variables, and/or other therapeutic activities that subjects may have engaged in while the study treatments were going on.

Relation between Attendance and Outcome

We first examined variables that might affect attendance. Regression analyses indicate that the number of weekly group sessions attended was not a function of client gender, education, or socioeconomic status, but was associated with the client's age, older clients attending more sessions ($t_{(95)} = 3.4$, $p < 0.001$, $R^2 = .11$). Attendance was not a function of any of the pretreatment drinking-related variables including days of heavy drinking, degree of dependence, complications of drinking, or family history of drinking, nor was attendance related to any of our hypothesized matching variables, namely, the severity of a client's psychological problems, the extent of his/her sociopathy, or neuropsychological impairment. Furthermore, no attendance differences were noted between the two types of treatment or between clients who were matched or mismatched according to our a priori hypotheses. However, attendance was a function of the client's social behavior, as assessed by the Psychosocial Functioning Inventory ($t_{(95)} = 2.15$, $p < 0.05$, $R^2 = .05$); those who functioned better socially at pretreatment attended more group sessions.

Attendance was related to several of the group process variables assessed by the GES, as rated by the clients at their third treatment session; scores on the Cohesion,

Leader Support, Expressiveness, Independence, Order/Organization, and Innovation subscales were all positively related to attendance (TABLE 1). However, attendance was not a function of type-of-treatment × GES-variable interactions.

Next we considered the impact of group attendance on various outcome measures. In a survival analysis utilizing a proportional hazards model (Cox regression[56]), attendance predicted time to first heavy drinking day ($p < 0.001$) (those who attended more went longer without relapsing), but there was no interaction of attendance with type of treatment. In a linear regression analysis, attendance was a significant negative predictor of the number of heavy drinking days during the 6-month treatment period, accounting for 9% of variance ($p < 0.01$), but the attendance × treatment-type interaction term was not significant. These results indicate that greater attendance at therapy sessions was predictive of positive outcome generally and was equally beneficial across both treatments.

Relation between Group Process and Outcome

As shown in TABLE 2, Coping Skills group members gave higher ratings to Leader Support, Leader Control, and Order/Organization at all three GES assessments (at the beginning, middle, and end of treatment) and higher ratings to Cohesion and Task Orientation at the first assessment. Interactional groups gave higher ratings to Expressiveness and Anger/Aggression at the third assessment only.

Survival analyses found main effects for the client ratings of group Cohesion and Leader Support as predictors of time to first heavy drinking day ($p < 0.05$), with higher ratings on the GES scales predicting longer survival. However, only one GES-rating × treatment-type interaction was significant; Expressiveness at the first GES assessment interacted with type of treatment to predict time to first heavy drinking day (improvement $\chi^2 = 5.4$, $p < 0.02$). Higher ratings of group Expressiveness were associated with longer time to relapse after interactional therapy and shorter time to relapse after cognitive-behavioral therapy.

Linear regression analyses were performed on all 10 GES subscales, at all 3 measurement points, and across 4 separate outcome intervals (6, 12, 18, and 24 months after the start of treatment), 120 analyses in all. The dependent variable was the number of heavy drinking days that occurred during one of the 6-month outcome

TABLE 1. Correlations between Group Environment Scale Subscales (Assessed at Session 3) and Attendance at Group Therapy Sessions

Variable	r
Cohesion	.43***
Leader Support	.41***
Expressiveness	.38***
Independence	.27**
Task Orientation	.26*
Self-Discovery	.14
Anger/Aggression	−.15
Order/Organization	.41***
Leader Control	.18
Innovation	.26*

*$p < 0.05$; **$p < 0.01$; ***$p < 0.001$.

TABLE 2. Comparison of Coping Skills and Interactional Group Ratings According to Group Environment Scale Subscales

GES Subscale	Session Number		
	3	12	23
Cohesion	CS>Int.		
Leader Support	CS>Int.	CS>Int.	CS>Int.
Expressiveness			Int.>CS
Independence			
Task Orientation	CS>Int.		
Self-Discovery			
Anger/Aggression			Int.>CS
Order/Organization	CS>Int.	CS>Int.	CS>Int.
Leader Control	CS>Int.	CS>Int.	CS>Int.
Innovation			

NOTE: Table entries indicate those GES subscales that were rated significantly higher (p <0.05) by the members of either Coping Skills (CS) or Interactional (Int.) groups at each of the three sessions when GES ratings were obtained.

measurement intervals. Of all the analyses that were run, only 12 were significant. Because so few were significant and no pattern was discernible, we conclude that in this study the GES variables were not robust predictors of treatment outcome.

Relation between Ancillary Treatments and Outcome

The percentages of subjects who participated in ancillary therapeutic activities at least once, during the 6 months that their groups were meeting, were as follows: self-help groups (e.g., AA) = 85%, Antabuse = 38%, and individual counseling = 23%. The numbers of times subjects used each of these modalities were uncorrelated with one another.

The number of participants in these additional therapeutic activities decreased for the most part over the course of the 2-year outcome period (TABLE 3). However, of those who participated in them, some continued their involvement at substantial levels over the course of the 2 years, as reflected in the median number of times participating.

No differences were noted between subjects assigned to the coping skills and interactional groups in their utilization of individual counseling or Antabuse (p >0.20), but those in the coping skills groups attended self-help meetings more often than did those assigned to interactional group therapy ($t_{(60)} = 2.01$, p <0.05), even though attendance at fellowship meetings was not actively encouraged in either type of treatment. No differences were found in the utilization of any of the ancillary treatments between subjects who were matched or mismatched to the two study treatments on the basis of the sociopathy characteristic.

The relation between the use of the ancillary treatments and a number of pre-treatment variables was tested. The variables examined were age, gender, education, socioeconomic status, marital status, days of heavy drinking, degree of dependence, complications of drinking, family history of drinking, ASI Psychiatric Severity, CPI-So score, DSM-III diagnosis of antisocial personality disorder, and extent of

TABLE 3. Number of Subjects Who Participated in Other Therapeutic Activities and Number of Times Participating

| | Outcome Assessment Points | | | | | |
| | 6 Months | | 12 Months | | 24 Months | |
Activity	Subjects (n)	Median Number of Times (range)	Subjects (n)	Median Number of Times (range)	Subjects (n)	Median Number of Times (range)
Individual counseling	20	11 (1–27)	11	25 (2–30)	14	14 (1–30)
Self-help groups/AA	74	20 (1–200)	43	13 (2–184)	32	30 (2–180)
Antabuse	33	45 (7–188)	3	72 (30–120)	2	29 (14–44)

neuropsychological impairment. The use of Antabuse was related to none of these pretreatment variables. Attendance at AA meetings was related only to CPI-So score ($r = .22$, $p < 0.05$), with those who were more sociopathic attending more meetings. Use of individual counseling was related to three pretreatment variables: gender, with females attending more counseling sessions ($r = .23$, $p < 0.05$); education, with the more educated using more counseling ($r = .24$, $p < 0.02$); and neuropsychological impairment, with those more impaired seeking more counseling ($r = .22$, $p < 0.05$).

There were no significant differences in heavy drinking days, at any of the outcome points, between those subjects who did and those who did not attend self-help meetings, take Antabuse, or receive individual counseling/therapy during the treatment phase. Neither were there any significant correlations between use of self-help meetings, Antabuse or individual counseling (reported at any of the follow-up points), and heavy drinking reported at those same follow-up points.

The interactions previously noted between CPI-So or ASI-Psychiatric Severity and treatment type were not reduced when the analyses (either linear regressions or survival analyses) controlled for the use of self-help meetings, Antabuse, or individual counseling.

It has been suggested that marital status may be an important factor in treatment outcome, through its impact on social support. In our sample, 44% of the subjects were married, and they were distributed approximately equally between the two treatment types. Marital status was not related to days of heavy drinking reported posttreatment, and there was no significant interaction between marital status and treatment type.

Accounting for Power. We have shown that several possible mediators of matching effects did not predict outcome, neither the time to the first heavy drinking day nor the number of heavy drinking days posttreatment. Power analyses were conducted to determine if our failure to find these effects was due simply to an insufficient number of subjects. In terms of survival analyses, the number of subjects reporting data was sufficient to have been able to detect even a small effect (i.e., $\mathbf{w} = .25$),[57] with a power of .80 and an alpha level of .05 if such an effect had been present. As for the regression analyses in which number of heavy drinking days was predicted, the N of the present study would have allowed us to detect an effect $\mathbf{f}^2 =$

.08 to .09, with a power of .80 and alpha set to .05. According to Cohen,[57] this effect size is in the small to medium range. We conclude that the failure of these potential mediating variables to account for the matching results is not due to a lack of power.

DISCUSSION

In the analyses presented here, we have examined several variables that might have played an intervening or moderating role in our matching findings. We found that attendance was related to subjects' age, level of social functioning, and several of the group process variables. Like others,[26,43,58] we found a main effect for attendance in predicting outcome (time to first heavy drinking day). However, the attendance variable did not interact with treatment type to affect treatment outcome. Thus, there was an impact of attendance, but it was equivalent across treatments and was not an intervening variable in the previously reported matching findings.

With respect to the GES variables, it was not surprising that the subjects in the coping skills treatment rated their groups higher on the factors of Leader Support and Control, Order/Organization, and Task Orientation or that the participants in the interactional therapy rated their groups higher on Expressiveness and on the degree to which the expression of Anger and Aggression are tolerated or encouraged. However, many of these rating differences were significant in only one of the three measurement periods; although the findings were in the expected direction, they were not as consistent as anticipated. Neither was there a consistent relationship of the GES variables to treatment outcome. Although the regression analyses revealed some significant findings, the chance of Type 1 error is very high because of the total number of analyses run. Only a single significant GES-variable × treatment-type interaction was found. If group process factors did indeed contribute to the significant matching findings in this research, those processes were not well captured by the subscales of the GES.

A few pretreatment variables were identified that were related to AA attendance and the use of individual counseling by our subjects. It was also found that subjects in the CB groups attended AA more than did subjects in the interactional groups, but the reason for this is not clear, because neither of the treatments took a position on AA attendance, either pro or con. No use of ancillary treatment activities, whether individual counseling, Antabuse, or AA, had a significant impact on treatment outcome or on the matching findings. Although it was not expected that these other therapeutic activities would interact with treatment type, it is curious that those subjects who made use of additional counseling or AA for a substantial length of time (most subjects discontinued Antabuse early) did not show improved drinking outcomes as a result of this involvement. However, it is difficult to interpret this null finding because subjects self-selected whether or not to engage in, and how much to attend, ancillary treatments.

In view of the inability of other variables to account for our matching findings, it seems reasonable to conclude that the observed benefits from patient-treatment matching are *not* a result of matched subjects making more use of additional treatment or supportive resources. Rather, the matching variables that were identified in the present study appear to have had a stronger effect than any of the ancillary treatment activities. It remains for further research to determine if the patient-treatment matching strategy suggested by our study will be an effective, practical method for assigning patients to treatment. Studies to evaluate this possibility should include the use of prospective matching rather than forming groups by random assignment, and the assignment of clients to their matched modality at the outset of their treatment experience rather than as aftercare.

REFERENCES

1. SNOW, R. E. 1991. Aptitude-treatment interaction as a framework for research on individual differences in psychotherapy. J. Consult. Clin. Psychol. **59:** 205–216.
2. KADDEN, R. M., N. L. COONEY, H. GETTER & M. D. LITT. 1989. Matching alcoholics to coping skills or interactional therapies: Posttreatment results. J. Consult. Clin. Psychol. **57:** 698–704.
3. MEEHL, P. E. 1954. Clinical Versus Statistical Prediction. University of Minnesota Press. Minneapolis.
4. LUBORSKY, L. & A. T. McLELLAN. 1978. Our surprising inability to predict the outcomes of psychological treatments—with special reference to treatments for drug abuse. Am. J. Drug Abuse **5:** 387–398.
5. SMART, R. G. 1978. Do some alcoholics do better in some types of treatment than others? Drug Alcohol Depend. **3:** 65–75.
6. BEUTLER, L. E. 1991. Have all won and must all have prizes? Revisiting Luborsky *et al.*'s verdict. J. Consult. Clin. Psychol. **59:** 226–232.
7. FINNEY, J. W. & R. H. MOOS. 1986. Matching patients with treatments: Conceptual and methodological issues. J. Stud. Alcohol **47:** 122–134.
8. ORFORD, J., E. OPPENHEIMER & G. EDWARDS. 1976. Abstinence or control: The outcome for excessive drinkers two years after consultation. Behav. Res. Ther. **14:** 409–418.
9. McLELLAN, A. T., G. E. WOODY, L. LUBORSKY, C. P. OBRIEN & K. A. DRULEY. 1983. Increased effectiveness of substance abuse treatment: A prospective study of patient-treatment "matching." J. Nerv. Ment. Dis. **171:** 597–605.
10. LEWIS, C. E., J. RICE & J. E. HELZER. 1983. Diagnostic interactions: Alcoholism and antisocial personality. J. Nerv. Ment. Dis. **171:** 105–113.
11. MANDELL, W. 1981. Sociopathic alcoholics: Matching treatment and patients. *In* Matching Patient Needs and Treatment Methods in Alcoholism and Drug Abuse. E. Gottheil, A. T. McLellan, & K. A. Druley, eds. : 325–369. Charles C Thomas. Springfield, IL.
12. HESSELBROCK, M. N., V. M. HESSELBROCK, T. F. BABOR, J. R. STABENAU, R. E. MEYER, & M. WEIDENMAN. 1983. Antisocial behavior, psychopathology and problem drinking in the natural history of alcoholism. *In* Longitudinal Research in Alcoholism. D. W. Goodwin, K. T. Van Dusen & S. A. Mednick, eds. :197–214. Kluwer-Nijhoff Publishing. Boston, MA.
13. CLONINGER, C. R. 1987. Neurogenetic adaptive mechanisms in alcoholism. Science **236:** 410–416.
14. BABOR, T. F., M. HOFMANN, F. K. DELBOCA, V. HESSELBROCK, R. E. MEYER, Z. S. DOLINSKY & B. ROUNSAVILLE. 1992. Types of alcoholics. I. Evidence for an empirically derived typology based on indicators of vulnerability and severity. Arch. Gen. Psychiatry **49:** 599–608.
15. WOODY, G. E., A. T. McLELLAN, L. LUBORSKY & C. P. OBRIEN. 1985. Sociopathy and psychotherapy outcome. Arch. Gen. Psychiatry **42:** 1081–1086.
16. ROUNSAVILLE, B. J., Z. S. DOLINSKY, T. F. BABOR & R. E. MEYER. 1987. Psychopathology as a predictor of treatment outcome in alcoholics. Arch. Gen. Psychiatry **44:** 505–513.
17. HELZER, J. E. & T. R. PRYZBECK. 1988. The co-occurrence of alcoholism with other psychiatric disorders in the general population and its impact on treatment. J. Stud. Alcohol **49:** 219–224.
18. WOODY, G. E., A. T. McLELLAN, L. LUBORSKY, C. P. OBRIEN, J. BLAINE, S. FOX, I. HERMAN & A. T. BECK. 1984. Severity of psychiatric symptoms as a predictor of benefits from psychotherapy: The Veterans Administration–Penn study. Am. J. Psychiatry **141:** 1172–1177.
19. McLELLAN, A. T., L. LUBORSKY, G. E. WOODY & C. P. OBRIEN. 1980. An improved

diagnostic evaluation instrument for substance abuse patients: The Addiction Severity Index. J. Nerv. Ment. Dis. **168:** 26–33.

20. McLELLAN, A. T., L. LUBORSKY, G. E. WOODY, C. P. OBRIEN & K. A. DRULEY. 1983. Predicting response to alcohol and drug abuse treatments: Role of psychiatric severity. Arch. Gen. Psychiatry **40:** 620–625.

21. RYAN, C. & N. BUTTERS. 1983. Cognitive deficits in alcoholics. *In* The Pathogenesis of Alcoholism: Biological Factors. B. Kissin & H. Begleiter, eds. Vol. 7 in The Biology of Alcoholism series. : 485–538. Plenum Press. New York, NY.

22. PARSONS, O. A. 1987. Neuropsychological consequences of alcohol abuse: Many questions—some answers. *In* Neuropsychology of Alcoholism: Implications for Diagnosis and Treatment. O. A. Parsons, N. Butters & P. E. Nathan, eds. : 153–175. Guilford. New York, NY.

23. GORDON, S. M., B. P. KENNEDY & J. D. McPEAKE. 1988. Neuropsychologically impaired alcoholics: Assessment, treatment considerations, and rehabilitation. J. Substance Abuse Treatment **5:** 99–104.

24. BECKER, J. T. & J. H. JAFFE. 1984. Impaired memory for treatment-relevant information in inpatient men alcoholics. J. Stud. Alcohol **45:** 339–343.

25. ERWIN, J. E. & J. J. HUNTER. 1984. Prediction of attrition in alcoholic aftercare by scores on the embedded figures test and two Piagetian tasks. J. Consult. Clin. Psychol. **52:** 354–358.

26. SUSSMAN, S., R. G. RYCHTARIK, K. MUESER, S. GLYNN & D. M. PRUE. 1986. Ecological relevance of memory tests and the prediction of relapse in alcoholics. J. Stud. Alcohol **47:** 305–310.

27. WALKER, R. D., D. M. DONOVAN, D. R. KIVLAHAN & M. R. O'LEARY. 1983. Length of stay, neuropsychological performance, and aftercare: Influences on alcohol treatment outcome. J. Consult. Clin. Psychol. **51:** 900–911.

28. DONOVAN, D. M., D. R. KIVLAHAN & R. D. WALKER. 1984. Clinical limitations of neuropsychological testing in predicting treatment outcome among alcoholics. Alcoholism: Clin. Exp. Res. **8:** 470–475.

29. ECKARDT, M. J., R. R. RAWLINGS, B. I. GRAUBARD, V. FADEN, P. R. MARTIN & L. A. GOTTSCHALK. 1988. Neuropsychological performance and treatment outcome in male alcoholics. Alcoholism: Clin. Exp. Res. **12:** 88–93.

30. McCRADY, B. S. & D. E. SMITH. 1986. Implications of cognitive impairment for treatment of alcoholism. Alcoholism: Clin. Exp. Res. **10:** 145–149.

31. CHANEY, E. F., M. R. O'LEARY & G. A. MARLATT. 1978. Skill training with alcoholics. J. Consult. Clin. Psychol. **46:** 1092–1104.

32. FREEDBERG, E. J. & W. E. JOHNSTON. 1978. The effects of assertion training within the context of a multi-modal alcoholism treatment program for employed alcoholics. Addiction Research Foundation. Toronto.

33. JACKSON, P. & T. P. S. OEI. 1978. Social skills training and cognitive restructuring with alcoholics. Drug Alcohol Dependence **3:** 369–374.

34. OEI, T. P. S. & P. JACKSON. 1980. Long-term effects of group and individual social skills training with alcoholics. Addict. Behav. **5:** 129–136.

35. JONES, S. L., R. KANFER & R. I. LANYON. 1982. Skill training with alcoholics: A clinical extension. Addict. Behav. **7:** 285–290.

36. ANNIS, H. M. & C. S. DAVIS. 1988. Self-efficacy and the prevention of alcoholic relapse: Initial findings from a treatment trial. *In* Assessment and Treatment of Addictive Disorders. T. B. Baker & D. Cannon, eds. Praeger. New York, NY.

37. ITO, J. R., D. M. DONOVAN & J. J. HALL. 1988. Relapse prevention in alcohol aftercare: Effects on drinking outcome, change process, and aftercare attendance. Br. J. Addict. **83:** 171–181.

38. SJOBERG, L. & V. SAMSONOWITZ. 1985. Coping strategies in alcohol abuse. Drug Alcohol Dependence **15:** 283–301.

39. YALOM, I. D., S. BLOCH, G. BOND, E. ZIMMERMAN & B. QUALLS. 1978. Alcoholics in interactional group therapy: An outcome study. Arch. Gen. Psychiatry **35:** 419–425.

40. VANICELLI, M. 1992. Removing the Roadblocks: Group Psychotherapy with Substance Abusers and Family Members. Guilford Press. New York, NY.

41. COONEY, N. L., R. M. KADDEN, M. D. LITT & H. GETTER. 1991. Matching alcoholics to coping skills or interactional therapies: Two-year follow-up results. J. Consult. Clin. Psychol. **59**: 598–601.
42. LITT, M. D., T. F. BABOR, F. K. DELBOCA, R. M. KADDEN & N. L. COONEY. 1992. Types of alcoholics. II. Application of an empirically derived typology to treatment matching. Arch. Gen. Psychiatry **49**: 609–614.
43. ITO, J. R. & D. M. DONOVAN. 1986. Aftercare in alcoholism treatment: A review. *In* Treating Addictive Behaviors: Processes of Change. W. R. Miller & N. Heather, eds. :435–456. Plenum Press. New York, NY.
44. KAUL, T. J. & R. L. BEDNAR. 1986. Experiential group research: Results, questions and suggestions. *In* Handbook of Psychotherapy and Behavior Change. S. L. Garfield & A. E. Bergin, eds. 3rd Ed. : 671–714. Wiley. New York, NY.
45. MOOS, R. H. 1981. Group Environment Scale Manual. Consulting Psychologists Press. Palo Alto, CA.
46. HOFFMANN, N. G., P. A. HARRISON & C. A. BELILLE. 1983. Alcoholics anonymous after treatment: Attendance and abstinence. Int. J. Addict. **18**: 311–318.
47. NACE, E. P. 1992. Alcoholics Anonymous. *In* Substance Abuse: A Comprehensive Textbook. 2nd Ed. J. H. Lowinson, P. Ruiz & R. B. Millman, eds. :486–495. Williams & Wilkins. Baltimore, MD.
48. DUCKERT, F. & J. JOHNSEN. 1987. Behavioral use of disulfiram in the treatment of problem drinking. Int. J. Addict. **22**: 445–454.
49. FULLER, R. K., L. BRANCHEY, D. R. BRIGHTWELL, R. M. DERMAN, C. D. EMRICK, F. L. IBER, K. E. JAMES, R. B. LACOURSIERE, K. K. LEE, I. LOWENSTAM, L. MAANY, D. NEIDERHISER, J. J. NOCKS & S. SHAW. 1986. Disulfiram treatment of alcoholism. JAMA **256**: 1449–1455.
50. ROUNSAVILLE, B. J. & K. M. CARROLL. 1992. Individual psychotherapy for drug abusers. *In* Substance Abuse: A Comprehensive Textbook. 2nd Ed. J. H. Lowinson, P. Ruiz & R. B. Millman, eds. :496–507. Williams & Wilkins. Baltimore, MD.
51. MONTI, P. M., D. B. ABRAMS, R. M. KADDEN & N. L. COONEY. 1989. Treating Alcohol Dependence: A Coping Skills Training Guide. Guilford. New York, NY.
52. GETTER, H. 1984. Aftercare for alcoholism: Short term interactional group therapy manual. Unpublished manuscript.
53. BROWN, S. & I. D. YALOM. 1977. Interactional group therapy with alcoholics. J. Stud. Alcohol **38**: 426–456.
54. VANICELLI, M. 1982. Group psychotherapy with alcoholics: Special techniques. J. Stud. Alcohol **43**: 17–37.
55. GETTER, H., M. D. LITT, R. M. KADDEN & N. L. COONEY. 1992. Measuring treatment process in coping skills and interactional group therapies for alcoholism. Int. J. Group Psychother. **42**: 419–430.
56. COX, D. R. 1972. Regression models and life tables. J. Roy. Statistical Soc. **34**: 187–220.
57. COHEN, J. 1988. Statistical Power Analysis for the Behavioral Sciences. 2nd Ed. Erlbaum. Hillsdale, NJ.
58. RYCHTARIK, R. G., D. M. PRUE, S. R. RAPP & A. C. KING. 1992. Self-efficacy, aftercare and relapse in a treatment program for alcoholics. J. Stud. Alcohol **53**: 435–440.

Implications of Typologies for Treatment

A Planner's Perspective

JEROME H. JAFFE

Center for Substance Abuse Treatment
Substance Abuse and Mental Health Services Administration
Rockville, Maryland 20857

Just 100 years ago, a Hartford physician, T. D. Crothers, Superintendent of Walnut Lodge Hospital and Editor of the *Journal of Inebriety,* published a brief article in the *Journal of the American Medical Association.* Titled "The law of periodicity in inebriety," the article described the periodic drinkers Crothers had observed and treated as falling into ". . . several distinct classes, with widely varying symptoms and conditions."[1]

Recent studies provide a firmer basis for subclassifying problem drinkers than those developed a century ago with less empirical approaches, but the implications of these studies for treatment are not absolutely clear. This paper discusses some of the questions and problems that people who are involved with funding and planning treatment programs must consider as work on typology of alcoholics progresses. These include:

Will the typologies now being proposed stand the test of time better than earlier ones? The typologies proposed by Crothers here in Hartford 100 years ago are now rarely cited.[2]

If the current typologies or subclassifications prove to be valid and durable, what additional evidence will be needed to show that individuals with different characteristics respond differently to different treatments? Will we accept the FDA's standard of two, pivotal, double-blind studies? Are double-blind studies appropriate or even possible in this area of research?

Assuming that people working in this field can agree on a standard of proof that different subtypes of alcoholics respond differently to different treatments, and can also agree on which research studies meet that standard, how can treatment providers be motivated to change the way they now provide treatment? Experience suggests that information alone will not be sufficient to bring about change, but additional information is necessary. For example, we need to know if any of the proposed changes will consume more or fewer resources. Depending on the answer to this question, the problems faced in implementing change will be very different, but it should not be assumed that they will be easier if the answer is fewer resources.

Will any decrease in cost be large enough to offset the increase in the cost of doing more detailed patient assessments, retraining staff, as well as any reductions in economy of scale now provided by delivering essentially the same treatment to a presumably homogeneous population?

If there are significant benefits to implementing treatments that consider typology, what mechanisms are available to foster the needed changes in treatment programs that now exist, and what obstacles impede such change?

THE WAY IT WAS

The idea that there are different types of alcohol and drug problems is not new, and neither is the idea that different types of patients might require different approaches to treatment. The American Association for the Cure of Inebriates, founded in the 1870s, included among its guiding principles the view that inebriety was a disease that should be treated—by legal compulsion if necessary. Furthermore, the Association advocated government funding to establish hospitals for this purpose.[3] Even then, among the dominant somatic wing of the Association, a crude typology was widely accepted, one not too different from some proposed in this volume. It was based in part on the differences observed among alcoholics with respect to antisocial behavior and deviance. One asylum president argued that unless an institution could be kept free from "dissolute and reckless characters" who would "demoralize a whole house," it would "soon degenerate into a school of intemperance and vice."[4]

As the debate evolved, it became clear that what was needed were places for just those indigent inebriates who outraged society by committing petty crimes, rotated through the jails, and created disruptions when cared for in hospitals for the insane. As then proposed, the public system would emphasize compulsory care and discipline, and would allow the private psychiatric facilities to deal with a "better class" of inebriates. However, a system of inebriate asylums was never set up, and antisocial inebriates remained mostly the burden of jails, prisons, almshouses, and insane asylums. Thus, the topic of the practical implications of typology for treatment is one that has been raised in the past, but has never been satisfactorily resolved.

OBSTACLES TO USING TYPOLOGY

Babor and Dolinsky[2] provided a scholarly review of the evolution of alcoholic typologies, beginning with the work in 19th century Europe and the United States and concluding with the post-Jellinek, empirically based efforts up to the present. They commented:

> Perhaps the greatest shortcoming of contemporary typology theory is its failure thus far to exert any influence on differential diagnosis, clinical practice, or treatment policy. . . . Few researchers have attempted to move beyond their theories to the development of diagnostic criteria, therapeutic interventions, or specialized treatment programs. The so-called matching hypothesis is practically impossible to test without some credible theory suggesting the types of alcoholics who should be matched to the variety of available treatments.[2]

From a planner's perspective, Babor and Dolinsky's observation cannot be overemphasized. To exert influence on clinical practice and treatment policy, those interested in the idea of typologies will have to demonstrate that there are differences that make a difference for treatment outcome, preferably to those who control the nature of the treatment provided—the providers and those who pay the providers. But even that may not be enough. Assume that researchers can persuasively show that when measured by cost-effectiveness and cost-benefit analyses, matching different patients to different treatments produces robust differences in outcome that are well worth the effort. What then? What incentives exist for treaters to adopt the new approach to differential diagnosis and differential treatment? What tools do policymakers have at their disposal to accelerate the adoption of the changes that would flow from the new findings? And what impediments might delay or prevent the introduction of such changes?

To answer such questions it is necessary to briefly describe both the American health care system and the system that now exists for providing treatment to individuals with alcohol and substance abuse problems. These topics are currently the subject of much debate in the public and political arena. The following excerpt from Schlesinger and Dorwart[5] aptly summarizes our present situation:

> By international standards, medical care in the United States is delivered and financed in an extraordinarily complicated fashion. . . . Health services are provided in an array of inpatient and outpatient settings, each with a different mix of for-profit, private nonprofit, and public facilities. Services are paid for by thousands of different private insurance companies, as well as federal, state, and local governments. . . . The current mix of providers and financing systems is portrayed as an embodiment of pluralistic American values, a mechanism for allowing different models of medical care to compete in the marketplace, and a bastion against overly intrusive interventions by government into what are appropriately private choices.[5]

PLURALISM HAS ITS COSTS

It is difficult to know how much treatment capacity exists, access is often limited, and those who seek treatment may not know the range of options available. Treatment providers may be selective and avoid patients who might prove difficult or might consume more than the ordinary level of resources. And the system may not adapt readily to changing needs or populations. For example, it has been slow to develop ways to provide services for pregnant women who use drugs.

But the most problematic aspect of pluralism in health care in the United States is that providers may respond to incentives and influences that, when judged from a public health perspective, are perverse. For example, the evidence suggesting that hospital-based care was not cost-effective did not prevent the massive expansion of hospital-based, private sector, chemical-dependency programs in the 1980s. Then, even as researchers concluded that time spent in treatment increases the likelihood that drug-dependent patients treated in public sector programs will recover successfully, economic forces motivated the private sector to sharply curtail the length of treatment offered in their programs. And as private providers have learned to exclude patients whose care is problematic, thereby raising their overall success rate and reducing their costs, they have shifted greater burdens onto the public sector, which makes it seem less effective, increases its costs, and makes it less attractive to anyone with a choice. According to Schlesinger and Dorwart:

> All these challenges to effective policy-making—conflicting goals, inadequate information, inconsistent responses by treatment providers—can, in our assessment, be traced to a more fundamental strategic error: an attempt to structure the drug treatment system along the pluralistic lines common to most health and human services in the United States.[5]

Thus, one fundamental impediment to changing the way treatment providers think about patient assessments and select modalities is that our current treatment arrangements vary markedly in how responsive they are to empirically based research on efficacy.

WAXING AND WANING INFLUENCE OF
THE FEDERAL GOVERNMENT

Twenty-five years ago, it would have been easier to correct this situation. At that time, most States received the bulk of funding for expanding drug abuse programs (and for many alcoholism programs as well) directly from Federal government sources. If the data had been available, it would have been possible to insist that patient assessments consider typology as a factor in treatment assignment. Indeed, although we did not then have as much evidence on the limited utility of inpatient detoxification as we do now, Federal policymakers were able to limit funding for costly hospital-based inpatient treatment for drug dependence to less than 5% of overall treatment capacity.

A policy started by the Federal government, in the early 1970s, of distributing money to the States to be used exclusively for drug and alcohol treatment and basing the amount each State got on a fixed formula tied to population ("block grants") was followed, in the early 1980s, by a policy of very limited supervision of how this block grant money was spent. This meant that the Federal government's capacity to influence, by controlling funding, the quality or types of treatments offered was virtually eliminated. Unfortunately, in too many cases, withdrawal of Federal supervision ("oversight") was not replaced by oversight and program evaluation at the State level. Many State administrators simply reallocated the block grant money among providers based on formulas that had evolved historically or were politically acceptable. The idea that a "treatment system" should be continually reshaped on the basis of treatment outcome was either forgotten or abandoned.

Within the last several years, the Federal government has again begun to require the States to develop more detailed plans to assess their needs for different types of treatment (e.g., women with children and intravenous drug users), so that funds can be distributed to agencies based on need. It remains to be seen how successful this effort will be. The idea of "accountability" for what treatment accomplishes is an element in the current National Strategy, but there is little definition of what "accountability" means and little mention of how the Federal government can ensure it.

It is possible that changes in the legislation that provides authority for block grants could also be used to compel treatment agencies to conduct more detailed intake assessments for alcoholics and other drug-dependent individuals. Such an assessment would permit differential assignment to treatment, if the needed array of treatments existed. Alternatively, it is possible that economic incentives could be developed that would induce treatment providers to use more detailed and standardized patient assessments. A number of discretionary grant programs now funded by the Center for Substance Abuse Treatment (CSAT), such as "Target Cities," require the use of standardized instruments as a condition of award and encourage thoughtful referral and placement of patients based on such assessments. What we have learned so far from these grant programs is that there are many barriers to changing the ways treaters assess patients, and that if those who do the assessments also provide treatment, there is substantial resistance to referring patients to another agency. Within the private sector, there are now so many separate payors that it is hard to see how any one of them acting alone could require that greater attention be paid to the notion of typology. However, this situation could change with a major restructuring of the health care system.

There is an important distinction between the notion of typology, as described by Babor and Dolinsky[2] and most of the contributors to this volume, and the idea of using patient assessments to justify decisions about level of care and setting of care. It is not necessary to postulate different types of alcoholism or different types of drug dependence in order to develop criteria for deciding about assignment to a level of care (e.g., inpatient or outpatient).

FACTORS NECESSARY FOR PROGRAM CHANGE

Relevant decision makers—those who operate alcohol or drug treatment programs or those who pay for them—must believe that part of the variance in treatment outcome is due to the nature and quality of treatment. This means that a bad outcome cannot always be attributed to the patients or to the environments to which they return. If statements such as "He/she didn't work the program" or "He wasn't ready" are accepted as accounting for all cases of poor outcome, there is little reason for treaters to change their programs or funders to require change. This does not imply that a patient's motivation is not critical. But treaters and payors must also be willing to ask not just, "Is there something else we might have done?," They must also ask, "What might have happened if we had done something different?"

In the medical model, in the broadest sense, treatment change is driven by technological advances. We can point to innumerable examples, from the introduction of anesthesia and antisepsis, to x-rays, insulin, antibiotics, and hosts of new agents to control a diversity of metabolic disorders. But there have been times when such advances have signaled not just incremental advances, but paradigm shifts. Such a conceptual shift occurred in the 1950s, when the introduction of antipsychotic and antidepressant agents challenged the fundamental premises of the psychoanalytic/psychodynamic school of psychiatry dominant at the time and fostered a major reemphasis on the importance of diagnosis. It is appropriate to ask if the idea of typology, with distinctly different treatments for different types of alcoholics, is merely a change in technology (comparable to the introduction of a new generation of antidepressants) or a new paradigm.

Another example can be found in the response of professionals to the introduction of therapeutic communities for treating drug addicts. Many of the early programs rejected professional involvement and used recovering drug addicts, exclusively, as staff. Did therapeutic communities represent a technology change or a fundamental challenge to the way the dominant treatment system viewed the problem of treating drug addiction? It might help to recall the very skeptical reception organizations such as Synanon were given in 1958 by the addiction-treatment establishment. There was, of course, a certain irony that professionals who had espoused the view that only the analyzed therapist could properly use psychoanalytic methods should be confronted by former drug addicts who claimed that only the formerly addicted could help the addict.

The distinction between paradigm shift and technology change is not a trivial one. In the drug abuse field, there have been, and still are, many people with advanced medical degrees who are committed to one or another specific treatment approach. To the degree that typology, and what it implies for changing some fundamental assumptions about existing treatment methods, demands a paradigm shift, there may be considerable resistance despite either economic incentives for change or government efforts to require more detailed patient assessments.

FACTORS THAT MAY RETARD CHANGE

The most important factor that may retard change is freedom of choice. If current research on typology leads to convincing evidence that robust differences in treatment outcome occur when patients are optimally matched to different treatments, issues of patient choice may still create barriers to such matching. Belief in freedom of choice runs deep in American society. Yet freedom of choice is really an overarching concept that refers to distinct areas of decision: the right to decide if any intervention will be accepted; the right to decide about the type of intervention; the right to select the specific individual or organization to provide the intervention; and the right to discontinue treatment at any time. Acceptance of typology concepts will influence all of these rights to one degree or another.

People who are dependent on drugs and/or alcohol can have some very strong opinions about what is likely to help them. In some cases, if patients believe that the information they provide will determine the kind of care they will receive, the information they provide may change. Eventually some patients will learn to say what they have to say to get the treatment they believe will help them. There is a lesson to be learned from the studies that assigned drug addicts to radically different treatments and found, to their dismay, that patients assigned to treatments other than the ones they initially stated a preference for dropped out at rates so high as to make interpretation of outcome almost meaningless.[6]

Although freedom of choice refers to the patient's right to choose, it does not, and need not, refer to society's obligation to pay for any choice the patient may make. For example, many individuals now choose a variety of nonapproved, alternative, or "new age" treatments for illnesses for which there are accepted and demonstrably effective treatments. For the most part, neither private insurance carriers nor government programs will pay for these nonrecognized treatments. If third-party payors become overly prescriptive about what treatments an individual with a drug or alcohol problem may select, more affluent patients may have a range of choices not available to the indigent (or even the less privileged). We are already seeing a constriction of choice based on the belief of certain third-party payors that inpatient treatment for alcoholism beyond brief detoxification cannot be justified by its results.

Also important as factors that may retard change are issues of political *versus* scientific advocacy. If the capacity of the various treatment approaches in the existing system were to correspond to the number of patients who might be expected to benefit from the particular programs, then widespread use of typologies for selecting treatment might involve only a reassortment of existing patient loads. However, it is unlikely that such a fortunate fit now exists. Basing treatment decisions on typology concepts might require that programs change the kinds of treatment they provide or that some programs might have less income. However, it is likely that before the publicly funded programs change or go out of business, they will exercise their prerogative to take their case directly to the legislators. And their arguments may fall on sympathetic ears. Politicians are human. They are often more impressed by a simple dramatic case than by mountains of statistics: a witness giving testimony; a congressman in recovery. What operates at the Federal level may operate even more persuasively at State and local levels.

ECONOMIC INCENTIVES HAVE LIMITATIONS

The current treatment system has built-in economic incentives that may frustrate efforts to stimulate treatment assignments based primarily on typology. Let me offer one or two examples. When regulators introduced diagnosis-related grouping (DRG) into clinical medicine as the basis for reimbursement, their objective was to remove any economic incentive to extend hospitalization or to carry out more medical tests and procedures than are absolutely necessary. The method grouped patients into a fixed number of diagnostic groups that took severity and complicating factors into consideration and established fixed payments for each diagnostic group, with the more complicated problems garnering substantially higher payment. If a hospital could render effective care with less than the amount provided, it would benefit; if treatment cost more, it would run a deficit. In essence, this was a reimbursement system based on a kind of typology. Within a surprisingly short time after DRGs were introduced, treatment providers were being offered new user-friendly computer software that could consider all of a patient's problems and complaints and select for the treater the DRG that would allow for the greatest monetary return. A second generation of DRG software considered the initial findings and suggested additional diagnostic tests that, if positive, might permit shifting the patient from one DRG to another with a higher rate of payment.

In the area of alcohol and drug abuse it is remarkable how quickly some providers became sensitive to the co-occurrence of psychiatric difficulties, when some payors began to limit payment for inpatient treatment only to those drug- and alcohol-using patients who were "dually diagnosed."

At present, most outpatient programs, especially in the public sector, have little incentive to conduct in-depth assessments. Currently, payment and case load are not linked to the complexity of problems encountered or, unfortunately, to the program's success in dealing with the patients it encounters. Any effort to induce such a system to adopt the concepts of differential diagnosis and differential treatment must consider the economic incentives operative in the current system, both public and private.

Despite these obstacles there are social and economic forces that could lead to a greater appreciation of the utility of concepts of typology within the general framework of the medical model. Among these are recent efforts by Federal agencies, such as CSAT, to require more comprehensive assessments of drug- and alcohol-dependent individuals as a condition of funding under its block grant and demonstration grant programs. To reduce the burden of the additional assessments, CSAT has been supporting the development of computer software that can serve both clinical and administrative functions. However, thus far it is not the concept of typology that is motivating the support for more detailed assessment. Rather, it is the belief that we must begin to look more carefully at the effectiveness of treatment (i.e., its outcome) in deciding which treatment programs will be funded or to which programs particular patients are best referred. But in order to compare programs fairly, there must be a way to adjust for the difficulty and complexity of the problems they are treating. A more thorough and standardized assessment is a necessary first step. Plans for a substantial change in the way health care is financed and organized are likely to create additional pressures to examine the effectiveness of treatment and, by implication, whether better matching will increase effectiveness. Yet, by placing initial emphasis on reducing costs for overall treatment through managed care, it is possible that conducting the more detailed assessments that are prerequisites for utilizing current typologies will not be seen as a priority.

There are other positive influences as well. As Enoch Gordis points out in this volume, there is a growing professionalism in the field of alcoholism treatment, as exemplified by organizations such as the American Society for Addiction Medicine (ASAM) and AMERSA. There may also be important distinctions between the drug treatment and alcoholism treatment fields. The latter, developed over the last few decades with closer links to hospitals, may be more receptive to the idea of assessment and differential treatment.

I don't know whether my comments here should be seen as pessimistic. I presented a paper 12 years ago, in Hartford, in which I raised essentially the same questions about how to get treatment programs to make full assessments standard practice, rather than an exceptional practice occurring only in research settings.[7]

Reginald Smart once observed that the use of small, daily doses of lemon juice was first noted, in 1601, to reduce the occurrence of scurvy. The finding was replicated about 150 years later, but it still took the British Navy another 48 years to provide citrus juice for its sailors on a regular basis, and still another 70 years, until 1865, before a citrus juice policy was applied in the British Mercantile Navy.[8] Judged by such standards, bringing about change with respect to a fuller assessment of alcoholics and drug addicts is occurring with breathtaking speed.

REFERENCES

1. CROTHERS, T. D. 1892. The law of periodicity in inebriety. JAMA **19:** 394. Reprinted in 1992. JAMA **268**.
2. BABOR, T. F. & Z. S. DOLINSKY. 1988. Alcoholic typologies: Historical evolution and empirical evaluation of some common classification schemes. *In* Alcoholism: Origins and Outcomes. R. M. Rose & J. Barrett, eds.: 245–266. Raven Press. New York, NY.
3. JAFFE, J. H. 1992. Current concepts of addiction. *In* Addictive States. Research Publications: Association for Research in Nervous and Mental Disease, Vol. **70**. C. P. O'Brien & J. H. Jaffe, eds.: 1–21. Raven Press. New York, NY.
4. BAUMOHL, J. 1990. Inebriate institutions in North America, 1840–1920. Br. J. Addict. **85:** 1187.
5. SCHLESINGER, M. & R. A. DORWART. 1992. Falling between the cracks: Failing National Strategies for the treatment of substance abuse. *In* Political Pharmacology: Thinking about Drugs. Daedalus. Proc. Acad. Arts & Sci. **121:** 195–237.
6. BALE, R. N., W. W. VAN STONE, J. M. KULDAU, T. J. J. ENGELSING, R. M. ELASHOFF & V. P. ZARCONE. 1980. Therapeutic communities *vs* methadone maintenance. A prospective controlled study of narcotic addiction treatment: Design and one-year follow-up. Arch. Gen. Psychiatry **37:** 179–193.
7. JAFFE, J. H. 1981. The evaluation of the alcoholic: General discussion. *In* Evaluation of the Alcoholic: Implications for Research, Theory, and Treatment. National Institute on Alcohol Abuse and Alcoholism. Research Monogr. No. 5: 395–409. DHHS Publ. No. (ADM) 81-1033. US Government Printing Office. Washington, DC.
8. SMART, R. 1993. Alcohol prevention. *In* Drugs, Alcohol, and Tobacco: Making the Science and Policy Connections. G. Edwards, J. Strang & J. H. Jaffe, eds.: 195–237. Oxford University Press. New York, NY.

Toward a Comprehensive Theory of Alcoholism

ROGER E. MEYER

The George Washington University
Washington, DC 20037

A comprehensive theory of alcoholism must take into account the heterogeneity of alcoholic symptomatology, the natural history of the disorder, and the presence of a core syndrome of alcohol dependence. Moreover, identified risk factors, such as a family history of alcoholism and/or the presence of antisocial personality disorder, are present in some alcoholic patients and not in others. The complexity of these issues has sometimes served to fuel the debate on the "disease concept." Opponents of the disease concept argued that alcoholism did not always follow a clearcut course with a characteristic train of symptoms, whereas advocates argued that many diseases in medicine present with the same variability of onset, progression, and severity as alcoholism. Indeed, the history of the concept of disease in alcoholism needs to be seen in the context of our evolving definitions of "disease" in clinical medicine, public health, and psychiatry.

If there is an explicit disease within the spectrum of symptoms and consequences of alcoholism, it is found within the formulation of alcohol dependence described in DSM-III-R[1] and ICD-10[2]. These criteria refer to a cluster of cognitive, behavioral, and physiological symptoms that indicate that the person has impaired control over alcohol and continues to drink despite adverse consequences. The criteria are based largely on the symptom complex proposed as the "alcohol dependence syndrome" by Edwards and Gross[3] in 1976. With some minor changes, the proposed criteria for alcohol dependence in DSM-IV[4] adhere closely to DSM-III-R. Although the dependence construct in these nosologies is explicitly categorical, severity of dependence may be quantified along a continuum. Severe dependence may be associated with increased disability, but alcohol-related disabilities may also be significantly influenced by genetic, demographic, and environmental factors as well as by the presence of co-morbid psychiatric, biomedical, or substance use disorders.[5] The outcome of alcoholism treatment may be affected by both severity of dependence and severity of alcohol-related disabilities.[6]

To a great degree, the boundaries of "disease" within medicine have been expanded to include a concern about risk factors within populations (heritable traits, the environment, human behavior) that might justify intervention with individuals to reduce the risk of subsequent pathology. This formulation involves definitions of "normal" or "abnormal" (e.g., cholesterol levels), about which there may be controversy, because these identify relative risk in a population and not individual risk. Within this framework, human behavior can also serve as a risk factor as well as a lever which can be influenced to reduce both heritable and environmental sources of risk of disease. Diagnoses such as alcohol abuse (DSM-III-R and DSM-IV) and harmful use (ICD-10) may warrant intervention on the basis of presumed risk, but for all intents and purposes, these categories stand apart from the disease construct.

If the disease construct has any validity, it should bear some predictive relationship to two important clinical phenomena: a tendency to use again after a period of abstinence and the rapid reinstatement of the elements of the dependence syndrome if alcohol use recurs. Most but not all studies have suggested that the severity of dependence (variably defined) has prognostic significance in clinical samples of alcoholic patients.[7,8] More severely dependent alcoholics are generally unable to sustain a moderate drinking outcome.[9] For these individuals, the recurrence of alcohol use leads rapidly to a recurrence of alcohol-dependent drinking patterns (e.g., loss of control drinking). Alcohol-related disabilities, which may be affected by co-morbid, psychiatric, or addictive disorders as well as environmental, demographic, and family environment issues, will affect the individual's ability to maintain abstinence.

THE HETEROGENEITY OF ALCOHOLIC PATIENTS IN CLINICAL SAMPLES

The heterogeneity of alcoholic patients in clinical samples can be described in a variety of ways including: demography (gender, age, ethnicity, marital status and quality of family support, presence or absence of significant friends, socioeconomic status, employment status); health status (presence or absence of specific biomedical consequences or other medical conditions); level of neuropsychological intactness; severity of alcohol dependence; level or severity of psychosocial disability; personality traits; presence, absence, and intensity of family history of alcoholism; presence or absence of specific co-morbid psychiatric or addictive disorders and/or global psychopathology; presenting symptom picture (e.g., symptoms of physical dependence); and number of previous treatment episodes. Because many of the characteristics influence the selection of treatment setting or program (as well as the response to treatment), it is difficult to generalize about the universe of alcohol-dependent individuals from data that have been collected from a single program. Community-based epidemiologic surveys offer the advantage of greater generalizability, but a substantial majority of individuals identified in these surveys never actually come to treatment. Data from community samples may therefore not be relevant to the specific characteristics of the universe of alcoholics in treatment. As one important example, severity of alcohol dependence predicted unstable moderate drinking outcomes in clinical samples, but failed to predict this outcome in a community sample.[10,11]

The heterogeneity of clinical samples of alcoholic patients has stimulated interest in the classification of subtypes of patients that might have some prognostic significance. Interest in typologies has also been stimulated by the fact that some of the aforementioned characteristics affect the age of onset and course of the disorder as well as the presenting symptom picture. Babor and Lauerman,[12] in reviewing typological constructs dating from 1850 to 1941, were able to identify 39 typologies based on four major criteria: dependence or addiction severity, consumption pattern, chronicity, and hypothesized etiology. Interest in classification has continued to the present with most of the typology literature (until recently) focusing on one-dimensional typologies (e.g., typologies based on personality characteristics of patients in one or more treatment programs). For the most part, these studies have employed factor analysis and other clustering techniques to statistically sort patients/subjects into more homogeneous groupings. As Brown has noted in this volume, false assumptions of homogeneity "wreak havoc" on the scientist's ability to test hypotheses by introducing unwanted error variance. Moreover, typologies developed solely for scientific inquiry may be of limited clinical value. Despite their limitations, single dimensional typologies have identified characteristics that are associated with greater severity and poorer treatment prognosis. These characteristics include antisocial personality disorder in males, strong family history of alcoholism, early age of onset in males, multiple previous treatment failures, little family cohesion, social isolation, and significant co-morbid psychopathology and/or substance abuse/dependence.[13–15] Depression, in several studies, appeared to confer a better prognosis in female alcoholics.[6,16]

What characteristics of a typology go beyond the statistical sorting of a single patient sample to confer clinical usefulness and validity? It should be readily apparent that the validity of subtype groupings derived from a cluster analysis of data from a specific clinical sample cannot be proven in the same sample. Yet, many cluster analytic studies of cross-sectional data from alcoholic patients have never been replicated in new patient samples. Replication is an important step in establishing validity and clinical usefulness. Clinical usefulness would also be well served by prospective patient/treatment matching designs in which the classification system suggested the greater usefulness of a specific treatment intervention; and treatment outcome data supported the hypothesis that patients matched to the treatment did better than did patients assigned to other treatments. *Validity* would be served by well-designed pedigree studies in which the family history of alcoholism was enlarged to identify the stability of subtypes within families. If these "types" conferred some predictive utility regarding age of onset, course, severity of dependence, and types of disability, the validity of the typology would also be strengthened.

Until recently, typologies derived from analyses of cross-sectional data from clinical samples of alcoholic patients have presented problems of reliability, validity, and clinical utility. The multidimensional typology proposed by Babor and colleagues[17] and the developmental model proposed by Zucker[18] represent recent efforts to address some of the problems of clinically derived typologies. Zucker's model (this volume) incorporates data on age of onset, gender, family history, ethnicity, drinking pattern, and the effects of alcohol in both cross-sectional and longitudinal databases. Babor and colleagues examined premorbid risk and vulnerability, severity of dependence and alcohol-related problems, chronicity and negative alcohol-related consequences, and co-morbid psychopathology in a cluster analysis that identified two types of alcoholics. Type A alcoholics were characterized by less premorbid risk and vulnerability, later onset, less severe dependence, fewer alcohol-related problems, a less chronic course with fewer negative consequences of alcohol consumption, and less co-morbid psychopathology. Conversely, type B alcoholics presented with more premorbid risk and vulnerability, earlier onset, more severe dependence and alcohol-related problems, a more chronic course with more negative consequences of alcohol consumption, and more co-morbid psychopathology. Their typology bears some resemblance to the early onset/late onset typology developed by von Knorring *et al.*[19] and Buydens-Branchey *et al.*[20] It also bears some resemblance to the typology developed from a sample of Swedish adoptees by Cloninger and colleagues.[21]

The Babor typology[5] predicted the treatment outcome at 12 and 36 months following treatment; type B alcoholics (compared with type A alcoholics) had more drinking problems and greater alcohol-related impairment. The typology developed by Babor *et al.* was also applied prospectively in a treatment-matching design by Litt *et al.*[22] These investigators found that the less severe, type A, alcoholics did better in interactional group therapy, whereas type B alcoholics had better outcomes with coping skills treatment. A significant effect was found for patient treatment matching. Since the single characteristic of antisocial personality disorder also served the patient treatment-matching function in the study by Litt *et al.*,[22] it is not clear which variables may be most useful in the Babor *et al.* classification. In the present volume, Brown *et al.* found that only 5 of the original 16 clustering variables were required to correctly classify almost 95% of an outpatient sample of male and female alcoholics in a discriminate function analysis in the type A/type B typology. For a statistically derived typology to be clinically useful, it should have predictive utility, treatment-matching implications, and ease of prospective classification. The typology developed by Babor *et al.* may be close to fulfilling these criteria.

EVIDENCE OF HETEROGENEITY IN ALCOHOLIC PEDIGREES, BIOLOGICAL MARKERS, AND CHILDHOOD ANTECEDENTS

Searles and Alterman (this volume) note that most individuals with a family history of alcoholism do not become alcoholic, and most individuals who are diagnosed as alcoholic do not have a biological history of the disorder. In their 30-year follow-up of men at high risk for alcoholism, Goodwin *et al.* (this volume) found that familial alcoholism was generally more severe than nonfamilial alcoholism. Cadoret (this volume) noted significant heterogeneity in the manifestations of a genetic effect (regarding the heritability of alcoholism) comparing one adoption agency to another in cohorts of adoptees from the State of Iowa. Cadoret identified three distinct putative risk factors that could each contribute to the risk for the development of alcoholism. These include a biological family history of alcoholism, the presence of alcohol problems in adoptee families, and the presence of antisocial personality disorder in the adoptee. Goodwin *et al.* (this volume) noted that antisocial personality disorder was found in both familial and nonfamilial forms of alcoholism in the cohort of Danish adoptees that had been followed for 30 years.

Results of these studies need to be compared with the results of the most influential adoptee study reported over the last 10 years by Cloninger, Bohman, and colleagues[21,23] which was based on adoptee records from Sweden. That study described two distinct heritable forms of alcoholism in males. Type I alcoholics could inherit alcoholism from the mother or the father. The severity of alcoholism was sensitive to demographic characteristics in the adoptive home. Adoptees with a positive family history of alcoholism who had been reared in lower socioeconomic environments tended to have more severe alcoholism, whereas adoptees with a positive family history of alcoholism who had been reared in more affluent homes had less severe forms of alcoholism. Severity was defined on the basis of the number of times the individual had been registered with the local temperance board. Type II alcoholics were individuals who inherited the risk of alcoholism only from their biological fathers, and these fathers had a history of criminality. Cloninger[23] has described the personality characteristics associated with the risk of alcoholism in these two groups of adoptees. Type I alcoholics tend to score low on novelty seeking and high on reward dependence and harm avoidance. Type II alcoholics are high on novelty seeking and low in reward dependence and harm avoidance. Cloninger has extended his observations about personality characteristics in these two adoptee groups by suggesting that these personality characteristics were related to heritable differences in neurotransmitter function. He has linked the heritable dimension of alcoholism risk to the explicit personality characteristics that he believes are inherited. Although some have argued that Cloninger's typology is consistent with typologies that differentiate alcoholics on the basis of the presence or absence of antisocial personality traits, in fact Cloninger's data do not refer to the presence or absence of antisocial personality traits in type I or Type II adoptees. (The biological fathers of Type II alcoholics had histories of criminality. No mention was made of histories of criminality in their offspring who became alcoholic.) Nevertheless, two papers on biological markers of vulnerability are of interest in regard to the Cloninger typology. Linnoila and his colleagues[24] in Finland identified low levels of 5-hydroxy indoleacetic acid in the cerebrospinal fluid of a cohort of Finnish alcoholics with histories of severe childhood conduct disorder and adult criminality. Devor and colleagues (this volume) studied platelet monoamine oxidase (MAO) levels in 21 families with 189 members in which alcoholism was ascertained through alcoholic probands, as part of the St. Louis Family Interview Study of Alcoholism. These authors reported lower activity levels of the enzyme MAO B in the platelets of alcoholics than of nonalcoholics. They found that low levels of MAO B were more prominent in Type II alcoholic families and that these families manifested greater risk for other psychiatric illnesses as well as alcoholism.

In a prospective study designed to characterize the developmental course of subtypes of alcoholism, Zucker and colleagues (this volume) found it necessary to discard some of the Type I/Type II characteristics described by Cloninger because of problems of classification among their alcoholic probands. They used a dimensional approach to characterize antisocial behaviors. Alcoholics were divided into antisocial and nonantisocial types without the requirement that the antisocial alcoholics fulfill all of the DSM-III-R criteria for an antisocial personality disorder diagnosis. They report that antisocial alcoholics have an earlier onset of alcohol problems and manifest greater severity of alcohol-related symptoms. The continuity of childhood conduct problems and adult antisocial behaviors made the most robust contribution to the typology. Indeed, children of antisocial alcoholics appear to manifest externalizing behavioral problems that were at least partly heritable. The emergence of psychopathological problems among children of alcoholics appear to be associated much more clearly with the antisocial typology than with the nonantisocial typology. Zucker's data seem compatible with earlier reports by Hesselbrock and colleagues[14,25] relative to the onset, course, and severity of alcohol dependence in antisocial (compared with nonantisocial) alcoholics. Although the early onset form of alcohol dependence and alcohol problems appears to be (for most studies) fairly homogeneous, Tarter (this volume) reports significant variability of childhood and adolescent psychiatric diagnoses among teenagers with alcohol problems. In his cohort, the diagnoses included childhood conduct disorder, anxiety disorder, and attention deficit hyperkinetic disorder. Importantly, these individuals are explicitly reporting "alcohol problems" and not the explicit behavioral criteria of alcohol dependence. Indeed, as this cohort of teenagers with alcohol problems is followed in longitudinal studies, it is likely that only the adolescents manifesting antisocial personality characteristics will develop diagnoses of alcohol dependence (of early onset).

Despite some controversy and lack of clarity in the pedigree and adoptee studies of alcoholism relative to the existence of distinct typologies, the early onset alcoholism in males with a history of childhood conduct disorder and adult antisocial behavior appear to constitute a relatively homogeneous subtype. Shuckit[26] questions if this is a form of alcoholism or of antisocial personality disorder. If the individual fulfills the criteria for alcohol dependence, then the diagnosis must include the latter. If the individual fulfills the criteria for antisocial personality disorder, this may be included as a co-morbid disorder, but (as described by Zucker) many of these individuals have antisocial personality traits but do not fulfill the criteria for the personality disorder. At this writing, there is support for a relatively antisocial subtype of alcoholism both within the clinical literature and in pedigree and adoptee studies. What is unclear is whether there are distinct subgroups of individuals with childhood conduct disorder that are more at risk for adult alcoholism.

Childhood conduct disorder may be related to a learning disability or violence in the home, or it may represent a type of atypical childhood depression. Some forms of childhood conduct disorder may indeed be heritable; some may be related to peer group involvement. Further work is clearly needed to define more precisely those characteristics of childhood conduct disorder that may be associated with the risk of alcoholism. Indeed, as Goodwin and his colleagues note (this volume), although antisocial personality disorder was associated with the development of alcoholism in the Danish adoptees, it was present in both family history positive and negative subjects. In other words, within the larger group of individuals with childhood conduct disorder and adult antisocial personality traits (or disorder), there may be a distinct subgroup with heritable biological features that differentially contribute to the risk for developing alcoholism. The identification of these characteristics should help to account for the earlier onset of alcohol problems and the more rapid progression to alcohol dependence in these individuals. One can speculate that these traits might involve either differential CNS responses to alcohol-related reinforcement, differential satiation to alcohol-related effects, differential tolerance to the aversive effects of higher doses of alcohol, and/or some nonspecific diathesis related to specific personality traits (e.g., novelty seeking, impulsivity, harm avoidance, reward dependence, "hypophoric" mood states). Because of the strength of childhood conduct disorder and adult antisocial personality traits (or disorder) as a risk factor for the development of alcohol dependence, the clarification of these questions in some well-designed longitudinal studies in well-characterized pedigrees should constitute an important research priority.

THE CORE SYNDROME AND THE PREDICTION OF RELAPSE

As just described, more recent formulations of alcohol dependence are firmly based on behavioral theory, with a focus on alcohol-seeking and alcohol-consuming behaviors. Both ICD-10 and DSM-III-R recognize the persistent risk of relapse in recommending that patients continue to receive the diagnosis of alcohol dependence even if they are "currently abstinent" (in the case of ICD-10) or in "full remission" for 6 months or longer (DSM-III-R). ICD-10 also notes that the return to alcohol use after a period of abstinence leads to a more rapid reappearance of other features of the (dependence) syndrome than occurs with nondependent individuals. Thus, alcohol-dependent individuals manifest a tendency to use again after they have been withdrawn from alcohol, and the elements of a dependence syndrome will recur in a matter of days or weeks following resumption of use.

Relapse to alcohol addiction is most likely to occur in the first 3–6 months of abstinence, a period that is characterized by physiological abnormalities, mood dysregulation, and a variety of somatic symptoms.[27] Abnormalities of stress response suggest homeostatic dysregulation during this period of "protracted abstinence."[28] Many patients complain of anxiety and depression, even in the absence of a diagnosable anxiety or mood disorder.[29] Insomnia and disturbances in sleep architecture may persist for many months,[30] suggesting the persistence of residual effects of chronic alcohol consumption on circadian rhythm. Abnormalities in hypthothalamic pituitary adrenal (HPA) axis function[31] as well as other endocrine abnormalities[32] have been documented in alcoholic patients up to 4 weeks postwithdrawal. There is persistent impairment of cognitive function, particularly short-term memory, judgment, and visuospatial relations.[33] Structural brain imaging studies demonstrate increased ventricular-brain ratios that recover over time,[34] and abnormalities in the electroencephalogram and evoked potential return to normal over the first 2–3 weeks of abstinence.[35] Begleiter et al.[36,37] demonstrated a state of latent CNS hyperexcitability persisting for at least 5 weeks following alcohol withdrawal in rats and monkeys. A challenge dose of alcohol in these animals elicited hyperexcitability in several brain regions. The hyperexcitability, which could not be elicited in control subjects, included some characteristics of acute alcohol withdrawal, suggesting that recovering alcoholics who resume drinking may trigger withdrawal-like CNS effects leading to a return to dependent drinking.

The acute withdrawal syndrome following abrupt discontinuation of dependent drinking is the most proximal example of homeostatic dysregulation consequent to chronic alcoholism. Persistent abnormalities following the initiation of abstinence suggest continued homeostatic dysregulation. Reports of abstinent patients that they need alcohol to "feel normal" have suggested to some observers that the physiological adaptations to chronic alcoholization may be reflected in changes in cellular function and molecular biology.[38] Like heroin addicts who do relatively well when maintained on high doses of oral methadone,[39] alcoholics may "require" continued high alcoholization (or its pharmacological equivalent). Evidence suggests that high doses of methadone, indeed, normalize neuroendocrine function in abstinent heroin addicts.[40] Analogous data on alcoholics being maintained on alcohol have not been systematically obtained.

On the basis of the success of methadone maintenance in treating the symptoms and signs of protracted opiate withdrawal, Kissen[41] proposed that benzodiazepines might serve the same functions for alcohol. Although his reasoning by analogy did not result in a new pharmacotherapy for alcoholism, it is possible that the GABA system is involved in some aspects of protracted alcohol withdrawal (as it appears to be involved in some aspects of the neuropharmacology of ethanol). Alcohol and benzodiazepines have similar acute affects on saccadic eye movements, respiration, mood, and evoked EEG response in human subjects.[42] Through different mechanisms, both alcohol and benzodiazepine agonists markedly potentiate muscimol-stimulated chloride ion flux in rat synaptoneurosomes.[43,44] With chronic exposure to alcohol, this effect is diminished. Cross-tolerance occurs between alcohol and benzodiazepines in humans and animal models, and benzodiazepines are the treatment of choice in acute alcohol withdrawal. Animal models of alcohol dependence demonstrate decreased sensitivity to GABAergic agonists and enhanced sensitivity to GABA inverse agonists.[45,46] The same change of sensitivity occurs with benzodiazepine dependence.[47] The enhanced sensitivity of the benzodiazepine/GABA receptor to endogenous inverse agonists, when coupled with decreased sensitivity to endogenous agonists, could account for the symptoms of anxiety and central nervous system hyperexcitability during alcohol or benzodiazepine withdrawal. Interestingly, changes in sensitivity of the benzodiazepine/GABA receptor consequent to chronic alcohol or benzodiazepine treatment will be reversed by single dose treatment with a benzodiazepine antagonist (flumazanil).[48] Alterations in sensitivity to alcohol and changes in preferential binding (agonist versus inverse agonist) with chronic alcohol administration may be a consequence of changes in gene expression of benzodiazepine/GABA receptor subunits. Chronic alcohol administration results in a 40–50% reduction in alpha$_1$ subunit mRNAs in the cerebral cortex of the rat as well as increases in mRNAs of the alpha$_6$ subunit and the beta$_2$ subunit (and cerebellum).[49]

In addition to its effects on the inhibitory neurons of the benzodiazepine/GABA system, alcohol also affects glutamate receptors in brain.[50] Glutamate is the major excitatory neurotransmitter in brain. Chronic alcohol administration results in up-regulation of the n-methyl D-aspartate (NMDA) receptor for at least 8 hours following acute withdrawal in mice.[51] Withdrawal seizure severity was reduced by the administration of an NMDA receptor antagonist. NMDA receptors are a subtype of glutamate receptor. In sum, the proconvulsant effects of alcohol withdrawal may be a result of enhanced glutamate activity along with diminished agonist sensitivity at the benzodiazepine/GABA receptor. At this writing it is unclear if changes produced by chronic alcohol treatment at the NMDA and benzodiazepine/GABA receptor could account for some of the symptoms and signs of protracted alcohol withdrawal as well as those of acute withdrawal. It will be important for molecular neurobiologists to define the changes in gene expression occurring at the NMDA and benzodiazepine/GABA receptors with chronic alcohol administration and to define the duration of changes in gene expression that affect receptor function. Could these changes account for the homeostatic dysregulation associated with "protracted abstinence?" These studies might also help to clarify the mechanism by which a return to alcohol use by abstinent alcoholics results in a rapid reinstatement of dependence symptoms and patterns of excessive drinking.

Another model that has been invoked to explain relapse and the rapid reinstatement of alcohol dependence involves operant and classical conditioning. Conditioning models postulate that relapse occurs in those environments or circumstances that have been associated with previous alcohol use. The model posits that "craving" for alcohol in abstinent individuals is a function of Pavlovian conditioning. Conditioning models have also been invoked to explain the pattern of relapse associated with other addictive disorders. In 1948, Wikler[52] proposed a conditioning model to explain heroin addiction. Because he viewed the development of physical dependence as the defining stage of opiate addiction, Wikler[53] believed that by a process of Pavlovian conditioning the symptoms and signs of opiate withdrawal (the unconditioned stimulus) are paired to environmental stimuli in the addict's home community. Over time, these stimuli elicit withdrawal symptoms that result in relapse to heroin use when the drug-free addict returns home, even after long periods of incarceration. O'Brien et al.[54] were able to confirm the conditioning of opiate withdrawal symptoms in human subjects. Ludwig extended Wikler's theoretical approach to explain relapse to alcohol addiction, linking craving for alcohol by alcoholics to conditioned withdrawal symptoms.[55] By the mid 1970s, Wikler[56] expanded his conditioning model to postulate that environmental stimuli could elicit "counter-adaptive interoceptive responses," mirror opposites of the effects of opiates, as individuals anticipated the administration of heroin. Siegel et al.[57] demonstrated a role for opponent process conditioning in the development of tolerance ("learned tolerance").

In addition to evidence of conditioning of the signs and symptoms of abstinence and of opponent process conditioning, there has long been evidence of conditioning of drug effects. Pavlov[58] noted conditioning of morphine's effects in the dog in 1926. Conditioned place preference represents a behavioral example of the conditioning of the rewarding properties of opiates, stimulants, and alcohol.[59] Meyer and Mirin[60] demonstrated evidence of conditioned heroin-like effects in drug-free addicts who continued to self-administer heroin (despite pharmacologically effective narcotic blockade) as long as they experienced conditioned narcotic effects. "Craving" for heroin was highest while subjects anticipated and experienced the conditioned or unconditioned effects of heroin as a type of priming effect. Priming effects have also been observed in alcoholics,[61–63] and conditioned drug-like effects have been observed in cocaine-dependent patients.[64] Clinicians have long been intrigued by the tendency of drug- or alcohol-free addicts to return to those settings associated with their addictive behavior despite counseling that encourages the individual to avoid "high risk" situations. This behavior seems strikingly analogous to conditioned place preference in animal models, where rats come to prefer a location in a maze that is associated with drug administration. It does not seem compatible with hypothetical models of relapsing behavior based on the conditioning of aversive (e.g., withdrawal symptoms) or opponent process effects. Stewart et al.[65] postulated that while opponent process conditioning is a parsimonious explanation for learned tolerance, it is the conditioned drug-like effects that appear to be responsible for relapse. Their model is consistent with the priming effects of low doses of a preferred drug (or alcohol) as well as recent data on sensitization which suggests an additive effect when repeated drug or alcohol use occurs over time in the same environmental setting.[66] Sensitization has been described for the locomotor stimulant effects of a variety of drugs of abuse.[67] Wise and Bozarth[68] linked the locomotor stimulant effects of drugs to their reinforcing properties. In their model, even drugs with depressant effects (such as alcohol) manifest locomotor stimulant properties at low doses and with repeated administration, as tolerance develops to the depressant effects and sensitization develops to the stimulant effects.

The locomotor activity associated with reinforcement has been linked to the dopaminergic system. Considerable evidence links the mesolimbic dopamine system (particularly the nucleus accumbens) to the reinforcing properties of alcohol and other drugs of abuse (e.g., see ref. 69). There is also some recent direct and indirect evidence linking this system to conditioned drug effects and sensitization. Two recent reports, using different methods to assess dopamine release in the nucleus accumbens, found dopamine release in association with *anticipated voluntary alcohol consumption* in the rat.[70,71] Studies with opiates and stimulants provide strong support for the mesolimbic dopamine system as essential for the manifestation of conditioned components of behavioral sensitization.[72,73] O'Brien et al.[74] reported dopamine release in nucleus accumbens following saline injections in rats previously exposed to multiple cocaine injections in the same setting.

With the development of biological techniques such as regionally specific intracerebral alcohol administration, *in vivo* voltametry, and *in vivo* microdialysis, it is now possible to examine the neurobiological correlates of conditioned drug effects and sensitization. Nestler et al. (this volume) offer evidence that changes in intracellular messenger pathways within neurons in the ventral tegmental area-nucleus accumbens pathway may underlie changes in drug or alcohol reward mechanisms associated with chronic use. They have extended these findings to contrast underlying differences between alcohol-preferring and non-preferring rat strains in these same intracellular messenger pathways. Their work suggests that either by inheritance or chronic alcohol exposure, molecular mechanisms within dopamine neurons of specific reinforcing pathways of the brain may account for the differential rewarding properties of alcohol (and other addictive drugs) in the rat.

Because reliable animal models of drug self-administration and alcohol consummatory behavior serve as reasonable homologs of substance dependence disorders (as defined in DSM-III-R or ICD-10), it should be possible to apply the technologies of molecular neurobiology to gain a greater understanding of the biological basis of alcohol dependence. Because of the existence of well-characterized pharmacogenetic differences in alcohol preference across rat strains, animal model data (using the new tools of molecular biology) may suggest possible biological mechanisms that contribute to heritable dimensions of risk for developing alcohol dependence. Will these animal models inform our perspective on the heritable dimensions of risk in human alcoholism? Will further research in human genetics clarify questions of alcoholic typologies that have intrigued clinicians and clinical investigators for more than 150 years? The danger that we face in applying these rich new methodologies to clinical investigation concerns the lack of reliability in many of our current typologies. In searching for a genetic basis for any medical disorder, the diagnosis of the disorder should be highly reliable. If there are, as Cadoret strongly suggests (this volume), heritable and nonheritable forms of alcoholism, genetic studies should focus on clearly heritable alcoholic subtypes and their family pedigrees. It is time to bring the literature on clinically derived typologies together with the literature on typologies derived from family pedigree data (e.g., in twin and adoptee samples) to better frame questions for molecular biologists and geneticists. This volume represents an important step in that direction!

REFERENCES

1. AMERICAN PSYCHIATRIC ASSOCIATION. 1987. Diagnostic and Statistical Manual of Mental Disorders, 3rd ed. Revised. Washington, DC.
2. WORLD HEALTH ORGANIZATION. 1992. The ICD-10 Classification of Mental and Behavioral Disorders: Clinical Descriptions and Diagnostic Guidelines. Geneva, Switzerland.
3. EDWARDS, G. & M. M. GROSS. 1976. Alcohol dependence: Provisional description of a clinical syndrome. Br. Med. J. 1: 1058–1061.
4. AMERICAN PSYCHIATRIC ASSOCIATION. 1993. DSM-IV Draft Criteria. Washington, DC.
5. BABOR, T. F. 1992. Nosological considerations in the diagnosis of substance use disorders. In Vulnerability to Drug Abuse. M. D. Glantz & R. Pickens, eds.:53–73. American Psychological Association Press. Washington, DC.
6. ROUNSAVILLE, B., Z. S. DOLINSKY, T. F. BABOR & R. F. MEYER. 1987. Psychopathology as a predictor of treatment outcomes in alcoholics. Arch. Gen. Psychiatry 44: 505–513.
7. BABOR, T. F., N. L. CONNEY & R. J. LAUERMAN. 1986. The drug dependence syndrome as an organizing principle in the explanation and prediction of relapse. In Relapse and Recovery in Drug Abuse. NIDA Research Monograph 72, DHHS Publication No. (ADM)86-1473. F. Tims & C. Lukefeld, eds.: 20–35. Government Printing Office. Washington, DC.
8. BABOR, T. F., N. L. COONEY & R. J. LAUERMAN. 1987. The drug dependence syndrome concept as a psychological theory of relapse behavior: An empirical evaluation. Br. J. Addict. 82: 393–405.
9. SOBELL, M. B. & L. C. SOBEL, Eds. 1987. Moderation as a Goal or Outcome of Treatment for Alcohol Problems: A Dialogue. The Hawthorne Press. New York, NY.
10. POLICH, A. M., D. J. ARMOR & H. B. BRAIKER. 1980. The course of alcoholism four years after treatment (R-2433-NIAAA). RAND Corporation.
11. HELZER, J. E., L. N. ROBINS, J. R. TAYLOR, K. CAREY, R. H. MILLER, T. COMBS-ORME & A. FARMER. 1985. The extent of long-term moderate drinking among alcoholics discharged from medical and psychiatric treatment facilities. N. Engl. J. Med. 312: 1678–1682.
12. BABOR, T. F. & R. J. LAUERMAN. 1986. Classification and forms of inebriety: Historical antecedents of alcoholic typologies. In Recent Developments in Alcoholism. M. Galanter, ed. Vol. 4: 113–144. Plenum Press. New York, NY.
13. BABOR, T. F. & Z. S. DOLINSKY. 1988. Alcoholic typologies: Historical evolution and empirical evaluation of some common classification schemes. In Alcoholism: Origins and Outcomes. R. M. Rose & J. Barrett, eds.:245–263. Raven Press, Ltd. New York, NY.
14. HESSELBROCK, M. N., V. M. HESSELBROCK, T. F. BABOR, R. E. MEYER, J. R. STABENAU & M. A. WEIDENMAN. 1984. Antisocial behavior, psychopathology and problem drinking

in the natural history of alcoholism. *In* Longitudinal Research in Alcoholism. D. W. Goodwin, K. T. Van Dusen & S. A. Mednick, eds. Kluwer-Nijhoff. Boston, MA.

15. FRANCES, R. J., S. BUCKY & G. S. ALEXOPOULOS. 1984. Outcome study of familial and non-familial alcoholism. Am. J. Psychiatry **141:** 1469–1471.

16. SCHUCKIT, M., F. N. PITTS, T. REICH, L. KING & G. WINOKUR. 1969. Alcoholism: Two types of alcoholism in women. Arch. Gen. Psychiatry **42:** 1050–1055.

17. BABOR, T. F., M. HOFMANN, F. DEL BOCA, V. HESSELBROCK, R. MEYER, Z. DOLINSKY & B. ROUNSAVILLE. 1992. Types of alcoholics. I. Evidence for an empirically-derived typology based on indicators of vulnerability and severity. Arch. Gen. Psychiatry **49:** 599–608.

18. ZUCKER, R. A. 1987. The four alcoholisms: A developmental account of the etiologic process. *In* Alcohol and Addictive Behaviors. P. C. Ribers, eds.: 27–83. University of Nebraska Press. Lincoln, NB.

19. VON KNORRING, L., A. L. VON KNORRING, L. SMIGAN, U. LINDBERG & M. EDHOLDM. 1987. Personality traits in subtypes of alcoholics. J. Stud. Alcohol **48:** 523–527.

20. BUYDENS-BRANCHEY, L., M. H. BRANCHEY & D. NOUMAIR. 1989. Age of alcoholism onset. I. Relationship to psychopathology. Arch. Gen. Psychiatry **46:** 231–236.

21. CLONINGER, C. R. 1987. Neurogenetic adaptive mechanisms in alcoholism. Science **236:** 410–416.

22. LITT, M. D., T. F. BABOR, F. K. DEL BOCA, R. M. KADDEN & N. L. COONEY. 1992. Types of alcoholics. II. Application of an empirically-derived typology to treatment matching. Arch. Gen. Psychiatry **49:** 609–614.

23. CLONINGER, C. R., M. BOHMAN & S. SIGARDSSON. 1981. Inheritance of alcohol abuse: Cross-fostering analysis of adopted men. Arch. Gen. Psychiatry **38:** 861–868.

24. LINNOILA, M., M. VIRKKUNEN, M. SCHEININ, R. RIMON & F. K. GOODWIN. 1983. Low cerebrospinal fluid 5-hydroxyindole acetic acid concentration differentiates impulsive from nonimpulsive violent behavior. Life Sci. **33:** 2609–2614.

25. HESSELBROCK, M. N., R. E. MEYER & J. J. KEENER. 1985. Psychopathology in hospitalized alcoholics. Arch. Gen. Psychiatry **42:** 1050–1055.

26. SCHUCKIT, M. 1985. The clinical implications of primary diagnostic groups among alcoholics. Arch. Gen. Psychiatry **42:** 1043–1049.

27. MEYER, R. E. 1989. Prospects for a rational pharmacotherapy of alcoholism. J. Clin. Psychiatry **50:** 403–412.

28. MULLER, N., M. HOEHE, H. E. KLEIN, G. NIEBERLE, H. P. KAPFHAMMER, F. MAY, O. A. MULLER & M. FICHTE. 1989. Endocrinological studies in alcoholics during withdrawal and after abstinence. Psychoneuroendocrinology **14:** 113–123.

29. KRANZLER, H. & N. LIEBOWITZ. 1985. Anxiety and depression in substance abuse: Clinical implications. Med. Clin. North. Am. **72:** 867–885.

30. GILLIN, J. C., T. L. SMITH & M. IRWIN. 1990. EEG sleep studies in "pure" primary alcoholism during subacute withdrawal: Relationships to normal controls, age, and other clinical variables. Biol. Psychiatry **27:** 477–488.

31. KHAN, A., D. A. CIRAULO, W. H. NELSON, J. T. BECKER, A. NIES & J. H. JAFFE. 1984. Dexamethasone suppression tests in recently detoxified alcoholics: Clinical implications. J. Clin. Pharmacol. **4:** 94–97.

32. LOOSEN, P. T., A. J. PRANGE & I. C. WILSON. 1979. TRH (protireline) in depressed alcoholic men: Behavioral changes and endocrine responses. Arch. Gen. Psychiatry **36:** 540–547.

33. BECKER, H. T. & R. F. KAPLAN. 1986. Neurophysiological and neuropsychological concomitance of brain dysfunction in alcoholics. *In* Psychopathology and Addictive Disorders. R. E. Meyer, ed.:262–292. Guilford Press. New York, NY.

34. CARLEN, P. L., G. WORTZMAN, R. C. HOLGATE, D. A. WILKINSON & J. C. RANKIN. 1978. Reversible cerebral atrophy in recently abstinent chronic alcoholics measured by computerized tomography scans. Science **200:** 1076–1078.

35. PORJESZ, B. & H. BEGLEITER. 1978. Human brain electrophysiology and alcoholism. *In* Alcohol and the Brain: Chronic Effects. R. E. Tartar & D. H. V. Thiel, eds.: 139–182. Plenum Press. New York, NY.

36. BEGLEITER, H. & B. PORJESZ. 1977. Persistence of brain hyperexcitability following chronic alcoholic exposure in rats. Adv. Exp. Med. Biol. **85:** 209–222.
37. BEGLEITHER, H., V. DENOBLE & B. PORJESZ. 1980. Protracted brain dysfunction after alcohol withdrawal in monkeys. *In* Biological Effects of Alcohol. H. Begleiter, ed.:231–239. Plenum Press. New York, NY.
38. BEITNER-JOHNSON, D., X. GUITART & E. J. NESTLER. 1992. Common intracellular actions of chronic morphine and cocaine in dopaminergic brain reward regions. *In* The Neurobiology of Drug and Alcohol Addiction. P. O. Kalivas & H. H. Sampson, eds. Ann. N.Y. Acad. Sci. **654:** 70–87.
39. DOLE, V. P. & M. E. NYSWANDER. 1965. A medical treatment for diacetylmorphine (heroin) addiction. J.A.M.A. **193:** 646–650.
40. KREEK, M. J. 1992. Rationale for maintenance pharmacotherapy of opiate dependence. *In* Addictive States, C. P. O'Brien & J. H. Jaffe, eds.:205–230. Raven Press. New York, NY.
41. KISSIN, B. 1975. The use of psychoactive drugs in the long term treatment of chronic alcoholics. Ann. N.Y. Acad. Sci. **252:** 385–395.
42. ERWIN, C. W., M. LINNOILA & J. HARTWELL. 1986. Effects of buspirone and diasepam, alone and in combination with alcohol, on skill performance and evoked potentials. J. Clin. Psychopharmacol. **6:** 199–209.
43. MEHTA, A. K. & M. K. TICKU. 1988. Ethanol potentiation of gamma amino butyric acid gated chloride channels. J. Pharmacol. Exp. Ther. **246:** 558–564.
44. SUZDAK, P. D., R. D. SCHWARTZ, O. SKOLNICK & S. M. PAUL. 1986. Ethanol stimulates gamma amino butyric acid receptor mediated chloride transport in rat brain synaptoneurosomes. Proc. Natl. Acad. Sci. USA **83:** 4071–4075.
45. BUCK, K. J. & R. A. HARRIS. 1990. Benzodiazepine agonist and inverse agonist actions on GABAa receptor-operated chloride channels I acute effects of ethanol. J. Pharmacol. Exp. Ther. **253:** 706–712.
46. BUCK, K. J. & R. A. HARRIS. 1990. Benzodiazepine agonist and inverse agonist actions on GABAa receptor-operated chloride channels II chronic effects of ethanol. J. Pharmacol. Exp. Ther. **253:** 713–719.
47. LITTLE, H. J., R. GALE, N. SELLARS, D. J. NUTT & S. C. TAYLOR. 1988. Chronic benzodiazepine treatment increased the effects of the inverse agonist FG7142. Neuropharmacology **27:** 383–389.
48. BUCK, K. J., J. HEIM & R. A. HARRIS. 1991. Reversal of alcohol dependence and tolerance by a single administration of flumazenil. J. Pharmacol. Exp. Ther. **257:** 984–989.
49. MORROW, A. L., P. MONTPIED & S. M. PAUL. 1991. Ethanol and the GABAa receptor gated chloride ion channel. *In* Neuropharmacology of Ethanol: New Approaches. R. E. Meyer, G. F. Koob, M. J. Lewis & S. M. Paul, eds.: 49–76. Birkhauser. Boston, MA.
50. TABAKOFF, R., C. S. RABIE & K. A. GRANT. 1991. Ethanol and NMDA receptor: Insights into ethanol pharmacology. *In* Neuropharmacology of Ethanol: New Approaches. R. E. Meyer, G. F. Koob, M. J. Lewis & S. M. Paul, eds.:93–106. Birkhauser. Boston, MA.
51. GRANT, K. A., P. VALVERIUS & M. HUDSPITH. 1990. Ethanol withdrawal seizures and NMDA receptor complex. Eur. J. Pharmacol. **176:** 289–296.
52. WIKLER, A. 1948. Recent progress in research on the neurophysiological basis of morphine addiction. Am. J. Psychiatry **105:** 329–338.
53. WIKLER, A. 1965. Conditioning factors in opiate addiction and relapse. *In* Narcotics. D. M. Wilmer & G. G. Kassebaum, eds.:85–100. McGraw Hill. New York, NY.
54. O'BRIEN, C. P., T. TESTA, T. J. O'BRIEN, J. P. BRADY & B. WELLS. 1977. Conditioned narcotic withdrawal in humans. Science **195:** 1000–1002.
55. LUDWIG, A. M. & A. WIKLER. 1974. Craving and relapse to drink. Q. J. Stud. Alcohol **35:** 108–130.
56. WIKLER, A. 1974. Dynamics of drug dependence: Implications of a conditioning theory for research and treatment in opiate addiction. *In* Opiate Addiction: Origins and Treatment. S. Fisher & A. M. Freedman, eds.:7–22. Winston Press. Washington, DC.
57. SIEGEL, S., R. E. HINSON & M. D. KRANK. 1987. Anticipation of pharmacological and

nonpharmacological events: Classical conditioning and addictive behavior. J. Drug Issues **17:** 83–110.
58. PAVLOV, I. P. 1926. Conditioned Reflexes. Dover Press. New York, NY.
59. BROWN, E. E., J. M. FINLAY, J. P. WONG, G. DAMSMA & H. C. FIBIGER. 1991. Behavioral and neurochemical interactions between cocaine and buprenorphine: Implications for the pharmacotherapy of cocaine abuse. J. Pharmacol. Exp. Ther. **256:** 119–126.
60. MEYER, R. E. & S. M. MIRIN, Eds. 1979. The Heroin Stimulus: Implications for a Theory of Addiction. Plenum Press. New York, NY.
61. KAPLAN, R. F., R. E. MEYER & C. F. STOREBEL. 1983. Alcohol dependence and responsibility to an ethanol stimulus as predictors of alcohol consumption. Br. J. Addict. **78:** 259–267.
62. LABERG, J. C. 1986. Alcohol and expectancies: Subjective, psychophysiological, and behavioral responses to alcohol stimuli in severely, moderately, and non-dependent drinkers. Br. J. Addict. **81:** 797–808.
63. HODGSON, R., H. RANKIN & T. STOCKWELL. 1979. Alcohol dependence and the priming effect. J. Behav. Res. Ther. **17:** 379–387.
64. O'BRIEN, C. P., A. R. CHILDRESS, A. T. MCLELLAN & R. EHRMAN. 1992. Classical conditioning in drug dependent humans. *In* The Neurobiology of Drug and Alcohol Addiction. P. W. Kalivas & H. H. Samson, eds. Ann. N.Y. Acad. Sci. **654:** 400–415.
65. STEWART, J., H. DEWITT & R. EIKELBLOOM. 1984. Role of unconditioned and conditioned drug effects in the self-administration of opiates and stimulants. Psychol. Rev. **91:** 251–268.
66. STEWART, J. 1992. Neurobiology of conditioning to drugs of abuse. *In* The Neurobiology of Drug and Alcohol Addiction. P. W. Kalivas & H. H. Samson, eds. Ann. N.Y. Acad. Sci. **654:** 335–344.
67. STEWART, J. & P. VEZINA. 1988. Conditioning and behavioral sensitization. *In* Sensitization in the Nervous System. P. W. Kalivas & C. Barnes, eds.:207–224. Telford Press. Caldwell, NJ.
68. WISE, R. A. & M. A. BOZARTH. 1987. A psychomotor stimulant theory of addiction. Psychol. Rev. **94:** 469–492.
69. DICHIARA, G. & A. IMPERATO. 1988. Drugs abused by humans preferentially increase synaptic dopamine concentrations in the mesolimbic system of freely moving rats. Proc. Natl. Acad. Sci. USA **85:** 5274–5278.
70. VAVROUSEK-JAKUBA, E., C. A. COHEN & W. J. SHOEMAKER. 1990. Ethanol effects of CNS dopamine receptors: *In vivo* binding following voluntary ethanol intake in rats. *In* Novel Pharmacological Interventions for Alcoholism. C. A. Naranjo & E. M. Sellers, eds.:372–374. Springer-Verlag. New York, NY.
71. WEISS, F., Y. L. HURD, U. UNGERSTEDT, A. MARKOU, P. M. PLOTSKY & G. F. KOOB. 1992. Neurochemical correlates of cocaine and ethanol self-administration. *In* The Neurobiology of Drug and Alcohol Addiction. P. W. Kalivas & H. H. Samsom, eds. Ann. N.Y. Acad. Sci. **654:** 220–241.
72. POST, R. M., S. R. B. WEISS, D. FONTANA & A. PERT. 1992. Conditioned sensitization to the psychomotor stimulant cocaine. *In* The Neurobiology of Drug and Alcohol Addiction. P. W. Kalivas & H. H. Samson, eds. Ann. N.Y. Acad. Sci. **654:** 386–399.
73. VEZINA, P. & J. STEWART. 1989. The effect of dopamine receptor blockade on the development of sensitization to the locomote activating effects of amphetamine and morphine. Brain Res. **499:** 108–120.
74. KALIVAS, P. & P. DUFFY. 1990. Effect of acute and daily cocaine treatment on extracellular dopamine in the nucleux accumbens. Synapse **5:** 48–58.

Subject Index

Index of Contributors